Aspects of Littorinid Biology

Developments in Hydrobiology 133

Series editor
H. J. Dumont

Aspects of Littorinid Biology

Proceedings of the Fifth International Symposium on Littorinid Biology, held in Cork, Ireland, 7–13 September 1996

Edited by

Ruth M. O'Riordan, Gavin M. Burnell, Mark S. Davies and Neil F. Ramsay

Reprinted from Hydrobiologia, volume 378 (1998)

Kluwer Academic Publishers

Dordrecht / Boston / London

Library of Congress Cataloging-in-Publication Data

A C.I.P. Catalogue record for this book is available from the Library of Congress.

ISBN 0-7923-5461-3

Published by Kluwer Academic Publishers,
P.O. Box 17, 3300 AA Dordrecht, The Netherlands

Sold and distributed in North, Central and South America
by Kluwer Academic Publishers,
101 Philip Drive, Norwell, MA 02061, U.S.A.

In all other countries, sold and distributed
by Kluwer Academic Publishers Group,
P.O. Box 322, 3300 AH Dordrecht, The Netherlands

Printed on acid-free paper

Cover illustration: Logo from the Fifth International Symposium on Littorinid Biology,
drawn by N. Buttimer, University College Cork

Contents

Short Communications

Hydrobiologia **378:** ix–x, 1998.
R. M. O'Riordan, G. M. Burnell, M. S. Davies & N. F. Ramsay (eds), Aspects of Littorinid Biology.

Preface

Members of the family Littorinidae are among the most widely studied gastropod molluscs and the more questions we answer about this group, the more questions are inevitably posed. Hence, despite decades of research on ecology, systematics, physiology and evolutionary biology, much still remains to be investigated. The epilogue to this volume, by M. S. Davies & G. A. Williams, highlights themes covered herein, but also suggests areas of focus for future research efforts.

The Fifth International Symposium on Littorinid Biology was held at University College Cork, Ireland from 7 to 13 September 1996. For the first time in this series of symposia, a workshop (on the importance of marine gastropods for aquaculture), was incorporated into the programme. This format allowed for valuable dialogue between the theoretical and applied aspects of littorinid research. Previous symposia met at The Natural History Museum in London, England (1986), Tjarnö Marine Biological Laboratory, Sweden (1988), Dale Fort Field Centre, Wales (1990) and the Marine Biological Station at Roscoff, France (1993). The Proceedings of the second and fourth symposia were published as volumes of *Hydrobiologia* (volumes 193 and 309, respectively), while the third was published by The Malacological Society of London.

These are listed below.

The Fifth Symposium was attended by more than fifty participants, from sixteen countries. Altogether thirty-three papers and seventeen posters were presented across a broad range of subjects. This volume includes twenty-four papers and three short communications from fifty-four authors, all but one of which were first presented at the symposium. Continuing a trend set in previous meetings, some contributions describe research on gastropods other than littorinids, and there are several on the effects of TBT pollution. All papers in this volume were refereed by at least two relevant specialists and we are very grateful to them all for their meticulous work. Biological discussion continued on the field excursion to Garrettstown (collection), Kenmare (coffee/collection) and Killarney National Park (culture). Despite inclement weather most participants scrambled on the beach for snails, especially for those elusive neglectoids.

We would like to thank a triumvirate of organisers of previous symposia, John Grahame and Peter Mill of the University of Leeds and David Reid of The Natural History Museum, London, for their advice and enthusiasm. We are indebted to University College Cork for providing the facilities for the conference and to our sponsors for their financial support. The International Science Foundation (Logovaz Travel Fund) generously covered the travel costs of our two Russian delegates.

The symposia continue to attract both long-standing and new researchers in littorinid biology and the success of the symposia is based on the participation and support of delegates, especially those who have contributed to organising and editing in the past. The stimulating and friendly atmosphere at these small, focused symposia is unique and is a particularly receptive environment for those researchers who are making their first presentations.

We thank all the participants of the symposium for their presentations, general discussions and lively company. In 1999, for the first time, the symposium will be held outside Europe and we wish the organisers (Joe Britton & Bob McMahon) of the sixth symposium in Jamaica every success.

References

Johannesson, K., D. G. Raffaelli & C. J. Hannaford Ellis (eds), 1990. Progress in Littorinid and Muricid Biology. Proceedings of the Second International Symposium on Littorinid Biology. Developments in Hydrobiology 56. Kluwer Academic Publishers, Dordrecht, 288 pp. Reprinted from Hydrobiologia 193.

Grahame, J., P. J. Mill & D. G. Reid (eds), 1992. Proceedings of the Third International Symposium on Littorinid Biology. The Malacological Society of London. 324 pp.

Mill, P. J. & C. D. McQuaid (eds), 1995. Advances in Littorinid Biology. Proceedings of the Fourth International Symposium on Littorinid Biology. Developments in Hydrobiology 111. Kluwer Academic Publishers, Dordrecht, 200 pp. Reprinted from Hydrobiologia 309.

RUTH O'RIORDAN
GAVIN BURNELL
MARK DAVIES
NEIL RAMSAY

Hydrobiologia **378:** xi, 1998.
R. M. O'Riordan, G. M. Burnell, M. S. Davies & N. F. Ramsay (eds), Aspects of Littorinid Biology.

List of participants

Front row, kneeling, left to right: Ruth O'Riordan, Gavin Burnell, Dan Minchin

Front row, standing, left to right: Lisa Johnson, A. Bruheim, Thierry Backeljau, B. Winnepenninckx, Sue Hull, Ulrike Schulte-Oehlmann, Pio Fioroni

Second Row, left to right: Raquel Cruz-Guerrero, Jorge Otero-Schmitt, Mike Johnson, Mary Barry, Francesco Toscano, Jörg Oehlmann, Esther Fioroni

Third Row, left to right: Craig Wilding, Diana Padilla, Dawn Dittmann, Cesare Sacchi, Gráinne Lynch

Back rows, left to right: Yiu Ming Mak, Bob McMahon, John Grahame, Anette Ekendahl, Richard Clarke, Jens Christensen, Iain Wilson, Derick Casey, John Knowles, Elizabeth Platts, Peter Mill, Gray Williams, Mark Davies, Susan Steele, Elizabeth Boulding, Hans De Wolf, Elizabeth Gosling, Peter Clarke, Joe Britton, Nadezhda Zaslavskaya, An Other, Gudrun Sundermann, Natalia Mikhailova, Peter Cook, Martin MacNulty, Anna Minchin, Sergei Sergievsky, Delmont Smith, Barbara Bauer, Mireille Sacchi, David Reid

Not present/visible in photo: J. Andrews, Becky Britton, John Crothers, Marilyn Crothers, Claudia De Magalhães, Johan Erlandsson, Carlos Garcia, Rupert Lewis, Máire Mulcahy, Luis Serpa

Hydrobiologia **378:** 1–10, 1998.
R. M. O'Riordan, G. M. Burnell, M. S. Davies & N. F. Ramsay (eds), Aspects of Littorinid Biology.

Genetic variability in *Littorina saxatilis* from different habitats on an island in Galway Bay

I. F. Wilson & E. M. Gosling
Regional Technical College, Galway, and Fermentation Laboratory, Microbiology Department, University College, Galway, Ireland

Key words: allozymes, natural selection, *AAT-1*, *Littorina saxatilis*, Ireland

Abstract

Starch-gel electrophoretic techniques were applied to the investigation of molecular genetic variation in populations of the rough periwinkle *Littorina saxatilis*. The investigation comprised two phases: (a) technique development to resolve as many as possible of the allozyme loci reported in the literature as having been screened in the genus *Littorina* and in *Melarhaphe neritoides*; (b) the use of these loci to assess levels of genetic variation in and patterns of genetic differentiation among populations of *L. saxatilis* from a relatively isolated group of populations from Galway Bay, Ireland.

More than 43 allozyme loci (of which four were screened for the first time here in this species), coding for 37 enzymes, were investigated and thirteen of these loci (including two loci screened for the first time here) were found to be variable and reliably scorable. Samples from five pairs of transects were collected from Inismór, Aran Islands, from sites with known exposure levels; one transect within each pair was collected from an exposed site and the other from a nearby, but relatively sheltered site. UPGMA for eleven loci, (*ARK* and *PGDH* were excluded from cluster and F_{ST} analysis as they were unscorable in a few samples), showed that the samples cluster mostly by pair, reflecting their geographic origin and is indicative of little gene flow between populations. Levels of population differentiation were high among samples from the top of the shore, but unusually so at *AAT-1* which showed nearly three times the mean F_{ST} value for the eleven loci. There was also a significant regression of frequency of $AAT\text{-}1^{100}$ against level of exposure. In addition, among midshore samples, there was a consistently higher frequency of $AAT\text{-}1^{100}$ in sheltered habitats. These results support the findings of others, indicating that this locus may be subject to natural selection.

Introduction

The rough periwinkle *Littorina saxatilis* (Olivi), inhabits the intertidal zone along Atlantic shores of western Europe. It is an ovoviviparous species with limited powers of migration, which lends itself to the study of evolution of genetic variation in isolated populations. Previous studies have found that this species shows high levels of genetic variability (see for example, Hull, 1994; Janson, 1987; Knight et al., 1987; Ward & Warwick, 1980; Wilkins & O'Regan, 1980), as do two other closely related species, *L. neglecta* Bean (Hull, 1994; Johannesson & Johannesson, 1990; Wilkins & O'Regan, 1980) and *L. arcana* Hannaford Ellis (Hull, 1994; Janson, 1985; Janson & Ward, 1984; Knight et al., 1987; Knight & Ward, 1991; Sundberg et al., 1990; Ward & Janson, 1985; Ward & Warwick, 1980; Ward et al., 1986, 1987). A fourth member of this group of rough periwinkles, *L. compressa* Jeffreys, is known to have comparatively reduced levels of variation (Beardmore & Morris, 1978; Knight & Ward, 1991; Wilkins & O'Regan, 1980).

Previous studies on *L. saxatilis* have shown that one allozyme locus, *AAT-1*, is influenced by natural selection. Johannesson & Johannesson (1989, 1990) found, in western Europe, that there was a cline in the frequency of $AAT\text{-}1^{100}$ down the shore, and Johannesson et al. (1995) reported evidence of strong selection

Table 1. The 37 enzymes and their EC numbers investigated in populations of *L. saxatilis*. Buffers: see text. The preferred buffer used to resolve (+, little activity; ++, some activity but not well resolved; +++, well resolved; O, not tested; –, no activity) each polymorphic enzyme is highlighted in grey. Also reported is the number of putative loci detected for each enzyme, and the superscript indicates those screened for the first time here

Enzyme	EC number	Buffer				Loci
		CAM	TC	TM	LiOH	
EST	3.1.1.-	O	+	+++	O	>1[1]
AAT	2.6.1.1	+++	+++	+	O	2
PGM	5.4.2.2	++	+++	+++	O	3
GPI	5.3.1.9	+++	+++	+++	O	1
MDH	1.1.1.37	+++	+++	+	O	2
MEP	1.1.1.40	+++	+++	+	O	1
IDHP	1.1.1.42	+++	+++	+	O	2
OPDH	1.5.1.15	+	+	+	O	1[1]
LDH	1.1.1.27	+	+	+	++	1
SOD	1.15.1.1	+	+	+	+	2
IDDH	1.1.1.14	+	+	+	O	1
MPI	5.3.1.8	+++	+++	+++	O	1
PNP	2.4.2.1	+++	+++	O	O	1
FBALD	4.1.2.13	++	++	++	O	1
G3PDH	1.1.1.8	+	-	+	O	1
GAPDH	1.2.1.12	+	-	+	O	1
XDH	1.1.1.204	O	+	+	O	1
DIA	1.8.1.4	O	++	+	O	1
PHE-PRO PEP	3.4.13.9	O	++	+	+++	1[1]
LEU-TYR PEP	3.4.-.-	O	++	+	+++	2
LEU-LEU-LEU PEP	3.4.-.-	O	+	+	++	3[1]
GLY-LEU PEP	3.4.-.-	O	O	O	+++	1
LEU-ALA PEP	3.4.-.-	O	O	O	+++	2
ODH	1.1.1.73	O	O	O	O	1
HDH	uncertain	O	O	O	O	1
ARK	2.7.3.3	-	+++	+++	-	1
CK	2.7.3.2	-	+	-	-	1
AK	2.7.4.3	-	-	-	-	-
AH	4.2.1.3	-	-	-	-	-
GLYD	1.1.1.29	-	-	-	-	-
IPP	3.6.1.1	-	+	-	-	1
ALP	3.1.3.1	++	+	-	-	1
PGDH	1.1.1.44	+++	+	++	-	1
G6PDH	1.1.1.49	-	-	-	-	-
LAP	3.4.1.1	++	+	+	+	>1
HBDH	1.1.1.30	-	+	O	O	1
FH	4.2.1.2	+	+	O	O	1

at this locus over time and space after recolonisation of a shore following extinction of populations of *L. saxatilis* caused by a bloom of a toxin-producing micro-flagellate.

This paper investigates further the association between variation at *AAT-1* and components of the environment as part of a larger ongoing study to investigate the evolution of this group of periwinkles on a group of islands in Galway Bay which are relatively isolated from populations on the mainland. Furthermore, it adds to the relatively restricted literature on genetic studies of Irish populations of this species (Knight et al., 1987; Knight & Ward, 1991; Wilkins & O'Regan, 1980; Wilkins et al., 1978).

Material and methods

Study area and collecting of littorinids

Samples were collected from the south Connemara coast, Co. Galway (Figure 1; Baile na hAbhann (Irish National Grid Reference: L 992 202), Ros an Mhíl (L 957 238) and Bearna (M 233 226)) during May 1996 for technique development. Allele fre-

Table 2. Allele frequencies among five pairs of *L. saxatilis* samples from the top of the shore on Inismór. Only variant alleles are shown, n, sample size

Locus	Allele	Sample																			
		Bun Gabhla-v		Cill Éinne-v		Iaráirne-v		Eoghanacht-v		Mainistir-v		Bun Gabhla-iii		Cill Éinne-iii		Iaráirne-iv		Eoghanacht-iii		Mainistir-iv	
		Exp	She	Exp	She	Exp	She	Exp	She	Exp	She	Exp	She	Exp	She	Exp	She	Exp	She	Exp	She
	(1)	(5.5)	(4.5)	(6.5)	(3)	(5)	(2)	(5.5)	(3)	(5)	(1)	(5.5)	(4.5)	(6.5)	(3)	(5)	(2)	(5.5)	(3)	(5)	
AAT-1	n	20	17	20	20	20	20	20	20	20	20	20	18	20	7	20	–	20	20	20	20
	†190	0.95	0.09	0.03	0.10	0.53	0.68	0.43	0.10	0.65	0.48	0.15	0.03	0.18	0.14	0.58	–	0.63	0.03	0.55	0.18
ARK	n	20	0	0	20	20	20	20	20	20	20	19	18	20	7	17	–	20	19	20	20
	290	0.00	–	–	0.00	0.00	0.00	0.00	0.00	0.00	0.03	0.00	0.00	0.03	0.00	0.00	–	0.00	0.00	0.03	0.03
	200	0.20	–	–	0.38	0.25	0.18	0.03	0.08	0.25	0.13	0.24	0.17	0.35	0.36	0.06	–	0.18	0.29	0.33	0.03
	20	0.13	–	–	0.10	0.13	0.03	0.10	0.05	0.08	0.08	0.11	0.22	0.05	0.00	0.12	–	0.05	0.08	0.08	0.18
AAT-2	n	20	17	20	20	20	20	20	20	20	20	20	15	20	7	20	–	20	17	20	20
	200	0.00	0.09	0.15	0.23	0.20	0.25	0.23	0.00	0.30	0.13	0.00	0.27	0.13	0.07	0.00	–	0.00	0.12	0.38	0.18
GPI	n	20	17	20	20	20	20	20	20	20	20	20	18	20	7	20	–	20	20	20	20
	115	0.00	0.00	0.00	0.00	0.00	0.00	0.00	0.00	0.00	0.00	0.00	0.00	0.03	0.00	0.00	–	0.00	0.00	0.00	0.00
	110	0.00	0.00	0.00	0.00	0.00	0.00	0.00	0.00	0.00	0.00	0.03	0.03	0.03	0.07	0.10	–	0.00	0.00	0.00	0.05
	†80	0.43	0.15	0.33	0.33	0.48	0.10	0.03	0.28	0.15	0.23	0.33	0.14	0.38	0.29	0.33	–	0.15	0.35	0.20	0.20
PGDH	n	20	0	0	20	20	0	11	20	6	0	20	5	20	7	10	–	19	17	20	20
	†150	0.38	–	–	0.13	0.18	–	0.09	0.05	0.42	–	0.15	0.00	0.13	0.14	0.10	–	0.13	0.21	0.30	0.18
	20	0.00	–	–	0.00	0.00	–	0.00	0.00	0.00	–	0.00	0.00	0.00	0.07	0.00	–	0.00	0.00	0.00	0.00
MPI	n	20	17	20	20	20	20	19	20	20	20	20	18	20	7	20	–	18	19	20	20
	†80	0.28	0.41	0.13	0.35	0.23	0.05	0.08	0.15	0.23	0.20	0.63	0.22	0.35	0.57	0.18	–	0.64	0.11	0.25	0.23
	60	0.08	0.03	0.08	0.13	0.05	0.03	0.00	0.03	0.13	0.18	0.03	0.03	0.05	0.21	0.15	–	0.00	0.03	0.13	0.08
LEU-TYR-1	n	20	17	20	20	20	20	20	20	20	20	20	18	20	7	20	–	20	17	20	20
	120	0.00	0.00	0.00	0.00	0.00	0.00	0.00	0.00	0.00	0.00	0.00	0.00	0.00	0.00	0.00	–	0.03	0.00	0.00	0.00
	110	0.03	0.00	0.08	0.03	0.03	0.03	0.05	0.08	0.13	0.10	0.08	0.03	0.10	0.21	0.00	–	0.08	0.00	0.03	0.00
	90	0.03	0.00	0.00	0.00	0.00	0.00	0.00	0.00	0.00	0.00	0.00	0.00	0.00	0.00	0.00	–	0.00	0.00	0.00	0.00
LEU-TYR-2	n	20	17	20	20	20	20	20	20	20	20	20	18	20	7	20	–	20	15	20	20
	†150	0.00	0.06	0.00	0.10	0.68	0.73	0.13	0.48	0.23	0.23	0.03	0.11	0.43	0.07	0.38	–	0.15	0.00	0.05	0.05
	50	0.33	0.44	0.15	0.28	0.03	0.00	0.28	0.13	0.13	0.23	0.25	0.17	0.08	0.36	0.00	–	0.18	0.18	0.25	0.28
PGM-1	n	20	17	20	20	20	20	19	20	20	20	20	18	20	7	20	–	20	19	20	20
	95	0.00	0.00	0.00	0.00	0.00	0.00	0.00	0.00	0.00	0.00	0.00	0.28	0.00	0.36	0.00	–	0.00	0.00	0.00	0.00
	†90	0.30	0.71	0.50	0.45	0.45	0.68	0.82	0.88	0.58	0.43	0.58	0.28	0.43	0.14	0.55	–	0.65	0.76	0.28	0.45
	85	0.00	0.00	0.00	0.00	0.00	0.00	0.00	0.00	0.00	0.00	0.00	0.00	0.00	0.00	0.03	–	0.00	0.00	0.00	0.00
	80	0.23	0.21	0.23	0.25	0.23	0.23	0.11	0.08	0.18	0.20	0.00	0.19	0.20	0.29	0.18	–	0.18	0.18	0.13	0.48

–, missing data; exposure levels are reported in brackets; †, alleles showing a range of at least 0.40 in allele frequencies across samples.

Table 2. (continued)

Locus	Allele	Sample																			
		Bun Gabhla-v		Cill Éinne-v		Iaráirne-v		Eoghanacht-v		Mainistir-v		Bun Gabhla-iii		Cill Éinne-iii		Iaráirne-iv		Eoghanacht-iii		Mainistir-iv	
		Exp	She	Exp	She	Exp	She	Exp	She	Exp	She	Exp	She	Exp	She	Exp	She	Exp	She	Exp	She
	(1)	(5.5)	(4.5)	(6.5)	(3)	(5)	(2)	(5.5)	(3)	(5)		(1)	(5.5)	(4.5)	(6.5)	(3)	(5)	(2)	(5.5)	(3)	(5)
PGM-2	n	20	17	20	19	20	20	20	20	20	19	20	18	20	7	20	–	20	16	20	20
	†*80*	0.28	0.24	0.33	0.50	0.43	0.30	0.58	0.28	0.48	0.45	0.05	0.44	0.50	0.36	0.30	–	0.13	0.06	0.40	0.20
	55	0.03	0.09	0.00	0.16	0.05	0.23	0.08	0.00	0.03	0.08	0.08	0.17	0.05	0.07	0.10	–	0.08	0.13	0.05	0.08
PNP	n	20	17	19	20	20	20	20	20	20	20	20	18	20	6	20	–	20	20	20	20
	125	0.00	0.00	0.03	0.03	0.18	0.03	0.00	0.00	0.03	0.00	0.00	0.00	0.08	0.08	0.03	–	0.00	0.00	0.05	0.00
	†*65*	0.60	0.38	0.55	0.60	0.35	0.50	0.20	0.23	0.65	0.65	0.65	0.75	0.43	0.42	0.73	–	0.65	0.58	0.65	0.18
	35	0.05	0.00	0.05	0.05	0.00	0.05	0.00	0.00	0.00	0.08	0.08	0.03	0.08	0.00	0.00	–	0.05	0.05	0.05	0.03
EST	n	19	17	20	19	20	20	20	19	20	20	20	16	20	7	20	–	12	16	19	20
	–50	0.08	0.06	0.05	0.05	0.23	0.25	0.10	0.03	0.30	0.23	0.23	0.06	0.00	0.07	0.20	–	0.17	0.06	0.11	0.03
	†*–150*	0.16	0.59	0.05	0.11	0.13	0.23	0.43	0.61	0.25	0.30	0.20	0.19	0.20	0.07	0.15	–	0.13	0.06	0.13	0.10
	†*–200*	0.16	0.12	0.10	0.21	0.43	0.23	0.25	0.00	0.08	0.08	0.13	0.03	0.20	0.21	0.33	–	0.21	0.22	0.34	0.03
PHE-PRO	n	20	17	18	20	20	20	20	20	20	20	20	18	20	7	20	–	20	20	20	20
	†*115*	0.33	0.35	0.28	0.40	0.35	0.30	0.08	0.28	0.28	0.40	0.13	0.08	0.23	0.43	0.03	–	0.23	0.25	0.20	0.03
	90	0.13	0.09	0.03	0.13	0.08	0.05	0.03	0.05	0.05	0.05	0.13	0.11	0.03	0.21	0.08	–	0.08	0.05	0.13	0.10
	80	0.00	0.00	0.00	0.00	0.00	0.00	0.00	0.00	0.00	.000	0.00	0.00	0.00	0.00	0.00	–	0.00	0.00	0.00	0.05

–, missing data; exposure levels are reported in brackets; †, alleles showing a range of at least 0.40 in allele frequencies across samples.

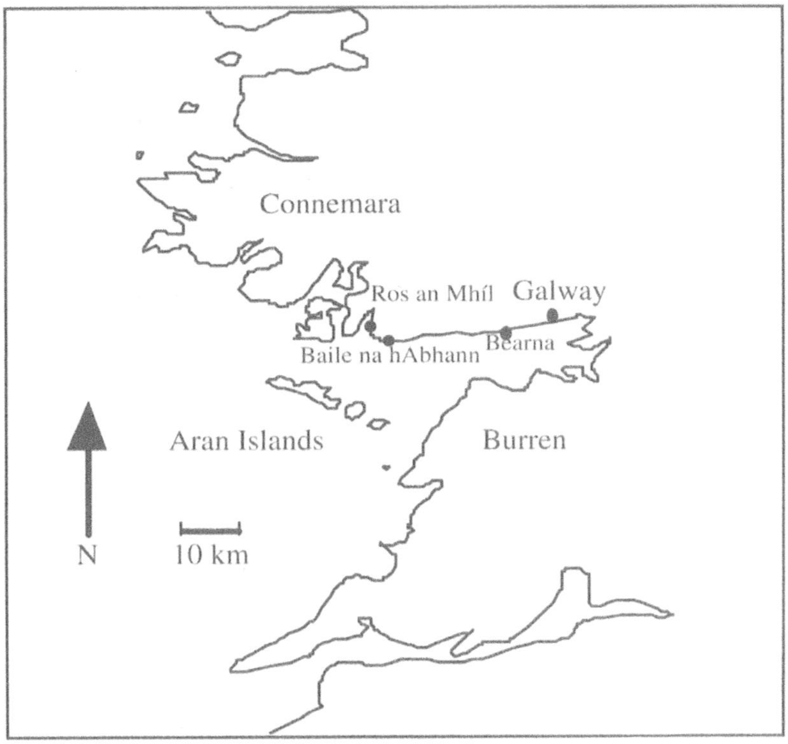

Figure 1. Map of the west of Ireland showing the position of the Aran Islands in Galway Bay and of the three sample sites used in technique development.

quencies from these samples are not reported here because, as they were used for technique development, not all individuals were screened for the same allozymes. In addition, samples from five pairs of transects from Inismór (Figure 2; exposed sites listed first; Bun Gabhla (L 771 113; L 771 115), Cill Éinne (L 883 080; L 879 083), Iaráirne (L 907 068; L 903 066), Eoghanacht (L 808 127; L 810 124) and Mainistir (L 873 108; L 873 105)) were collected between May and July 1996 for an investigation of the genetics of *L. saxatilis* populations in relative isolation. At each site samples were collected from two transects some 200 m to 300 m apart and, within each transect, from five stations: bottom, middle and top of the barnacle zone, top of the shore and midway between there and the top of the barnacle zone. Stations were numbered i to v from the bottom of the barnacle zone. Each pair of transects comprised samples from an exposed and a sheltered site. The transects were then assigned a level of wave exposure (Ballantine, 1961) by Michael O'Connell who had investigated some of our sites for his own research (O'Connell et al., 1992); the units of Ballantine's scale are integers, but to include sites intermediate between two different exposure levels half units were also used.

Individuals were classified to species on the basis of shell morphology and internal anatomy (Reid, 1996). Using shell morphology, *L. saxatilis* and *L. neglecta* can be successfully separated (Caley et al., 1995; but see Johannesson & Johannesson, 1990). However, *L. arcana* and *L. compressa* have a range of forms that can overlap with those of *L. saxatilis*. In addition, *L. arcana* and *L. compressa* have similar pallial oviduct morphology while *L. saxatilis* and *L. arcana* have similar penis morphology. If a morphologically distinct group of snails is present in a sample and the female pallial oviduct is similar to *L. compressa* while the males lack the penis shape of *L. compressa* and are more like *L. saxatilis*, then these individuals are almost certainly *L. arcana*. Only adults were analysed because of the difficulty imposed by overlap between species in shell morphology which is accentuated in juveniles.

Twenty individual *L. saxatilis* were analysed from the upper-most station (v) from each transect to enable comparisons to be made between allele frequencies at

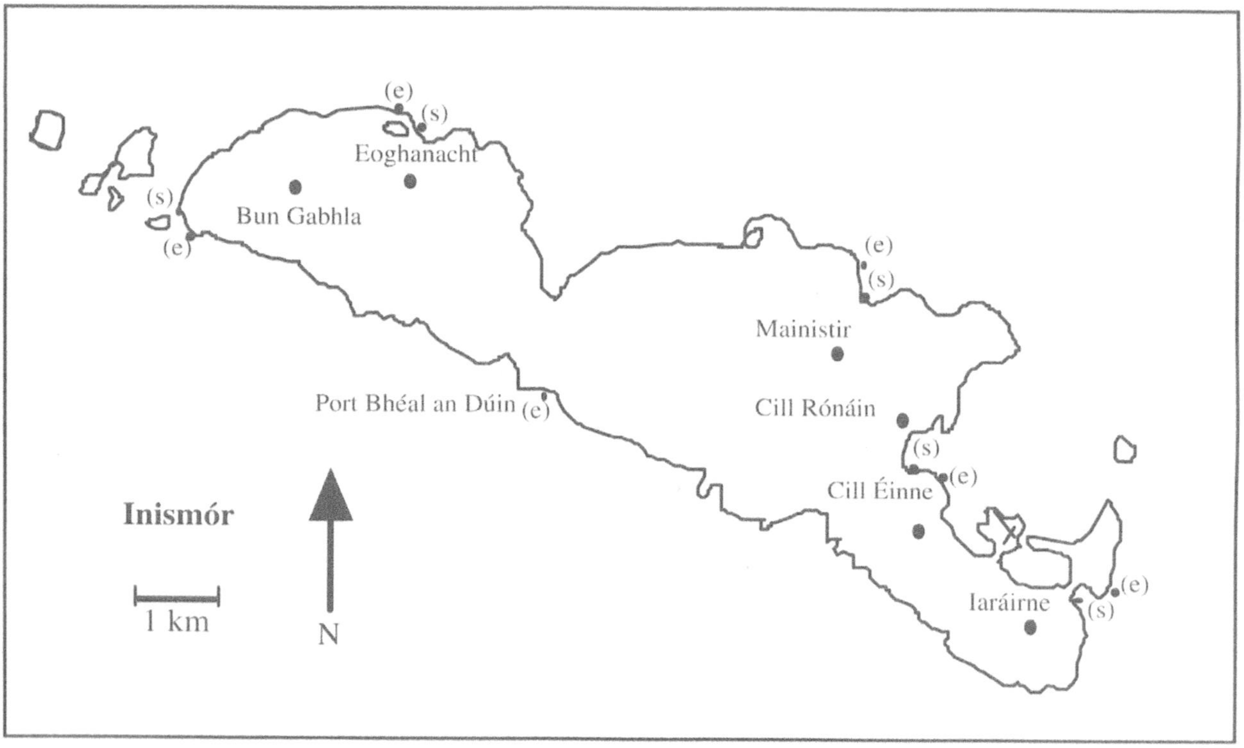

Figure 2. Map of Inismór showing the position of the sample sites used to investigate genetic differentiation among populations of *L. saxatilis*; e – exposed; s – sheltered.

sites with different levels of exposure. In addition, after having analysed the results for those five pairs of samples, up to 20 individual *L. saxatilis* from samples midway up the shore (either stations iii or iv) were analysed; because of the mixture of species lower down the shore, adequate sample sizes of *L. saxatilis* could not always be collected at each site. For example, we thought that we had collected an adequate number of *L. saxatilis* from Iaráirne-iii (sheltered transect), but only under the microscope did it become clear that all but two of these snails were *L. arcana*. Thus, for this sample, although the two individuals were analysed, the allele frequencies are not reported. This problem with sampling also prevented us from giving an analysis of the *F*-statistics and UPGMA from the midshore samples.

Starch gel electrophoresis and statistical analysis

Thirty-seven enzymes (Table 1) were investigated using four buffer systems: CAM, amine-citrate pH 6.6 (Clayton & Tretiak, 1972), TC, tris-citrate pH 8.0 (Ward & Warwick, 1980), TM, tris-maleate pH 7.4 and LiOH, discontinuous LiOH-borate (buffers XVIII and X respectively of Shaw & Prasad, 1970). In keeping with the convention for littorinid analysis using this technique, loci were numbered from the anodal end of the gel.

Analysis of gene frequencies, observed levels of heterozygosity (H_o) and percentage polymorphism (P) was performed by using BIOSYS-1 (Swofford & Selander, 1981) and GENEPOP (Raymond & Rousset, 1995), which was also used to perform an analysis of two-locus linkage disequilibrium coefficients. Because of the small number of expected genotypes in certain cases, samples were tested for deviation from Hardy-Weinberg expectations using Fisher's exact test. Monomorphic loci were excluded from the analysis. Heterogeneity among samples at each locus was estimated from the *F*-statistics of Wright (1978). Discrimination between pairs of samples and cluster analysis (UPGMA) was performed with BIOSYS-1 using Nei's (1978) unbiased genetic distance. Possible associations between allele frequencies and level of wave exposure were investigated by regression analysis.

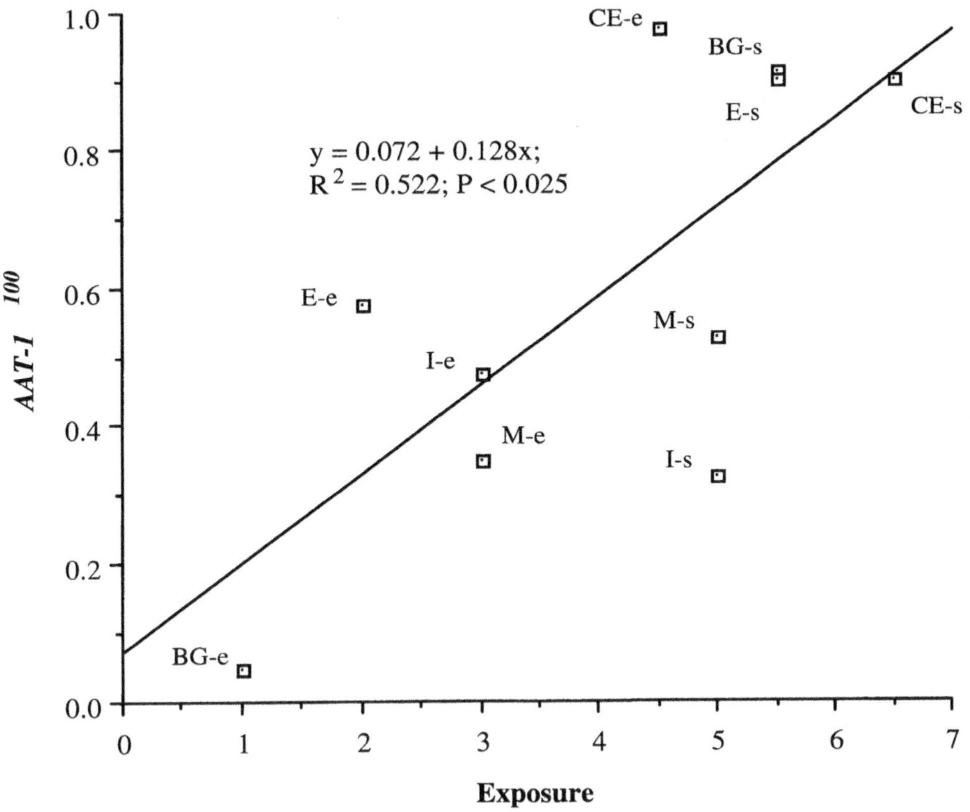

Figure 3. Regression of *AAT-1*^{100} frequency among *L. saxatilis* samples from the top of the shore against level of exposure. BG – Bun Gabhla; CE – Cill Éinne; I – Iaráirne; E – Eoghanacht; M – Mainistir; e – exposed site; s – sheltered site.

Results

Activity was detected at more than 43 putative loci (Table 1); two enzymes (EST and LAP) both showed more than one region of activity, but we are uncertain of the actual number of genetic loci involved. Fifteen of the enzymes, encoded by more than 19 loci (including at least one EST locus), were well resolved (Table 1), and of these four were investigated for the first time in this study. Thirteen allozyme loci were found to be variable, including two novel polymorphic loci (*EST* and *PHE-PRO*). The single GLY-LEU peptidase had the same mobility as the more anodal of the two LEU-TYR loci, while the two LEU-ALA loci had the same mobility as the two LEU-TYR loci. Of the three LEU-LEU-LEU loci the two less anodal loci had the same mobility as the two LEU-TYR loci while the uppermost locus was specific to LEU-LEU-LEU.

Exposure levels at exposed sites ranged from 1 to 4.5, while those at sheltered habitats ranged from 5 to 6.5 (Table 2). Each sample of *L. saxatilis* showed high levels of variability ($H_o = 0.27$ to 0.52; $P = 77\%$ to 100%) with an enormous range in allele frequencies at most loci (Table 2: at least 0.40 for alleles at ten of the thirteen loci). Not only were there large differences between sites and between transects within sites in allele frequencies at a number of loci, there were also a number of differences between stations within transects. There were differences in allele frequencies larger than 0.3 at eight of the thirteen loci between stations only 10 to 40 m apart. For instance, *PNP* showed large differences in allele frequencies between stations iii and v in the sheltered transect at Bun Gabhla, between stations iv and v in the exposed transect at Iaráirne, between stations iii and v in both transects at Eoghanacht, and between stations iv and v in the sheltered transect at Mainistir. However, the distribution of gene frequencies at this locus was not associated with habitat type. Twelve out of 234 tests showed deviation from Hardy-Weinberg expectations which is to be expected given the number of tests carried out.

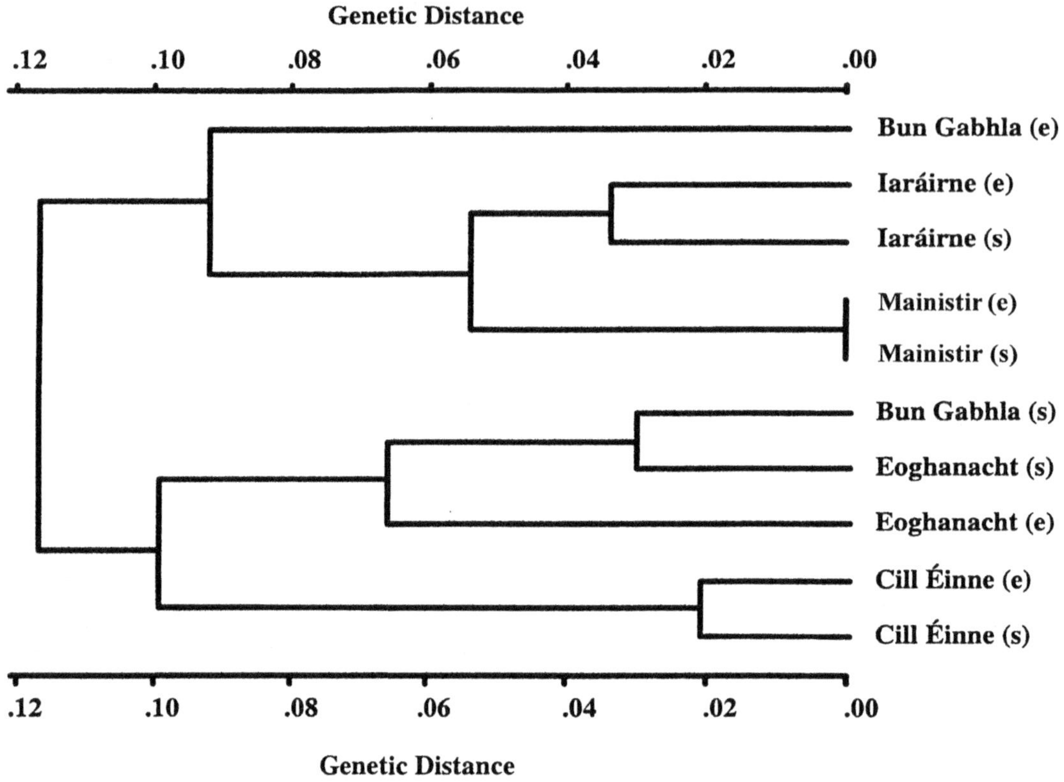

Figure 4. UPGMA, based upon allele frequencies at eleven polymorphic loci and using Nei's unbiased genetic distance (1978), of five pairs of samples of *L. saxatilis* from the top of the shore on Inismór. (e) – exposed; (s) – sheltered.

F-statistics analysis for the samples from the top of the shore (Table 3a) showed that levels of population differentiation varied among loci. (Note that only eleven loci were subjected to F_{ST} and cluster analysis as there are missing data for *ARK* and *PGDH*.) Although F_{ST} values were small for a few loci (including *LT-1, PHE-PRO* and *PGM-2*) others showed marked levels of differentiation, most noticeably *AAT-1*, where the F_{ST} value was more than three times the mean value (Tables 3a and 3b). In addition, for that locus, the partitioning of the variance attributable to differences between habitats (exposure levels) was very high (Table 3b), indicating that habitat has a large effect on the distribution of alleles at *AAT-1*. Furthermore, regression analysis showed that the frequency of $AAT\text{-}1^{100}$ increased significantly ($p < 0.025$) with decreasing level of exposure (Figure 3), bearing in mind that, using the Ballantine scale, exposed sites have a low exposure value. Although formal F-statistics analysis could not be carried out on the incomplete data from the midshore samples (some samples were from station iii and others from station iv), analysis showed that there was a non-significant regression of $AAT\text{-}1^{100}$ against level of exposure ($y = 0.413 + 0.079x$; $R^2 = 0.36$; $p > 0.05$). However, when the data for stations iii, iv and v were combined the regression was highly significant ($y = 0.244 + 0.103x$; $R^2 = 0.41$; $p < 0.005$). In addition, in six of nine transects the frequency of $AAT\text{-}1^{100}$ increased down the shore; the tenth middle shore site (Iaráirne-iv sheltered) only had two *saxatilis* which were both homozygous for $AAT\text{-}1^{100}$.

UPGMA for the eleven loci for which there are complete data showed that the five pairs of *L. saxatilis* samples from the top of the shore mostly clustered by geography (Figure 4); even after removing from the analysis *AAT-1*, which may be subject to selection.

There was no systematic linkage disequilibrium between any of the loci investigated.

Table 3. a: F-statistics for eleven allozyme loci among the five pairs of samples of *L. saxatilis* from the top of the shore; b: variance components and F-statistics combined across loci

a

Locus	F_{IS}	F_{IT}	F_{ST}
AAT-1	–0.093	0.308	0.366
AAT-2	–0.073	0.004	0.072
GPI	–0.178	–0.060	0.101
MPI	–0.128	–0.042	0.076
LT-1	0.014	0.040	0.027
LT-2	0.006	0.191	0.186
PGM-1	–0.021	0.076	0.094
PGM-2	0.039	0.099	0.063
PNP	0.071	0.176	0.113
EST	0.157	0.268	0.132
PHE-PRO	–0.037	0.007	0.043

b

Comparison		Variance	F_{XY}	
X	*Y*		Mean	*AAT-1*
Sample	Pair	0.47491	0.084	0.358
Sample	Total	0.58428	0.101	0.350
Pair	Total	0.10937	0.019	–0.011

Discussion

This study adds to the known number of allozyme loci screened in the genus *Littorina*. Although the data are not presented here for the other species of rough periwinkle, the buffer systems used here enabled most of the polymorphic loci reported in Table 2 to be resolved in juvenile specimens and in the minute rough periwinkle, *L. neglecta* (Wilson & Gosling, unpublished).

It is clear from this study and from the results of others in the literature (Hull, 1994; Janson, 1987; Knight et al., 1987; Ward & Warwick, 1980; Wilkins & O'Regan, 1980) that *L. saxatilis* is genetically a highly diverse species. Despite the relative proximity of populations of *L. saxatilis* on Inismór (the island is approximately 13.5 km long), there is great genetic differentiation among them and an enormous range in allele frequencies at most loci. UPGMA analysis of samples from the top of the shore showed that populations mostly cluster by pair (geographically, 200 m to 300 m apart and between 2 km and 13 km between pairs), indicating that there is limited gene flow between them which is to be expected given the low dispersal rate in this species (Johannesson et al., 1995). The fact that the populations cluster by pair indicates that the allele frequencies estimated from the sample sizes used here are probably representative of the populations from which they were taken. This leads to another interesting pattern of allele frequency distributions. The stations up the shore were between 10 m and 20 m apart, and although the allele frequencies are often similar between stations within a transect, a number of loci show quite large differences between stations, indicating that there is large micro-geographic population differentiation.

F-statistics analysis shows that the majority of the loci investigated here reflect the larger scale genetic differentiation. However, one locus (*AAT-1*) shows a remarkable pattern of differentiation. This paper represents an early stage in our study of this phenomenon, but nevertheless it is apparent that there are differences between habitats in the frequencies of alleles at this locus. The high F_{ST} value for this locus and the significant regression of *AAT-1*^{100} against exposure level suggest that the pattern of allelic differentiation at *AAT-1* is very different from that at other enzyme loci, indicating that natural selection may be shaping its distribution. The distribution of *AAT-1*^{100} is interesting and requires further explanation. The frequency is higher in the more sheltered sites and is fairly consistent in this distribution within pairs of samples. In addition, it also increases in frequency down the shore. These result support the findings of other researchers who have found evidence of selection at this locus (Grahame et al., 1992; Johannesson & Johannesson, 1989, 1990; Johannesson et al., 1995).

Acknowledgements

This research was supported by the MAST 3 programme of the European Commission under contract number MAS3-CT95-0042 (AMBIOS). We would also like to thank Michael O'Connell, Alba Dakoli and Jim Gosling for help with collecting samples, and John Patching for providing facilities in the Microbiology Department and the Martin Ryan Institute, University College, Galway.

References

Ballantine, W. J., 1961. A biologically-defined exposure scale for the comparative description of rocky shores. Field Stud. 1: 1–19.

Beardmore, J. & S. R. Morris, 1978. Genetic variation and species coexistence in *Littorina*. In B. Battaglia & J. Beardmore (eds), Marine Organisms: Genetics, Ecology and Evolution. Plenum Press, New York & London.

Caley, K. J., J. Grahame & P. J. Mill, 1995. A geographically-based study of shell shape in small rough periwinkles. Hydrobiologia 309: 181–193.

Clayton, J. W. & D. N. Tretiak, 1972. Amine-citrate buffers for pH control in starch gel electrophoresis. J. Fish Res. Bd Can. 29: 1169–1172.

Gosling, E. M., I. F. Wilson & J. Andrews, 1998. A preliminary study on genetic differentiation in *Littorina saxatilis* from Galway Bay, Ireland: *Littorina tenebrosa* Montagu – a valid species or ecotype? Hydrobiologia 378: 21–25.

Grahame, J., P. J. Mill, M. Double & S. Hull, 1992. Patterns of variation in Aat-1 allele frequencies in rough periwinkles (*Littorina*) suggest similar selection regimes rather than conspecificity. J. mar. biol. Ass. U. K. 72: 499–502.

Grahame, J., P. J. Mill, S. Hull & K. J. Caley, 1995. *Littorina neglecta* Bean: ecotype or species? J. nat. Hist., Lond. 29: 887–899.

Hull, S., 1994. Polymorphism, physiological ecology and the systematics of *Littorina*. Unpublished Ph.D. thesis, University of Leeds.

Janson, K., 1985. A morphologic and genetic analysis of *Littorina saxatilis* (Prosobranchia) from Venice, and on the problem of *saxatilis-rudis* nomenclature. Biol. J. linn. Soc. 24: 51–59.

Janson, K., 1987. Genetic drift in small and recently founded populations of the marine snail *Littorina saxatilis*. Heredity 58: 31–37.

Janson, K. & R. D. Ward, 1984. Microgeographic variation in allozyme and shell characters in *Littorina saxatilis* Olivi (Prosobranchia: Littorinidae). Biol. J. linn. Soc 22: 289–307.

Johannesson, K. & B. Johannesson, 1989. Differences in allele frequencies of *Aat* between high- and mid- rocky shore populations of *Littorina saxatilis* (Olivi) suggest selection in this enzyme locus. Genet. Res. 54: 7–11.

Johannesson, K. & B. Johannesson, 1990. Genetic variation within *Littorina saxatilis* (Olivi) and *Littorina neglecta* Bean: Is *L. neglecta* a good species? Hydrobiologia 193: 89–97.

Johannesson, K., B. Johannesson & U. Lundgren, 1995. Strong natural selection causes microscale allozyme variation in a marine snail. Proc. natn. Acad. Sci. U.S.A. 92: 2602–2606.

Knight, A. J., R. N. Hughes & R. D. Ward, 1987. A striking example of the founder effect in the mollusc *Littorina saxatilis*. Biol. J. linn. Soc. 32: 417–426.

Knight, A. J. & R. D. Ward, 1991. The genetic relationships of three taxa in the *Littorina saxatilis* species complex (Prosobranchia: Littorinidae). J. moll. Stud. 57: 81–91.

Nei, M., 1978. Estimation of heterozygosity and genetic distance from a small number of individuals. Genetics 89: 583–590.

O'Connell, M., J. M. Fives & P. Ó Céidigh, 1992. Ecological studies of littoral fauna and flora on Inishmore, Aran Islands, Co. Galway. Proc. R. I. A. 92B: 91–107.

Raymond, M. & F. Rousset, 1995. GENEPOP (version 1.2): A population genetics software for exact tests and ecumenicism. Users Manual (version 2). J. Hered. 86: 248–249.

Reid, D. G., 1996. Systematics and evolution of *Littorina*. The Ray Society, London. 463 pp.

Shaw, C. R. & R. Prasad, 1970. Starch gel electrophoresis of enzymes – a compilation of recipes. Biochem. Genet. 4: 297–320.

Sundberg, P., A. J. Knight & R. D. Ward & K. Johannesson, 1990. Estimating the phylogeny in the mollusc *Littorina saxatilis* (Olivi) from enzyme data: methodological considerations. Hydrobiologia 193: 29–40.

Swofford, D. L. & R. B. Selander, 1981. Biosys-1. A Computer Program for the Analysis of Allelic Variation in Genetics. Users Manual, 1.7. Department of Genetics and Development, University of Illinois, Urbana-Champaign, IL.

Ward, R. D. & K. Janson, 1985. A genetic analysis of sympatric subpopulations of the sibling species *Littorina saxatilis* (Olivi) and *Littorina arcana* Hannaford Ellis. J. moll. Stud. 51: 86–94.

Ward, R. D. & T. Warwick, 1980. Genetic differentiation in the molluscan species *Littorina rudis* and *Littorina arcana* (Prosobranchia: Littorinidae). Biol. J. linn. Soc. 14: 417–428.

Ward, R. D., T. Warwick & A. J. Knight, 1986. Genetic analysis of ten polymorphic enzyme loci in *Littorina saxatilis* (Prosobranchia: Mollusca). Heredity 57: 233–241.

Ward, R. D., T. Warwick & A. J. Knight, 1987. Further genetic analysis of polymorphic enzyme loci in *Littorina saxatilis* (Prosobranchia: Mollusca). Heredity 66: 151–158.

Wilkins, N. P. & D. O'Regan, 1980. Genetic variation in sympatric sibling species of *Littorina*. The Veliger 22: 355–359.

Wilkins, N. P., D. O'Regan & E. Moynihan, 1978. Electrophoretic variability and temperature sensitivity of phosphoglucose isomerase and phosphoglucomutase in littorinids and other marine molluscs. In B. Battaglia & J. A. Beardmore (eds), Marine Organisms: Genetics, Ecology and Evolution. Plenum Press, New York & London.

Wright, S., 1978. Variability Within and Among Natural Populations. Vol. 4, Evolution and the Genetics of Populations. University of Chicago Press, Chicago, 580 pp.

Hydrobiologia **378:** 11–19, 1998.
R. M. O'Riordan, G. M. Burnell, M. S. Davies & N. F. Ramsay (eds), Aspects of Littorinid Biology.

Electrophoretic heterogeneity within and between flat periwinkles (Mollusca: Gastropoda) along an intertidal transect at Ria Ferrol, northwest Spain

C. Olabarria[1], J.-M. Timmermans[2] & T. Backeljau[2,*]
[1] *Dept. Biology, Faculty of Biology, University of Santiago de Compostela, E-15706 Santiago de Compostela, Spain*
[2] *Royal Belgian Institute of Natural Sciences, Vautierstraat 29, B-1000 Brussels, Belgium*
(author for correspondence)*

Key words: esterases, Littorinidae, myoglobin, population genetics, Spain, systematics

Abstract

Using isoelectric focusing of esterases (EST), general proteins (GP) and myoglobin (Mb), we surveyed intra- and interspecific differentiation in flat periwinkles along a vertical intertidal transect in the Ensenada do Baño at Ria Ferrol, N.W. Spain. In this region, *L. obtusata* occurs in four algal belts, although it is rare in the lowest zone defined by *Fucus serratus*. *L. fabalis* is common in the *F. serratus* and *F. vesiculosus* belt, but is absent higher up on *Ascophyllum nodosum* and *F. spiralis*. Our data show that (1) EST and GP consistently differentiate between *L. obtusata* and *L. fabalis*, without however providing useful diagnostic markers, (2) *L. fabalis* is the less variable (heterozygous), but more heterogeneous species, (3) Mb patterns show significant heterogeneity in *L. obtusata* between the *F. serratus* zone and the other algal belts, but not in *L. fabalis*, and (4) the data on littorinid Mb appear inconsistent with a dimeric protein controlled by a single locus. Yet, assuming two loci coding for a monomeric (or dissociated dimeric) protein produces for the flat periwinkles a data set in which no significant deviations from Hardy-Weinberg expectations were detected. Nevertheless, this speculative interpretation fails to explain all littorinid Mb data. Hence the genetics and structure of littorinid Mb need further study.

Introduction

The closely related and highly similar flat periwinkles *Littorina obtusata* (Linnaeus, 1758) and *L. fabalis* (Turton, 1825) [formerly *L. mariae* Sacchi & Rastelli, 1966] are common and widely distributed in the eulittoral of the European Atlantic coasts, where they live sympatrically on fucoids (e.g. Reid, 1996). This obligatory macrophyte-association is unique within the genus *Littorina*, but at the same time differentiates both flat periwinkles since *L. obtusata* usually lives in the middle to upper eulittoral on *Ascophyllum nodosum* and *Fucus vesiculosus*, whereas *L. fabalis* prefers *F. serratus* in the lower to middle eulittoral. In addition, *L. fabalis* is an annual species that feeds on the micro-epiphytes growing on the macroalgae, while *L. obtusata* is a perennial species that lives upon the macrophytes themselves (e.g. Reid, 1996).

Besides these ecological differences, *L. obtusata* and *L. fabalis* show a number of other diagnostic features involving morphology, life history (both reviewed by Reid, 1996), and allozymes (reviewed by Tatarenkov, 1995a). Some of these features are subject to considerable geographic variation and interspecific overlap so that their value as species markers can vary between populations (Reid, 1996). This may be partly due to the fact that both species have direct development, without a planktonic dispersal stage. Under such conditions, gene flow between populations is expected to be insufficient to counteract differentiation provoked by genetic drift at neutral loci (e.g. Ward, 1990; Tatarenkov & Johannesson, 1994). In contrast, natural selection can impose substantial differentiation in any species, even those with high dispersal ability (e.g. Koehn et al., 1983; Johnson & Black, 1982, 1984; Hilbish, 1985, 1996).

Several allozyme studies have investigated the effects of drift and selection in *Littorina* spp., including flat periwinkles, by describing population differentiation and structuring at 'horizontal' micro- and macrogeographic scales (e.g. Vuilleumier & Matteo, 1972; Berger, 1977; Newkirk & Doyle, 1979; Janson & Ward, 1984; Janson, 1987a; Mill & Grahame, 1988; Dytham et al., 1992; Johannesson, 1992; Mill & Grahame, 1992; Tatarenkov & Johannesson, 1994; Tatarenkov, 1995b). Yet, relatively few papers deal with genetic differentiation between vertical intertidal zones (e.g. Johannesson et al., 1993, 1995a), although at least in rough periwinkles selection seems to maintain sharp allozyme frequency differences at an aspartate aminotransferase (AAT-1) locus between intertidal levels (Johannesson & Johannesson, 1989; Johannesson et al., 1995b).

Against this background we present here preliminary data on the population structure and differentiation of *L. obtusata* and *L. fabalis* along a vertical intertidal transect in N.W. Spain. We particularly aimed at (1) assessing whether the zonation pattern of both species could be correlated with electrophoretic variation in three protein markers for which a functional relationship with habitat, feeding and diet might be assumed, *viz.* esterases (EST), radular myoglobins (Mb) and general proteins (GP) (e.g. Berger et al., 1975; Oxford, 1975, 1978; Alyakrinskaya, 1989, 1994; Medeiros et al., this volume), (2) testing whether these markers support the allozyme-based observation that *L. obtusata* is the genetically more variable of the two flat periwinkles (e.g. Janson, 1987b; Zaslavskaya et al., 1992; Backeljau & Warmoes, 1992; Rolán-Alvarez et al., 1995; Tatarenkov, 1995a), and (3) evaluating to what extent these markers can be used to diagnose both species.

Because EST, Mb and GP are complex systems, requiring a high-resolution separation technique, we applied isoelectric focusing (IEF) as described by Backeljau et al. (1994) and Medeiros et al. (1998). The latter authors were also the first to report on the presence of Mb in *L. fabalis*.

Materials and methods

A total of 88 specimens of *L. obtusata* and 54 specimens of *L. fabalis* was collected along a vertical intertidal transect in the Ensenada do Baño at Ria Ferrol, N.W. Spain. The transect comprised four algal levels, with the uppermost zone defined by *F. spiralis*, and the three lower levels by, in descending order, *Ascophyllum nodosum, F. vesiculosus* and *F. serratus*. In each level an area of 0.5 m^2 was sampled. Specimens were transported alive to the laboratory in Brussels, where they were stored at –80 °C until prepared for electrophoresis. Specimens were identified by the penis morphology for the males (e.g. Sacchi & Rastelli, 1966; Warmoes et al., 1988; Reid, 1996) or the structure of the pallial oviduct for the females (Reid, 1990, 1996).

Individual tissue homogenates were prepared by crushing the shells and dissecting the radular muscles and digestive gland. Visibly parasitised animals were discarded. Tissues were kept separately and were homogenised in a 20% (w/v) aqueous sucrose solution (a fixed volume of 5 μl sucrose solution for radular muscles and a proportional volume of 5 μl sucrose solution per mg digestive gland tissue). Crude homogenates were centrifuged for 30 min at $\pm$19 000$\times$ $\boldsymbol{g}$ and at 4 °C. Digestive gland supernates were directly stored at –80 °C, radular muscle supernates were first further diluted by adding 4 μl sucrose solution per μl supernate.

Horizontal IEF in 3–9 pH gradients for digestive gland EST and 4–6.5 pH gradients for radular muscle proteins was performed as outlined by Medeiros et al. (1998). EST were stained with the recipe of Backeljau et al. (1994). Mb/GP were revealed by Coomassie and silver staining as described by Medeiros et al. (1998) and Backeljau et al. (1994), respectively.

EST and Mb/GP patterns were analysed by means of band counting and calculating Jaccard's matching coefficients (S_J) between adjacent protein profiles (Backeljau et al., 1994). S_J values were used to construct UPGMA trees with the program NTSYS (Rohlf, 1993). Mean numbers of protein bands per individual (N_B) were calculated and tested for intra- and interspecific differences with Kruskal-Wallis tests. Intraspecific band heterogeneity was assessed with the Shannon-Wiener index (H') as described by Mill & Grahame (1988, 1992).

Mb profiles were also used for a population genetic treatment involving allele frequency estimation, Hardy-Weinberg (HW) equilibrium testing with a

Table 1. Distribution of the Mb patterns (A–G) defined in Figure 3 over the algal belts in *L. obtusata* and *L. fabalis*. N = total number of specimens

Population	N	A	B	C	D	E	F	G
L. obtusata	68	2	54	7	1	1	2	1
F. spiralis	20	1	18	1	–	–	–	–
A. nodosum	21	1	19	–	–	1	–	–
F. vesiculosus	19	–	12	6	–	–	1	–
F. serratus	8	–	5	–	1	–	1	1
L. fabalis	40	5	30	5	–	–	–	–
F. vesiculosus	20	3	14	3	–	–	–	–
F. serratus	20	2	16	2	–	–	–	–
Total	108	7	84	12	1	1	2	1

Fisher exact test corrected for multiple testing by the sequential Bonferroni procedure (Rice, 1989), contingency chi-square analysis of allele frequency differences between populations, Cavalli-Sforza & Edwards' (1967) chord distance calculation and UPGMA tree reconstruction. All analyses were performed with the program BIOSYS (Swofford & Selander, 1981), except for the construction of the UPGMA dendrogram which was done with NTSYS.

Results

L. obtusata occurred in the four algal levels, even though it was rare on *F. serratus* ($N = 8$). *L. fabalis* was only found on *F. vesiculosus* and *F. serratus*. The overall EST differentiation between both species was obvious, but only the K-band was diagnostic (present in *L. obtusata*, absent in *L. fabalis*; Figure 1A). No other species-specific bands had a frequency of 100%.

The mean number of EST bands per individual was significantly larger *in L. obtusata* ($N_B = 10$) than in *L. fabalis* ($N_B = 9$) ($p < 0.05$), but within each species there were no significant differences between algal levels or sexes. Yet, *L. fabalis* was more heterogeneous ($H' = 3.58$) than *L. obtusata* ($H' = 2.57$). This was also reflected by the lower intraspecific S_J value *of L. fabalis* ($S_J = 0.55$) compared to that of *L. obtusata* ($S_J = 0.70$). However, UPGMA clustering of the interpopulation S_J values separated the two species (Figure 2A).

The silver stained Mb/GP profiles yielded only two diagnostic bands, viz. band F for *L. fabalis* and band O for *L. obtusata* (Figure 1B). However, some rarer bands were also species-specific. The mean number of Mb/GP bands per individual was significantly larger in *L. obtusata* ($N_B = 22$) than in *L. fabalis* ($N_B = 18$) ($p < 0.05$), but within species there were no significant differences between algal levels or sexes. Because the Mb/GP profiles were too complex, no H' values were calculated. Yet, the mean S_J value between individuals of *L. fabalis* ($S_J = 0.61$) was lower than the corresponding value for *L. obtusata* ($S_J = 0.71$). Once again clustering of the interpopulation S_J values separated both flat periwinkles (Figure 2B). Hence the Mb/GP and EST results were highly consistent, even though they yielded different topologies for the *L. obtusata* populations (Figure 2A–B).

The Mb profiles revealed five bands (a–e) and seven patterns (A–G) (Figure 3). Pattern D (one individual of *L. obtusata* from *F. serratus*) consisted of a single band. All other profiles comprised two, three or four bands (Figure 3). The distribution of Mb patterns (Table 1) shows that (1) both species share profiles A–C, with the two-banded B profile being the most common one, (2) profiles D–G are rare (frequencies of less than 3% in *L. obtusata*) and do not occur in *L. fabalis*, and (3) the proportion of rare profiles is highest in *L. obtusata* from *F. serratus* (38%).

Our tentative genetic interpretation of the Mb profiles assumes a monomeric protein coded by two loci (MbE and MbA), one of which (MbE) possibly carries a null allele (Figure 3). With this interpretation and after sequential Bonferroni correction, none of the 12 HW tests (two loci in six populations) showed significant differences between the observed and the expected genotype frequencies (Table 2).

The chi-square contingency table analysis of the allele frequencies over both loci jointly, revealed no heterogeneity among the two *L. fabalis* populations ($p = 0.715$), but a highly significant heterogeneity among the four *L. obtusata* populations ($p < 0.001$). However, this heterogeneity disappeared after excluding the *L. obtusata* population from *F. serratus* ($p = 0.116$). The differentiation of *L. obtusata* from *F. serratus* also followed from the UPGMA tree of Cavalli-Sforza & Edwards' (1967) chord distances (Table 3), which placed this population at the basis of the tree, far away from all other populations (Figure 4). The same analysis clustered both periwinkles from the *F. vesiculosus* belt as closest (i.e. most similar) pair. These results were not affected by the in- or exclusion of the specimen with the supposed null allele (pattern D).

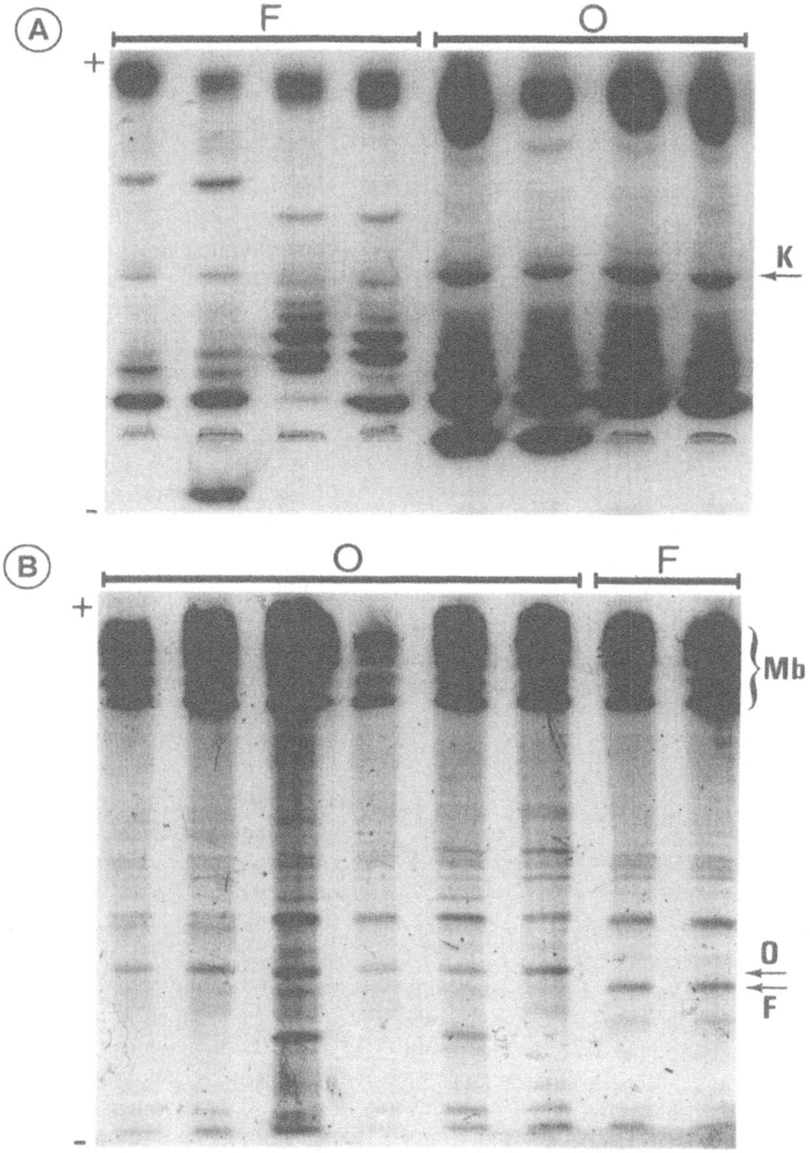

Figure 1. IEF patterns of *L. obtusata* (O) and *L. fabalis* (F). A. Esterases in a 3–9 pH gradient (the diagnostic K-band is indicated with an arrow). B. Silver stained radular GP in a 4–6.5 pH gradient (the diagnostic O and F bands are indicated by arrows; the heavily stained anodal zone is Mb).

Discussion

Although the EST and silver stained GP profiles differentiated *L. obtusata* from *L. fabalis* (Figures 1–2), both protein systems yielded only a very limited number of diagnostic bands, whose applicability in other areas is still uncertain. Therefore, IEF of EST and GP is not a useful tool to identify flat periwinkles. Nevertheless, our EST and GP results provide a new argument to support that *L. obtusata* and *L. fabalis* are well-defined, but closely related species, as was previously shown by several allozyme studies (reviewed by

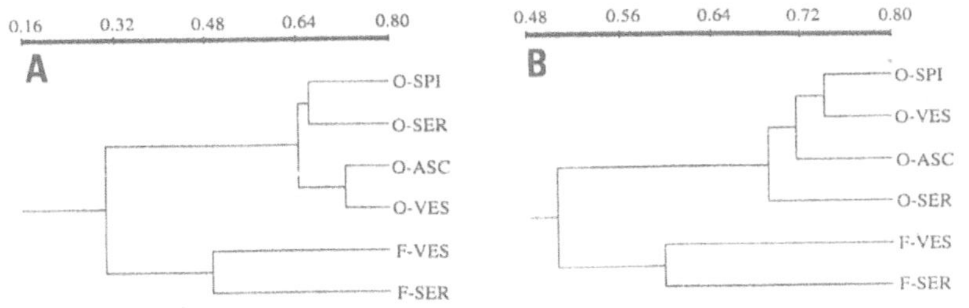

Figure 2. UPGMA dendrograms derived from mean interpopulation S_J values between *L. obtusata* and *L. fabalis* based on EST (A) and GP (B). Populations are identified by a species prefix (O = *L. obtusata*; F = *L. fabalis*) followed by: ASC = *Ascophyllum nodosum*; SER = *Fucus serratus*; SPI = *Fucus spiralis*; VES = *Fucus vesiculosus*.

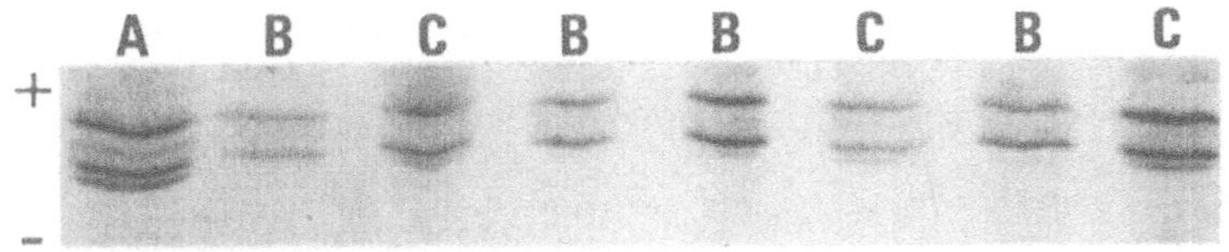

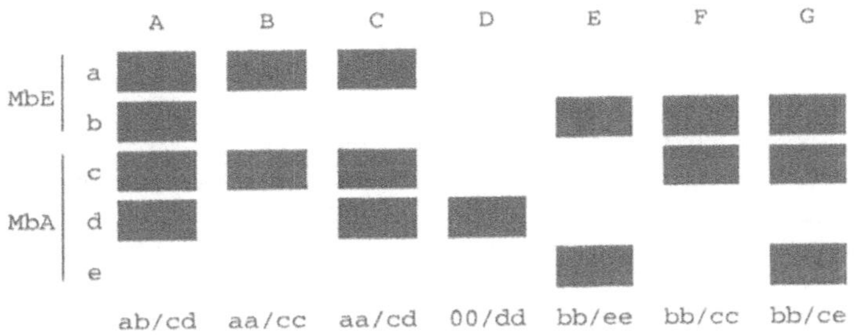

Figure 3. Examples and tentative interpretation of Mb profiles of *L. obtusata* and *L. fabalis* as resolved by IEF in a 4–6.5 pH gradient and stained by Coomassie Brilliant Blue R-250. Profiles are indicated by capitals (A–G), alleles by lower case letters (a–e; 0 for the null allele).

Tatarenkov, 1995a). Similarly, because protein band numbers are related to heterozygosity levels (the more heterozygous an individual is, the more bands it is expected to reveal), our EST and GP data also support the observation that *L. obtusata* tends to be more heterozygous than *L. fabalis* (Janson, 1987b; Zaslavskaya et al., 1992; Backeljau & Warmoes, 1992; Rolán-Alvarez et al., 1995; Tatarenkov, 1995a). Yet, the higher H' value for EST in *L. fabalis*, as well as the lower EST and GP similarity among the two *L. fabalis* populations, suggest that this species is more heterogeneous than *L. obtusata*. This agrees in a way with the lower F_{ST} values of *L. obtusata* compared to *L. fabalis* (Rolán-Alvarez et al., 1995). The lower heterozygosity, but stronger intraspecific differentiation of *L. fabalis* has been related to the species' annual life cycle, resulting in a smaller effective population size (e.g. Rolán-Alvarez et al., 1995; Reid, 1996).

In contrast, the Mb profiles did not differentiate between *L. obtusata* and *L. fabalis*, but revealed some remarkable population characteristics. The high incidence of rare Mb variants in *L. obtusata* from the *F. serratus* belt, perhaps reflects that this is a marginal environment for *L. obtusata* (e.g. Reid, 1996). On the other hand, the high Mb similarity between *L. obtusata* and *L. fabalis* in the *F. vesiculosus* belt (Figure 4), possibly indicates a similar response to an environment in which both species are common and show maximum overlap. Clearly, it remains to be investigated whether these Mb zonation patterns persist in time and space, and if so, whether they are adaptive (i.e. the result of selection) or ecophenotypic (i.e. the result of any non-selection based factor like gene regulation, post-translational modifications, etc.).

Such investigations require a correct interpretation of the Mb profiles, but Medeiros et al. (1998) have remarked that the genetic background of littorinid Mb is

Table 2. Allele frequencies, expected and observed heterozygosities (H_e, H_o) and exact probabilities of conformance between observed and HW expected genotype frequencies (p_{exact}) at two supposed Mb loci (MbE and MbA) in *L. obtusata* (O) and *L. fabalis* (F). Significance level after sequential Bonferroni correction: 0.004. Alleles and loci are defined in Figure 3. Abbreviations as in Figure 2 but with SED = *F. serratus* without the specimen with the D pattern

Locus		O-SPI	O-ASC	O-VES	O-SER	O-SED	F-VES	F-SER
MbE								
	N	20	21	19	8	7	20	20
	a	0.975	0.929	0.947	0.625	0.714	0.925	0.950
	b	0.025	0.071	0.053	0.250	0.286	0.075	0.050
	0	–	–	–	0.125	–	–	–
H_e		0.050	0.136	0.102	0.567	0.440	0.142	0.097
H_o		0.050	0.048	–	–	–	0.150	0.100
p_{exact}		1.000	0.073	0.027	0.007	0.021	1.000	1.000
MbA								
	N	20	21	19	8	7	20	20
	c	0.950	0.928	0.842	0.688	0.786	0.850	0.900
	d	0.050	0.024	0.158	0.125	–	0.150	0.100
	e	–	0.048	–	0.188	0.214	–	–
H_e		0.097	0.138	0.273	0.508	0.363	0.262	0.185
H_o		0.100	0.048	0.316	0.125	0.143	0.300	0.200
p_{exact}		1.000	0.073	1.000	0.077	0.231	1.000	1.000

Table 3. Cavalli-Sforza & Edwards' (1967) chord distances between populations of *L. obtusata* and *L. fabalis*, based on the data of Table 2. Abbreviations as in Figure 2

	O-SPI	O-ASC	O-VES	O-SER	F-VES	F-SER
O-SPI	0.000					
O-ASC	0.115	0.000				
O-VES	0.089	0.150	0.000			
O-SER	0.320	0.252	0.295	0.000		
F-VES	0.094	0.145	0.021	0.286	0.000	
F-SER	0.053	0.124	0.039	0.298	0.041	0.000

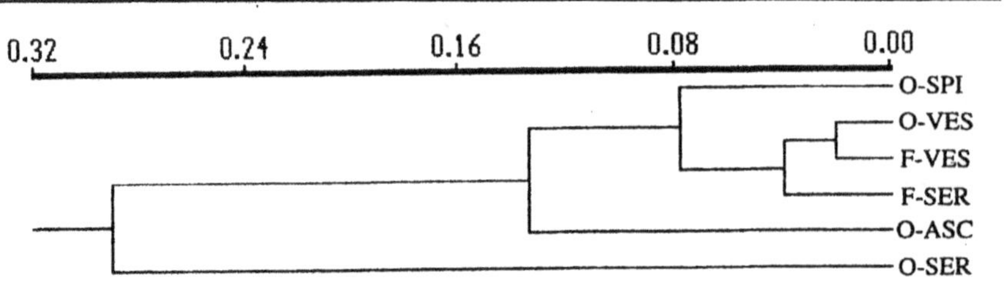

Figure 4. UPGMA dendrogram derived from Cavalli-Sforza & Edwards' (1967) chord distances between populations of *L. obtusata* and *L. fabalis*, based on the data of Table 3. Abbreviations as in Figure 2.

unknown. Yet, in rough periwinkles Wium-Andersen (1970) postulated a single Mb locus with two alleles, producing two-banded heterozygotes, thus implying a monomeric protein. With this model, Wium-Andersen (1970) found good fits between observed and HW expected genotype frequencies in the 13 populations studied. However, Wium-Andersen (1970) also observed that *L. littorea* and *L. striata* were monomorphic for a two-banded Mb profile that differed between both species, but was shared between *L. littorea* and the rough periwinkle. These two-banded patterns are difficult to reconcile with the Mendelian variation observed in the rough periwinkle, unless (1) *L. littorea* and *L. striata* show fixed heterozygosity at the Mb locus, (2) both species have a different genetic basis of Mb expression, and/or (3) the quaternary structure of Mb is different in *L. littorea* and rough periwinkles. In this context Read (1968) and Terwilliger & Read (1969) reported that the Mb of *L. littorea* is a dimer that at low concentrations mainly exists as a dissociated monomer. This situation could account for Wium-Andersen's (1970) interpretation, if it were not that a fixed heterozygosity in *L. littorea* (and *L. striata*) Mb is very unlikely. It would suggest namely an extreme selection against Mb homozygotes in a wide-ranging, high-dispersal species which shows considerable distributional and ecological overlap with the rough periwinkles that have the same Mb alleles.

Despite the fact that our interpretation of flat periwinkle Mb as a monomer coded by two loci is also at variance with the dimeric Mb structure of *L. littorea* (Read, 1968; Terwilliger & Read, 1969), it is possible that our experimental conditions may have dissociated the Mb (e.g. as a consequence of the dilution of the samples). On the other hand, if dissociation is not complete, heterozygotes of a dimeric protein controlled by a single locus, are expected to produce three equally-distant bands, a pattern that was neither observed by Wium-Andersen (1970), nor by us. Therefore we suspect that, if littorinid Mbs are truly dimeric, they must have been completely dissociated in the electrophoretic analyses hitherto published. Otherwise we have to assume that Mb quaternary structures may vary between littorinid species or that still other phenomena are involved. In this respect littorinid Mb is comparable with fish haemoglobins and crustacean haemocyanins, whose complex multi-band profiles may be very different between closely related species and/or may be related to environmental conditions (e.g. Bonaventura et al., 1975; Mangum, 1996; Mangum & Greaves, 1996; Mangum & McKenney, 1996). Yet, just as for crustacean haemocyanins, it is currently impossible to provide a reliable and consistent structural and genetic model to account for the littorinid Mb data observed by us and previous authors. Hence our genetic interpretation is highly speculative and is at least flawed by the assumption of a null allele [null alleles are usually considered to be rare (e.g. Buth, 1990), but are not infrequent in bivalves (e.g. Gaffney, 1994)], as well as by the arbitrary assignment of Mb bands to alleles and loci (Figure 3). Moreover, our model does not consider the many artefactual factors that may be involved (e.g. Medeiros et al., this volume). For a more general account of problems that may occur while interpreting electrophoretic protein profiles we refer to Buth (1990). However, even in the likely event that our genetic interpretation is incorrect, none of the HW tests showed significant deviations (Table 2). Hence, in the absence of other genetic information, HW testing may be misleading if used as a means to assess the reliability of population genetic data in outcrossing species, i.e. conformance to HW conditions is not *a posteriori* proof of a correct scoring and interpretation of electrophoretic profiles.

In conclusion, our EST and GP data corroborate previous allozyme reports on the specific differentiation, genetic variability and heterogeneity of *L. obtusata and L. fabalis*. The Mb data, on the other hand, reveal a significant heterogeneity between the *F. serratus* and other algal belts in *L. obtusata*, but not in *L. fabalis*. This suggests that besides 'horizontal' microgeographic differentiation, *L. obtusata* may also show heterogeneity among vertical intertidal zones. Yet, the biological meaning of this heterogeneity is unknown, particularly since a consistent genetic interpretation of littorinid Mb is still lacking. Therefore, in depth studies of the genetics and quaternary structure of littorinid Mb are needed, even if a number of speculative genetic models can be advanced to explain Mb variation in single species.

Acknowledgements

We are indebted to two anonymous referees for their constructive comments. This research was supported by the MAST 3 programme of the European Commission under contract number MAS3-CT95-0042 (AMBIOS). C. Olabarria was supported by the Xunta de Galicia.

References

Alyakrinskaya, I. O., 1989. Biochemical adaptations of some littoral molluscs to feeding. Zh. Evol. Biokhim. Fisiol. 25: 112–115. [in Russian].

Alyakrinskaya, I. O., 1994. Haemoglobin concentration and its functional meaning in White Sea *Littorina*. Gidrobiol. Zh. 30: 69–78. [in Russian].

Backeljau, T. & T. Warmoes, 1992. The phylogenetic relationships of ten Atlantic littorinids assessed by allozyme electrophoresis. In J. Grahame, P. J. Mill & D. G. Reid (eds), Proceedings of the Third International Symposium on Littorinid Biology. The Malacological Society of London, London: 9–24.

Backeljau, T., K. Breugelmans, H. Leirs, T. Rodriguez, D. Sherbakov, T. Sitnikova, J.-M. Timmermans, J. L. Van Goethem & E. Verheyen, 1994. Application of isoelectric focusing in molluscan systematics. The Nautilus, Suppl. 2: 156–167.

Berger, E. M., 1977. Gene-enzyme variation in three sympatric species of *Littorina*. II. The Roscoff populations, with a note on the origin of North American *L. littorea*. Biol. Bull. 153: 255–264.

Berger, V. Y., A. N. Pakhomov & A. G. Mukhlenov, 1975. Study of isozyme composition of esterases and lactate dehydrogenase during adaptation of *Littorina littorea* (L.) (Gastropoda) to the changes in water salinity. Zh. Obsh. Biol. 36: 579–584. [in Russian].

Bonaventura, J., C. Bonaventura & B. Sullivan, 1975. Hemoglobins and hemocyanins: comparative aspects of structure and function. J. exp. Zool. 194: 155–174.

Buth, D. G., 1990. Genetic principles and the interpretation of electrophoretic data. In D. H. Whitmore (ed.), Electrophoretic and Isoelectric Focusing Techniques in Fisheries Management. CRC Press, Boca Raton: 1–21.

Cavalli-Sforza, L. L. & A. W. F. Edwards, 1967. Phylogenetic analysis: models and estimation procedures. Evolution 21: 550–570.

Dytham, C., J. Grahame & P. J. Mill, 1992. Esterase variation in *Littorina arcana* and *L. saxatilis* at Robin Hood's bay, Yorkshire. In J. Grahame, P. J. Mill & D. G. Reid (eds), Proceedings of the Third International Symposium on Littorinid Biology. The Malacological Society of London, London: 39–44.

Gaffney, P. M., 1994. Heterosis and heterozygote deficiencies in marine bivalves: more light? In Beaumont, A. R. (ed.), Genetics and Evolution of Aquatic Organisms. Chapman & Hall, London: 146–153.

Hilbish, T. J., 1985. Demographic and temporal structure of an allele frequency cline in the mussel *Mytilus edulis*. Mar. Biol. 86: 163–171.

Hilbish, T. J., 1996. Population genetics of marine species: interaction of natural selection and historically differentiated populations. J. exp. mar. Biol. Ecol. 200: 67–83.

Janson, K., 1987a. Genetic drift in small and recently founded populations of the marine snail *Littorina saxatilis*. Heredity 58: 31–37.

Janson, K., 1987b. Allozyme and shell variation in two marine snails (*Littorina*, Prosobranchia) with different dispersal abilities. Biol. J. linn. Soc. 30: 245–256.

Janson, K. & R. D. Ward, 1984. Microgeographic variation in allozyme and shell characters *in Littorina saxatilis* Olivi (Prosobranchia: Littorinidae). Biol. J. linn. Soc. 22: 289–307.

Johannesson, K., 1992. Genetic variability and large scale differentiation in two species of littorinid gastropods with planktotrophic development, *Littorina littorea* (L.) and *Melarhaphe* (*Littorina*) *neritoides* (L.) (Prosobranchia: Littorinacea), with notes on a mass occurrence of *M. neritoides* in Sweden. Biol. J. linn. Soc. 47: 285–299.

Johannesson, K. & B. Johannesson, 1989. Differences in allele frequencies of *Aat* between high- and mid-rocky shore populations of *Littorina saxatilis* (Olivi) suggest selection at this enzyme locus. Genet. Res. Cambridge 54: 7–11.

Johannesson, K., B. Johannesson & E. Rolán-Alvarez, 1993. Morphological differentiation and genetic cohesiveness over a microenvironmental gradient in the marine snail *Littorina saxatilis*. Evolution 47: 1770–1787.

Johannesson, K., E. Rolán-Alvarez & A. Ekendahl, 1995a. Incipient reproductive isolation between two sympatric morphs of the intertidal snail *Littorina saxatilis*. Evolution 49: 1180–1190.

Johannesson, K., B. Johannesson & U. Lundgren, 1995b. Strong natural selection causes microscale allozyme variation in a marine snail. Proc. Natl. Acad. Sci. U.S.A. 92: 2602–2606.

Johnson, M. S. & R. Black, 1982. Chaotic genetic patchiness in an intertidal limpet, *Siphonaria* sp. Mar. Biol. 70: 157–164.

Johnson, M. S. & R. Black, 1984. Pattern beneath the chaos: the effect of recruitment on genetic patchiness in an intertidal limpet. Evolution 38: 1371–1383.

Koehn, R. K., A. J. Zera & J. G. Hall, 1983. Enzyme polymorphism and natural selection. In M. Nei & R. K. Koehn (eds), Evolution of Genes and Proteins. Sinauer, Sunderland, Massachusetts: 115–136.

Mangum, C. P., 1996. Subunit composition of polymorphic hemocyanins in the decapod crustaceans: differences between sibling species. Physiol. Zool. 69: 568–585.

Mangum, C. P. & J. Greaves, 1996. Hemocyanins of the genus *Uca*: structural polymorphisms and native oligomers. J. exp. mar. Biol. Ecol. 199: 1–15.

Mangum, C. P. & A. L. McKenney, 1996. Subunit composition of the crustacean hemocyanins: divergence in incipient speciation. Biol. Bull. 191: 33–41.

Medeiros, R., L. Serpa, C. Brito, H. De Wolf, K. Jordaens, B. Winnepenninckx & T. Backeljau, 1998. Radular myoglobin and protein variation within and among some littorinid species (Mollusca: Gastropoda). Hydrobiologia 378: 43–51.

Mill, P. J. & J. Grahame, 1988. Esterase variability in the gastropods *Littorina saxatilis* (Olivi) and *L. arcana* Ellis. J. moll. Stud. 54: 347–353.

Mill, P. J. & J. Grahame, 1992. Clinal changes in esterase variability in *Littorina saxatilis* (Olivi) and *L. arcana* Hannaford Ellis in southern Britain. In J. Grahame, P. J. Mill & D. G. Reid (eds), Proceedings of the Third International Symposium on Littorinid Biology. The Malacological Societyof London, London: 31–38.

Newkirk, G. F. & R. W. Doyle, 1979. Clinal variation at an esterase locus in *Littorina saxatilis* and *L. obtusata*. Can. J. Genet. Cytol. 21: 505–513.

Oxford, G. S., 1975. Food induced esterase phenocopies in the snail *Cepaea nemoralis*. Heredity 35: 361–370.

Oxford, G. S., 1978. The nature and distribution of food-induced esterases in helicid snails. Malacologia 17: 331–339.

Read, K. R. H., 1968. The myoglobins of the gastropod molluscs *Busycon contrarium* Conrad, *Lunatia heros* Say, *Littorina littorea* L. and *Siphonaria gigas* Sowerby. Comp. Biochem. Physiol. 25: 81–94.

Reid, D. G., 1990. Note on the discrimination of females of *Littorina mariae* Sacchi & Rastelli and *L. obtusata* (Linnaeus). J. moll. Stud. 56: 113–114.

Reid, D. G., 1996. Systematics and evolution of *Littorina*. The Ray Society, London, 463 pp.

Rice, W. R., 1989. Analyzing tables of statistical tests. Evolution 43: 223–225.

Rohlf, F. J., 1993. NTSYS-pc: numerical taxonomy and multivariate analysis system, version 1.80. Applied Biostatistics Inc., Setauket, New York.

Rolán-Alvarez, E., C. Zapata & G. Alvarez, 1995. Distinct genetic subdivision in sympatric and sibling species of the genus *Littorina* (Gastropoda: Littorinidae). Heredity 74: 1–9.

Sacchi, C. F. & M. Rastelli, 1966. *Littorina mariae*, nov. sp.: les differences morphologiques et écologiques entre 'nains' et 'normaux' chez l''espèce' *L. obtusata* (L.) (Gastr. Prosobr.) et leur signification adaptive et évolutive. Atti Soc. Ital. Sci. Nat. Mus. Civ. Stor. Nat. Milano 105: 351–369.

Swofford, D. L. & R. K. Selander, 1981. BIOSYS-1: a FORTRAN program for the comprehensive analysis of electrophoretic data in population genetics and systematics. J. Hered. 72: 281–283.

Tatarenkov, A. N., 1995a. Genetic divergence between sibling species *Littorina mariae* Sacchi & Rastelli and *L. obtusata* (L.) (Mollusca: Gastropoda) from the White Sea. Ophelia 40: 207–218.

Tatarenkov, A. N., 1995b. Genetic heterogeneity in populations of *Littorina brevicula* (Philippi) (Mollusca: Gastropoda) in the northern part of Peter the Great Bay (Sea of Japan). Veliger 38: 85–91.

Tatarenkov, A. & K. Johannesson, 1994. Habitat related allozyme variation on a microgeographic scale in the marine snail *Littorina mariae* (Prosobranchia: Littorinacea). Biol. J. linn. Soc. 53: 105–125.

Terwilliger, R. C. & R. H. Read, 1969. The radular muscle myoglobins of the amphineuran mollusc, *Acanthopleura granulata* Gmelin. Comp. Biochem. Physiol. 29: 551–560.

Vuilleumier, F. & M. B. Matteo, 1972. Esterase polymorphisms in European and American populations of the periwinkle *Littorina littorea* (Gastropoda). Experientia 28: 1241–1242.

Ward, R. D., 1990. Biochemical genetic variation in the genus *Littorina* (Prosobranchia: Mollusca). Hydrobiologia 193: 53–69.

Warmoes, T., T. Backeljau & L. De Bruyn, 1988. The littorinid fauna of the Belgian coast (Mollusca, Gastropoda). Bull. Inst. r. Sci. nat. Belg. 58: 51–70.

Wium-Andersen, G., 1970. Haemoglobin and protein variation in three species of *Littorina*. Ophelia 8: 267–273.

Zaslavskaya, N. I., S. O. Sergievsky & A. N. Tatarenkov, 1992. Allozyme similarity of Atlantic and Pacific species of *Littorina* (Gastropoda: Littorinidae). J. moll. Stud. 58: 377–384.

Hydrobiologia **378:** 21–25, 1998.
R. M. O'Riordan, G. M. Burnell, M. S. Davies & N. F. Ramsay (eds), *Aspects of Littorinid Biology.*

A preliminary study on genetic differentiation in *Littorina saxatilis* from Galway Bay, Ireland: *Littorina tenebrosa* Montagu – a valid species or ecotype?

E. M. Gosling[1], I. F. Wilson[1] & J. Andrews[2]
[1] *School of Science, Regional Technical College, Galway and Fermentation Laboratory, Microbiology Department, University College Galway, Ireland*
[2] *School of Biological Sciences, University College of Swansea, Swansea SA2 8PP, U.K.*

Key words: Littorina saxatilis, L. tenebrosa, allozymes, genetic differentiation, genetic distance

Abstract

Littorina tenebrosa is a small fragile-shelled periwinkle which lives on permanently submerged algae in coastal lagoons and non-tidal brackish pools. This periwinkle is a member of the rough periwinkle group which also comprises *Littorina saxatilis*, *L. arcana*, *L. compressa* and *L. neglecta* and is most closely related to *L. saxatilis* although its exact systematic status is in some doubt. Based on its unique ecology many believe *L. tenebrosa* to be a valid species. However, shell morphometric and allozyme analyses on Scottish and Swedish populations of *L. tenebrosa* and *L. saxatilis* have indicated that the two periwinkles are virtually identical. Preliminary results on five allozyme loci (AAT-1, GPI, PGM-1, PGM-2 and PNP) in samples of *L. tenebrosa* and *L. saxatilis* from Golam Head, Lettermullen, and other locations on the west coast of Ireland show *L. tenebrosa* to be genetically differentiated from *L. saxatilis*. At Golam Head, where opportunities for gene flow occur between the two taxa, *L. tenebrosa* is as genetically differentiated from local *L. saxatilis* as it is from *L. saxatilis* from more geographically distant locations.

Introduction

Rocky shore populations of rough periwinkles are presently thought to consist of four species, often referred to as the *saxatilis* complex: *Littorina saxatilis* (Olivi), *L. compressa* Jeffreys, *L. arcana* Hannaford Ellis and *L. neglecta* Bean. Species separation is on the basis of shell characteristics and features of the soft parts, in particular the male and female reproductive systems (Reid, 1996). The species are closely related and interspecific crosses between some species pairs produce fertile hybrids (Ward et al., 1986). Starch gel electrophoresis of enzymes has revealed no diagnostic alleles to date although Random Amplified Polymorphic DNA analysis has been successfully used to separate UK populations of *L. saxatilis, L. arcana* and *L. compressa* (Crossland et al., 1993, 1996).

Some authors have suggested that *Littorina tenebrosa* Montagu should also be included in the *saxatilis* complex (Fretter & Graham, 1980; Smith, 1982). *L. tenebrosa* is a small ovoviviparous periwinkle with a very fragile, smooth, black or dark brown shell. Unlike other members of the *saxatilis* group, *L. tenebrosa* does not live on the substrate, but is found at high densities (> 30 000 m^{-2}) on permanently submerged macrophytes in coastal lagoons or non-tidal brackish pools. Because of the rarity of the lagoon habitat *L. tenebrosa* is an uncommon and little studied winkle (Barnes, 1989).

Despite its unique ecology there are conflicting views as to whether *L. tenebrosa* is a valid species or simply an ecotype of *L. saxatilis*. Barnes (1993) and others believe that they are good species based on morphological and ecological differences between the two. However, shell morphometric and allozyme analyses of Scottish and Swedish populations of *L. tenebrosa* and *L. saxatilis* have shown them to be morphologically and genetically almost identical (Janson & Ward, 1985). These authors concluded that *L. tenebrosa* is an ecotype of *L. saxatilis*, rather than a species in its own right.

L. saxatilis is a highly variable species which occupies a wide range of habitats. On European shores, snails living only a few metres apart are strongly polymorphic for morphological characters such as size, shape, colour and ornamentation of the shell, as well as embryo numbers (Johannesson et al., 1993, 1997; Johannesson & Johannesson, 1996; Hull et al., 1996). Limited dispersal together with habitat-specific selection are the main reasons cited for the observed heterogeneity. Also, individuals sharing the same morph are more genetically similar to each other than to alternative morphs, thus indicating a partial barrier to gene flow between the morphs (Johannesson et al., 1993).

Since Janson and Ward's (1985) paper there have been no further genetic comparisons of *L. tenebrosa* and *L. saxatilis*. Therefore, the aim of our study was to carry out a preliminary genetic analysis of *L. tenebrosa* from a lagoon at Golam Head, Co. Galway and to compare *L. tenebrosa* there with *L. saxatilis* from the same location, and from other shores in the Galway Bay area. Unfortunately, the lagoon at Golam Head is the only confirmed site for *L. tenebrosa* on Irish shores (Fretter & Graham, 1980: Figure 214 and S. Smith, pers. comm., 1996) and therefore, it was not possible to collect more than this one sample of *L. tenebrosa*.

Table 1. Allele frequencies in samples of *Littorina tenebrosa* and *L. saxatilis* from Galway Bay. $\bar{H}$, mean heterozygosity per locus (unbiased estimate); T, *L. tenebrosa*; S, *L. saxatilis*; s, sheltered gully; e, exposed; GH, Golam Head; BH, Baile na hAbhann; Asterisks indicate significant deviation from Hardy-Weinberg equilibrium, *, $p =< 0.05$; **, $p =< 0.01$

	Population				
		GH		BH	ARAN
Locus	T	S (s)	S (e)	S (e)	S (s)
AAT-1					
190	0.089	0.679	0.685	0.684	0.395
100	0.911	0.321	0.315	0.316	0.605
N	95	42	62	19	19
GPI					
110	0.230	0.282	0.410	0.342	0.125
100	0.633	0.500	0.377	0.395	0.600
80	0.137	0.218	0.213	0.263	0.275
N	98**	39	61	19**	20
PGM-1					
100	0.432	0.224	0.169	0.029	0.025
95	0.042	0.053	0.000	0.294	0.625
90	0.525	0.434	0.371	0.412	0.300
80	0.000	0.289	0.460	0.265	0.050
N	59	38	62	17	20
PGM-2					
100	0.450	0.256	0.183	0.583	0.763
80	0.250	0.423	0.567	0.417	0.132
55	0.300	0.321	0.250	0.000	0.105
N	40	39	52*	18	19
PNP					
125	0.070	0.000	0.008	0.026	0.000
100	0.529	0.220	0.339	0.421	0.417
65	0.360	0.720	0.581	0.500	0.583
35	0.041	0.060	0.073	0.053	0.000
N	86*	41	62*	19	18
$\bar{H}$	0.495	0.571	0.569	0.580	0.497

Materials and methods

Samples were collected from Golam Head, Co. Galway (Irish National Grid Reference (INGR): L 826 214) which is an exposed rocky shore with the unique feature of a non-tidal marine lagoon (area 10,000 m^2) on the landward side of the shore. A sample of *L. saxatilis* was collected from large boulders and pools on the upper shore and from small rocks and pools in a sheltered gully leading to the lagoon; both sites were more than 50 m from the lagoon site. *L. tenebrosa* was collected from submerged *Chaetomorpha* sp. at the edge of the lagoon. The animals corresponded with those described by Fretter & Graham (1980) and with the form *Turbo tenebrosus* described by Reid (1996).

For comparison, additional samples of *L. saxatilis* were collected from an exposed site at Baile na hAbhann, (INGR L 992 202) about 18 km east of Golam Head, and from a sheltered site at Mainstir on Inismór (INGR L 873 105), one of the Aran Islands in Galway Bay, situated about 20 km south-west of Golam Head.

All samples were stored at –80 °C for a maximum period of five weeks prior to analysis. Because of its small size only the larger *L. tenebrosa* were

Table 2. Cavalli-Sforza & Edwards (1967) chord distance values between samples of *Littorina tenebrosa* and *L. saxatilis* from Galway Bay. Asterisks indicate levels of significance for pair-wise comparisons of allele frequencies; *, $p =< 0.05$; **, $p =< 0.01$; ***, $p < 0.001$. Mean $\bar{D}$ values across five loci are also included. Abbreviations as in Table 1

	AAT-1	GPI	PGM-1	PGM-2	PNP	$\bar{D}$
T × GH S (s)	0.415***	0.089	0.365***	0.142	0.285***	0.288
T × GH S (e)	0.420***	0.165***	0.483***	0.226***	0.183***	0.323
T × BH	0.419***	0.155	0.502***	0.366**	0.108	0.345
T × ARAN	0.238***	0.127	0.522*	0.211	0.241	0.299
GH S (s) × GH S (e)	0.005	0.091	0.178*	0.093	0.108	0.110
GH S (s) × BH	0.004	0.067	0.277*	0.396***	0.179	0.232
GH S (s) × ARAN	0.183*	0.126	0.461***	0.334***	0.201	0.288
GH S (e) × BH	0.001	0.049	0.392***	0.390***	0.076	0.251
GH S (e) × ARAN	0.188*	0.213*	0.604***	0.394***	0.185	0.356
BH × ARAN	0.187	0.174	0.251***	0.283*	0.182	0.220

selected for electrophoretic analysis and the whole animal was homogenised. In the case of *L. saxatilis* only the soft parts were used for analysis after discarding developing embryos.

Five loci, representing four enzymes, AAT (EC 2.6.1.1), GPI (EC 5.3.1.9), PGM (EC 5.4.2.2) and PNP (EC 2.4.2.1), were investigated using 0.1M Tris-maleic, pH 7.4 (Selander et al., 1971) for GPI and PGM and Tris-citrate, pH 8.1 (Ward & Warwick, 1980) for AAT and PNP. Loci were numbered from the anodal end of the gel except for PGM where there is a third faster locus (PGM-F in Janson & Ward, 1985). Allelic nomenclature follows Wilson & Gosling (1998). Whether variant alleles at loci in this paper are the same as those in, for example, Ward & Warwick (1980) remains to be seen.

Data were analysed using BIOSYS-1 (Swofford & Selander, 1981) and GENEPOP (Raymond & Rousset, 1995) for pair-wise comparisons. Genetic differentiation was assessed using the Wagner procedure (Farris, 1972) and the Cavalli-Sforza & Edwards (1967) chord distance measure. Sidák's multiplicative inequality equation was applied to correct for type-1 errors arising from multiple tests (Weir, 1990).

Results

Allele frequencies at the five polymorphic loci in *L. tenebrosa* and the four *L. saxatilis* samples are reported in Table 1. All alleles present in *L. saxatilis,* with the single exception of $PGM\text{-}1^{80}$, were also present in the *L. tenebrosa* sample. Similar levels of heterozygosity were observed in all samples (Table 1). Significant deficiencies of heterozygotes relative to Hardy-Weinberg expectations were observed in *L. tenebrosa* (GPI, $p < 0.01$; PNP, $p < 0.05$), in *L. saxatilis* from the exposed shore at Golam Head (PGM-2, $p < 0.05$; PNP, $p < 0.05$) and in *L. saxatilis* from Baile na hAbhann (GPI, $p < 0.01$).

In pair-wise comparisons, using a χ^2 contingency test, significant differences in allele frequencies at three or more loci were observed between *L. tenebrosa* and each sample of *L. saxatilis* (Table 2). Three loci, AAT-1, PGM-1 and PGM-2, contributed most to the genetic differentiation, although significant differences in allele frequencies for some pair-wise comparisons were also observed at GPI and PNP. It is noteworthy (Table 2 and Figure 1) that *L. tenebrosa* is as genetically different from the two samples of *L. saxatilis* from Golam Head ($\bar{D}$ = 0.288 and 0.323) as it is from the more geographically distant Baile na hAbhann ($\bar{D}$ = 0.345) and Aran Island ($\bar{D}$ = 0.299) samples.

Significant differences were also observed between samples of *L. saxatilis*. The two Golam Head samples were more similar to each other ($\bar{D}$ = 0.110) than they were to the Baile na hAbhann ($\bar{D}$ = 0.232 and 0.251) and Aran Islands ($\bar{D}$ = 0.288 and 0.356) samples (Table 2 and Figure 1).

Discussion

The results indicate that *L. tenebrosa* at Golam Head is genetically distinct from populations of *L. saxatilis*

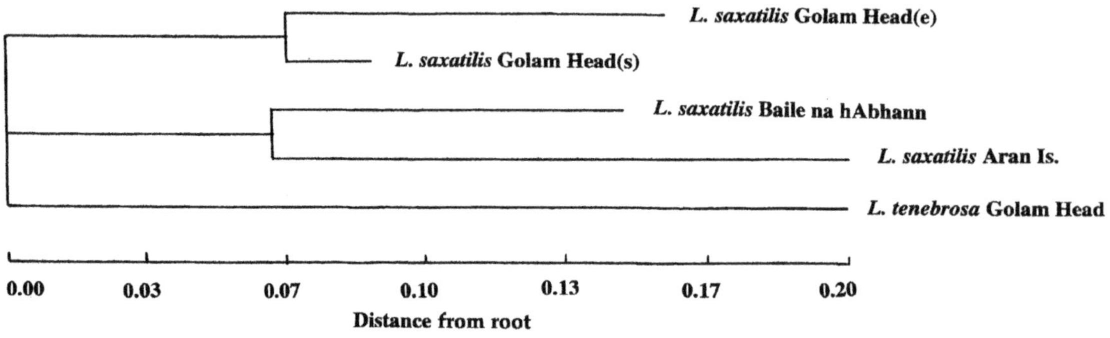

Figure 1. Wagner tree using Cavalli-Sforza & Edwards (1967) chord distance from allele frequencies at five polymorphic loci in populations of *Littorina tenebrosa* and *L. saxatilis* from Galway Bay, Ireland. (e), exposed, (s), sheltered.

around Galway Bay. Janson & Ward (1985), however, have shown Scottish and Swedish populations of *L. tenebrosa* to be genetically and morphologically similar to *L. saxatilis*. Their UPGMA cluster analysis showed populations of Swedish *tenebrosa* and *saxatilis* to be more similar to each other than either were to their Scottish counterparts, which also clustered together. Their mean genetic distance value based on 15 polymorphic loci (four of which were analysed in the present study) was 0.01, a value considerably smaller than that reported here. Janson & Ward (1985) concluded that, since there was no evidence of any intrinsic barrier to gene flow between *saxatilis* and *tenebrosa*, the latter should be regarded as an ecotype of *L. saxatilis*. There is, however, some doubt as to whether Janson & Ward (1985) actually sampled *L. tenebrosa* in their study. Their *L. tenebrosa* samples were not obtained from lagoons (i.e. animals were not permanently submerged), but from sediment, stones and fucoid algae and on or amongst *Zostera*, in sheltered marine inlets. Barnes (1993) points out that authors who have argued for the synonymy of the two species have used the name *tenebrosa* for a largely intertidal brackish/marine mudflat winkle, while all those who have argued for the validity of a separate *L. tenebrosa* have applied the name specifically to a winkle living permanently submerged in lagoons.

In our study significant differences in allele frequencies were also observed between samples of *L. saxatilis* (Table 2). Differentiation is not unexpected in an ovoviviparous species with a dispersal rate of a few metres per month (Johannesson & Johannesson, 1995). Not surprisingly, therefore, the largest genetic differentiation was observed between those samples of *L. saxatilis* which were more geographically distant from each other. In samples of *L. saxatilis* collected from Filey Brigg, UK, mean genetic distance based on seven loci (four of which have been analysed in our study) was 0.055 (Ward & Janson, 1985), while mean genetic distance between South African and North Atlantic populations was 0.203 (Knight et al., 1987).

What is unexpected in our study is the considerable genetic differentiation between *L. saxatilis* and *L. tenebrosa* at Golam Head. At this site there is ample opportunity for gene flow to occur between the two taxa through crawling individuals, and through contact with the sea on the spring tides and during storms. Despite opportunities for gene flow the two gene pools are significantly differentiated. There are several possible explanations for this phenomenon. We could be dealing with a single panmictic population of *L. saxatilis* which is subject to differential selection at the analysed (or closely-linked) loci in the pool and intertidal habitat; evidence for selection has been reported for a number of allozyme loci in the *saxatilis* group (reviewed by Reid, 1996). Alternatively, the significant differentiation may reflect partial or complete reproductive isolation between the gene pools of *L. tenebrosa* and *L. saxatilis* at this site.

This preliminary study does not resolve the systematic status of *L. tenebrosa*. We have after all examined only a small number of polymorphic genes in a single population of *L. tenebrosa*. Using additional allozyme loci and DNA analysis it is planned to extend our survey to other sites in Ireland (B. Healy, pers. comm. 1996) and the UK where putative *L. tenebrosa* has been reported, and where we may be able to throw further light on the systematics of this little-studied and intriguing periwinkle.

Acknowledgements

This research was supported by the MAST 3 programme of the European Commission under contract number MAS3-CT95-0042 (AMBIOS). Thanks to J. Patching for provision of facilities in the Microbiology Department and the Martin Ryan Institute, University College, Galway. The authors would also like to thank two anonymous reviewers for valuable comments and suggestions.

References

Barnes, R. S. K., 1989. The coastal lagoons of Britain: an overview and conservation appraisal. Biol. Cons. 49: 295–313.

Barnes, R. S. K., 1993. On the nature of the coastal lagoon winkles attributed to *Littorina tenebrosa* and *Littorina saxatilis*. Cah. Biol. mar. 34: 477–495.

Cavalli-Sforza, L. L. & A. W. F. Edwards, 1967. Phylogenetic analysis: models and estimation procedures. Am. J. hum. Genet. 19: 233–257.

Crossland, S., D. Coates, J. Grahame & P. J. Mill, 1993. Use of random amplified polymorphic DNAs (RAPDs) in separating two sibling species of *Littorina*. Mar. Ecol. Prog. Ser. 96: 301–305.

Crossland, S., D. Coates, J. Grahame & P. J. Mill, 1996. The *Littorina saxatilis* species complex – interpretation using random amplified polymorphic DNAs. In J. Taylor (ed.), Origin and Evolutionary Radiation of the Mollusca, Oxford University Press, Oxford: 205–209.

Farris, J. S., 1972. Estimating phylogenetic trees from distance matrices. Am. Nat. 106: 645–668.

Fretter, V. & A. Graham, 1980. The Prosobranch Molluscs of Britain and Denmark. V. Marine Littorinacea. J. moll. Stud. Suppl. 7: 243–283.

Hull, S. L., J. Grahame & P. J. Mill, 1996. Morphological divergence and evidence for reproductive isolation in *Littorina saxatilis* (Olivi) in northeast England. J. moll. Stud. 62: 89–99.

Janson, K. & R. D. Ward, 1985. The taxonomic status of *Littorina tenebrosa* Montagu as assessed by morphological and genetic analysis. J. Conch. 32: 9–15.

Johannesson, K. & B. Johannesson, 1995. Dispersal and population expansion in a direct developing marine snail (*Littorina saxatilis*) following a severe population bottleneck. Hydrobiologia 309: 173–180.

Johannesson, B. & K. Johannesson, 1996. Population differences in behaviour and morphology in the snail *Littorina saxatilis*: phenotypic plasticity or genetic differentiation? J. Zool., Lond. 240: 475–493.

Johannesson, K., B. Johannesson & E. Rolán-Alvarez, 1993. Morphological differentiation and genetic cohesiveness over a microenvironmental gradient in the marine snail *Littorina saxatilis*. Evolution 47: 1770–1787.

Johannesson, K., E. Rolán-Alvarez & J. Erlandsson, 1997. Growth rate differences between upper and lower shore ecotypes of the marine snail *Littorina saxatilis* (Olivi) (Gastropoda). Biol. J. linn. Soc. 61: 267–279.

Knight, A. J., R. N. Hughes & R. D. Ward, 1987. A striking example of the founder effect in the mollusc *Littorina saxatilis*. Biol. J. linn. Soc. 32: 417–426.

Raymond, M. & F. Rousset, 1995. A population genetics software for exact tests and ecumenicism. J. Hered. 86: 248–249.

Reid, D. G., 1996. Systematics and evolution of *Littorina*. The Ray Society, London. 463 pp.

Selander, R. K., M. H. Smith, S. Y. Yang, W. E. Johnson & J. R. Gentry, 1971. Biochemical polymorphism and systematics in the genus *Peromyscus*. I. Variation in the old-field mouse (*Peromyscus polionotus*). Studies in Genetics VI. University of Texas Publication 7103: 49–90.

Smith, S., 1982. A review of the genus *Littorina* in British and Atlantic waters (Gastropoda: Prosobranchia). Malacologia 22: 535–539.

Swofford, D. L. & R. K. Selander, 1981. BIOSYS-1: a FORTRAN program for the comprehensive analysis of electrophoretic data in population genetics and systematics. J. Hered. 72: 281–283.

Ward, R. D. & K. Janson, 1985. A genetic analysis of sympatric populations of the sibling species *Littorina saxatilis* (Olivi) and *Littorina arcana* Hannaford Ellis. J. moll. Stud. 51: 86–94.

Ward, R. D. & T. Warwick, 1980. Genetic diferentiation in the molluscan species *Littorina rudis* and *Littorina arcana* (Prosobranchia: Littorinidae). Biol. J. linn. Soc. 14: 417–428.

Ward, R. D., T. Warwick & A. J. Knight, 1986. Genetic analysis of ten polymorphic enzyme loci in *Littorina saxatilis* (Prosobranchia: Mollusca). Heredity 57: 233–241.

Weir, B. S., 1990. Genetic Data Analysis, Sinauer Associates, Sunderland, MA, USA, 377 pp.

Wilson, I. F. & E. M. Gosling, 1998. Genetic variability in *Littorina saxatilis* from different habitats on an island in Galway Bay. Hydrobiologia 378: 1–10.

Hydrobiologia **378:** 27–32, 1998.
R. M. O'Riordan, G. M. Burnell, M. S. Davies & N. F. Ramsay (eds), *Aspects of Littorinid Biology*.

Lack of significant esterase and myoglobin differentiation in the periwinkle, *Littorina striata* (Gastropoda, Prosobranchia)

H. De Wolf[1], T. Backeljau[2] & R. Verhagen[1]
[1] *Department of Biology, University of Antwerp (RUCA), Groenenborgerlaan 171, B-2020 Antwerp, Belgium*
[2] *Royal Belgian Institute of Natural Sciences, Vautierstraat 29, B-1000 Brussels, Belgium*

Key words: esterases, isoelectric focusing, Littorinidae, Macaronesia, myoglobin, population genetics

Abstract

The relationship between gene flow and the maintenance of geographic or morphology-related variation in the polymorphic Macaronesian periwinkle, *Littorina striata*, was investigated by means of isoelectric focusing of esterases (EST) and myoglobin (Mb). This revealed that: (1) individual EST variation is very high, (2) there is no EST differentiation between sexes, shell morphotypes or wave-exposure regimes, (3) there is no clear macrogeographic patterning of EST variability, although both Cape Verde Islands have the highest, with exception of Flores, mean number of EST bands, and (4) there is no Mb variation, not even between islands separated by more than 2000 km. These results indicate that *L. striata* shows a high degree of genetic homogeneity among geographic populations and that the morphological patterning in this species persists in the presence of intense gene flow.

Introduction

Littorina striata King & Broderip, 1832 is a periwinkle with planktonic development, which occurs in the eulittoral zone and the littoral fringe of Macaronesian rocky shores (Azores, Madeira, Canary Islands and Cape Verde Islands) (Reid, 1996). Like many other littorinids, *L. striata* displays a remarkable variation in shell size, shape and sculpture (Reid, 1996). Yet, despite the species' planktonic development, and thus supposedly high dispersal capacity (i.e. gene flow), it seems that at least part of its shell variability is not randomly distributed, but rather shows geographical/ecological correlates (Britton, 1995; Reid, 1996; De Wolf et al., 1997, 1998a). Similar sculptural variation has been described by Struhsaker (1968) in another planktonic developing littorinid *L. picta* (now *Nodilittorina hawaiiensis*) and has been shown to be genetically determined. This raises the question as to whether the patterning and functionality of shell variation in *L. striata* is adaptive (i.e. due to differential selection) or whether it is the result of some ecophenotypic effect (i.e. phenotypic plasticity).

The resolution of this question requires, amongst others, a population genetic analysis of the degree of gene flow and differentiation between populations and shell morphotypes at different spatio-temporal scales. Such analyses were performed by Backeljau et al. (1995) and De Wolf et al. (1998b), who applied native polyacrylamide gel electrophoresis (PAGE) to assess allozyme variation in *L. striata*. These data revealed neither a population genetic structuring other than a possible 'isolation by distance' model (i.e. gene flow between geographic populations was very high and only a function of distance), nor any genetic differentiation between shell morphotypes (De Wolf et al., 1994; De Wolf et al., 1998b). However, these results were based on a survey of only one highly polymorphic and four moderately to weakly polymorphic enzyme loci (34 other loci were monomorphic or not interpretable). Hence, the available allozyme data need support from other polymorphic markers in order to achieve a more reliable and representative genetic sampling.

In the present contribution we assess geographic and morphotype-related variation in *L. striata* with two additional markers, viz. non-specific esterases

(EST) and radular myoglobins (Mb). Backeljau et al. (1995) has shown that EST are highly variable in *L. striata*, so that this enzyme system could potentially provide the same kind of population genetic data as it did in other littorinids (e.g. Vuilleumier & Matteo, 1972; Gaines et al., 1974; Newkirk & Doyle, 1979; Mill & Grahame, 1988; Dytham et al., 1992; Mill & Grahame, 1992). In contrast, preliminary surveys of Mb in *L. striata* revealed hitherto no variation in this species (Wium-Andersen, 1970; Medeiros et al., 1998). Nevertheless, since these surveys were based on very limited sample sizes, and since Mb yielded interesting polymorphisms in other littorinids (Wium-Andersen, 1970; Medeiros et al., 1998; Olabarria et al., 1998), we decided that a more comprehensive analysis of Mb variation in *L. striata* could still be relevant.

As electrophoretic technique we preferred isoelectric focusing (IEF), because this method provides a better resolution of littorinid EST and Mb, than other techniques (Mill & Grahame, 1988; Dytham et al., 1992; Mill & Grahame, 1992; Backeljau et al., 1994; 1995; Medeiros et al., 1998).

Materials and methods

Three sets of experimental animal groups were surveyed. The first set comprised 240 specimens collected at Ilheu de Vila Franca do Campo (further referred to as 'Ilheu'), a drowned volcanic crater along the south coast of the island of São Miguel, Azores (Morton, 1990; De Wolf et al., 1997). At this site two morphotypes of *L. striata* occur microsympatrically, viz. one with a 'smooth' shell and one with a 'nodulous' shell (Reid, 1996; De Wolf et al., 1997). Ilheu furthermore comprises both very wave-exposed (outside of the crater) and very sheltered (lagoon inside the crater) rocky shores. Therefore, the material from this site was used for assessing microscale variation of EST between wave-exposure regimes (sheltered vs exposed), morphotypes (smooth vs nodulous) and sexes (males vs females).

A second series of samples consisted of 470 specimens from wave-exposed coasts in nine islands representing the four Macaronesian archipelagos, viz. the Azores ($N = 197$), Madeira ($N = 90$), the Canary Islands ($N = 37$) and the Cape Verde Islands ($N = 146$) (For detailed descriptions of sample sites see Table 1). This material was used for a macrogeographical survey of EST variation.

Finally, a third set of 160 individuals from the Azores (São Miguel: Ilheu de Vila Franca $N = 40$, Vila Franca $N = 40$) and the Cape Verde Islands (Sal: Jose Fonseca $N = 40$; São Vicente: Baia das Gatas $N = 40$) served to analyse Mb.

IEF sample preparation followed Backeljau et al., 1994, 1995) for EST and Medeiros et al. (1998) for Mb. IEF was performed with LKB-Pharmacia's PhastSystem® (Olsson et al., 1988; Backeljau et al., 1994), using 5–8 pH ranges for EST (running conditions in Table 2) or 3–9 pH ranges for Mb (Medeiros et al., 1998). EST were stained according to Backeljau et al. (1994) and Mb were revealed by Coomassie Brilliant Blue R-250 following Medeiros et al. (1998).

The EST profiles were analysed by means of simple band counting. Differences in mean numbers of bands between males (M) and females (F), between smooth (S) and nodulose (N) morphs and between wave exposed (We) and wave sheltered (Ws) specimens were analysed by means of a three-way analysis of variance (ANOVA) with unequal sample size (.We: $MN = 9$, $MS = 50$, $FN = 11$, $FS = 85$; Ws: $MN = 17$, $MS = 16$, $FN = 23$, $FS = 29$) and with all three factors being fixed.

Macrogeographic differences in mean EST band counts were tested, using a one factor ANOVA model (random factor = population) with unequal sample size (Table 1), followed by *post-hoc* Sheffé tests to evaluate pairwise population means.

Normality was analysed by means of a Kolmogorov-Smirnov test, while the Levenes test for homogeneity was used to test for variance homogeneity.

All statistical analyses were performed with the Statistica v. 5.0 (Statsoft, 1995) package. A 5% significance level was used throughout. The Mb profiles were compared qualitatively, without further statistical treatment.

Results

The EST profiles appeared to be highly variable (Figure 1), with numbers of bands ranging from 10 to 25 per individual. In all populations, except for Vila de Nordeste (São Miguel, Azores) ($p = 0.0047$), band counts were normally distributed. The statistical analysis of band count differences between sexes, morphotypes and wave-exposure regimes at Ilheu, showed that none of these factors was associated with a significant EST heterogeneity (Table 3). Hence, there were no significant differences in EST band

Table 1. Mean number of EST bands and standard deviations for each population, each island and each archipelago

Archipelago	Island	Population	*N*	Mean	Mean	Mean
Azores	Flores	Faia Grande	30	18.40		
	(FLO)			(2.99)		
	Faial	Caphelinhos	29	16.62		
	(FAI)			(3.42)		
	São Miguel	Ilheu	46	15.41		
	(SMI)			(2.82)		
		Vila Franca	32	18.31		
				(2.70)		
		Vila de Nordeste	30	15.20	16.21	
				(2.83)	(3.09)	
	Santa Maria	Anjos	30	17.56		16.81
	(SMA)			(3.85)		(3.28)
Madeira	Madeira	Caniço de Baixo	60	15.98		
	(MAD)			(2.78)		
	Porto Santo	Porto Santo	30	16.26		16.07
	(PSA)			(2.73)		(2.75)
Canary Is.	G. Canaria	Augustin	37	17.35		17.35
	(GCA)			(2.85)		(2.85)
Cape Verde	Sal	Josè Fonseca	45	17.75		
	(SAL)			(2.98)		
		Santa Maria	45	18.44	18.10	
				(2.09)	(2.58)	
	São Vicente	Baia das Gattas	31	18.91		
	(SVC)			(3.19)		
		Mindelo	25	18.12	18.16	18.12
				(2.99)	(3.07)	(2.77)

Table 2. Programmed conditions for IEF separations of Mb/GP in polyacrylamide gels (pH gradient = 5–8) using PhastSystem®

Sample applicator down at				1.2		0 Vh
Sample applicator up at				1.3		0 Vh
Extra alarm sound at				1.1		73 Vh
SEP 1.1	2000 V	2.0 mA	3.5 W		5 °C	75 Vh
SEP 1.2	200 V	2.0 mA	3.5 W		5 °C	15 Vh
SEP 1.3	2000 V	5.0 mA	3.5 W		5 °C	510 Vh

counts between sexes, morphotypes or wave-exposure regimes.

The macrogeographic EST analysis showed a significant difference between mean band numbers among the 13 populations compared ($p < 0.0001$; Table 4). However, this result was entirely due to a significant difference between one Cape Verde (Santa Maria) and two Azorean populations (Ilheu and Vila de Nordeste). None of the other pairwise comparisons yielded significant values (Table 5). Hence, there was no clear pattern of macrogeographic EST heterogeneity. Nevertheless, the highest mean numbers of EST bands were observed in Flores and at the two Cape Verde islands (Sal and São Vicente) (Table 1).

Because of the complexity of the EST profiles (Figure 1), it was impossible to assign the bands to

Table 3. Results of the three-way ANOVA, for possible microgeographical differences in mean number of EST bands, between wave exposure (1), sex (2) and morphotype (3)

Effect	df effect	MS effect	df error	MS error	F	p
1	1	5.8789	232	6.5502	0.8975	0.3444
2	1	18.1688	232	6.5502	2.8806	0.0909
3	1	4.2432	232	6.5502	0.6478	0.4217
1 × 2	1	24.5919	232	6.5502	3.7543	0.0539
1 × 3	1	7.0249	232	6.5502	1.0725	0.3615
2 × 3	1	14.0734	232	6.5502	2.1485	0.1441
1 × 2 × 3	1	0.0805	232	6.5502	0.0123	0.9118

Table 4. Results of the one-way ANOVA, for possible macrogeographical differences in EST variability

Effect	df effect	MS effect	df error	MS error	F	p
1	12	50.9103	456	8.2991	6.1344	<0.0001

specific loci or to characterise the bands in an unambiguous way. Consequently, no reliable similarity indices could be calculated.

In contrast to the EST, the Mb profiles of *L. striata* revealed no detectable variation or heterogeneity, not even between samples separated by as much as 2000 km. The Mb patterns were identical to those reported by Medeiros et al. (1998) and consisted of several Mb fractions with relatively low isoelectric points (Figure 2). Evidently, since Mb was revealed by Coomassie staining it is likely that the patterns we obtained, also include other, non-specific proteins (GP = general proteins) (Medeiros et al., 1998).

Discussion

The high degree of individual EST variation in *L. striata* observed during this work, is in line with the preliminary data of Backeljau et al. (1995). Similarly, the lack of EST differentiation between sexes, morphotypes and wave-exposure regimes is in complete agreement with the genetic homogeneity that *L. striata* shows at four other enzyme loci (De Wolf et al., 1998b). Yet, it contrasts with the sex-related EST differences reported in rough periwinkles (Dytham et al., 1992). Our EST data provide additional evidence for panmixis among morphotypes or wave-exposure regimes, thus reinforcing, the fact that shell variation in *L. striata* persists, despite the presence of a high degree of gene flow.

Also at macrogeographic scale our data suggest a high degree of EST homogeneity between *L. striata* populations (except for two comparisons). This is somewhat surprising as (macro)geographic differentiation and clinal variation have been commonly reported in other littorinids for which EST were surveyed and this for both species with or without planktonic dispersing larvae (e.g. Vuilleumier & Matteo, 1972; Gaines et al., 1974; Berger, 1977; Newkirk & Doyle, 1979; Mill & Grahame, 1988; Dytham et al., 1992; Mill & Grahame, 1992). Yet, the macrogeographic EST homogeneity in *L. striata* is to a large extent concordant with the results of a parallel population genetic analysis of five enzyme loci. This study suggested that although a number of unique, but rare, alleles separated the Cape Verde *L. striata* populations from all others, gene flow estimates always remained $Nm \gg 1$, so that population differentiation due to random genetic drift would be prevented (De Wolf et al., unpubl. data). However, the same allozyme work revealed a significant correlation between geographic latitude and the mean number of alleles per locus, such that numbers of alleles decreased with increasing latitude. The observation of the high mean number of EST bands at both Cape Verde Islands, fits into this picture.

Finally, the Mb data presented here suggest an extreme macrogeographic homogeneity and thus con-

Table 5. Results of *post hoc* Sheffé tests, for interpopulational comparisons of EST variability at a macrogeographical scale

SMI1												
SMI2	0.09											
SMI3	1.00	0.12										
SAM1	0.60	1.00	0.61									
FAI1	0.99	0.95	0.99	1.00								
FLO1	0.08	1.00	0.11	1.00	0.93							
MAD1	1.00	0.33	1.00	0.91	1.00	0.30						
PSA1	1.00	0.80	0.99	0.99	1.00	0.76	1.00					
GCA1	0.68	1.00	0.68	1.00	1.00	1.00	0.95	1.00				
SAL1	0.02	1.00	0.03	1.00	0.85	1.00	0.10	0.59	1.00			
SAL2	0.24	1.00	0.29	1.00	1.00	1.00	0.37	0.96	1.00	1.00		
SAV1	0.14	1.00	0.18	1.00	0.97	1.00	0.45	0.87	1.00	1.00	1.00	
SAV2	0.31	1.00	0.32	1.00	1.00	1.00	0.66	0.94	1.00	1.00	1.00	1.00

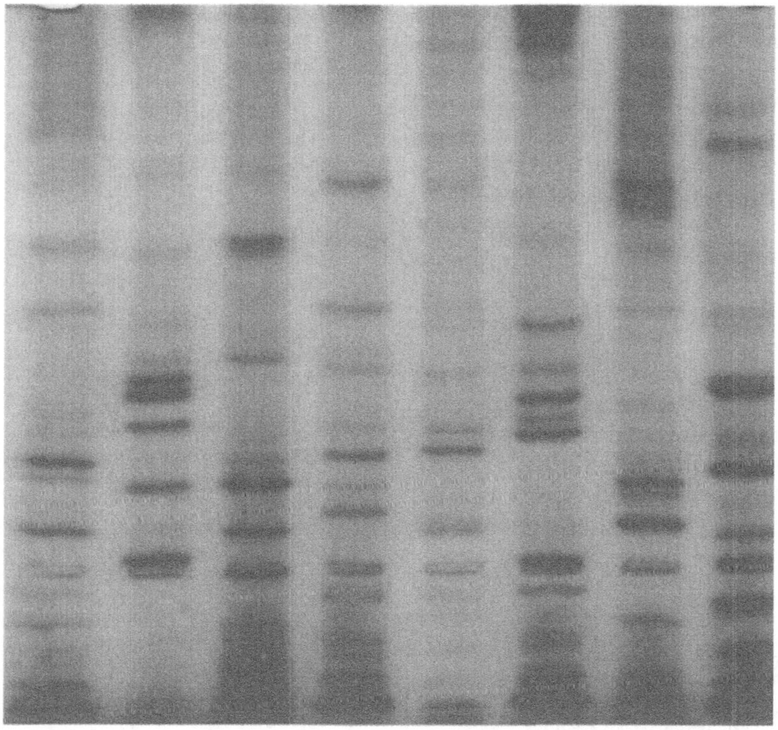

Figure 1. IEF patterns (pH gradient = 5–8) of EST in *L. striata*, all eight specimens were taken from Ilheu de Vila Franca do Campo.

firm the monomorphism reported by Wium-Andersen (1970) and Medeiros et al. (1998). Yet, whether this lack of Mb variation is real or merely due to the inability of the employed techniques (resp. PAGE and IEF) to resolve hidden variation, will be further investigated by means of two-dimensional electrophoresis.

In summary, the present data on EST and Mb variation in *L. striata* support the results of previous allozyme analyses in two ways: (1) although there is a high degree of genetic homogeneity among populations, there is a tendency of decreasing genetic variation with increasing latitude, and (2) the lack of EST differentiation between sexes, morphotypes and

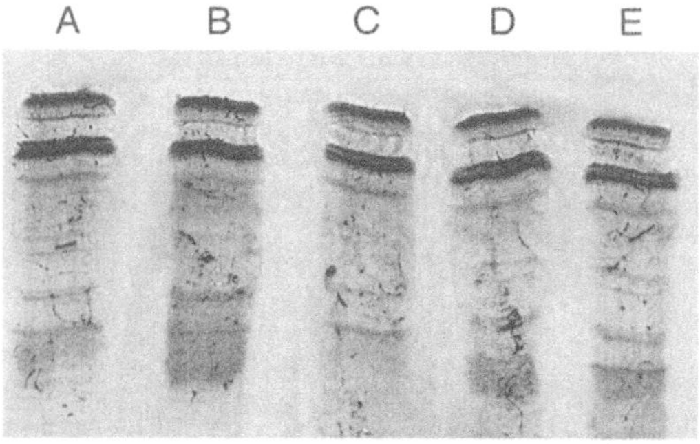

Figure 2. IEF patterns (pH gradient = 3–9) of Mb in *L. striata*. (A, C, E = Ilheu de Vila Franca do campo, Azores; B, D = Jose Fonseca, Cape Verde).

wave-exposure regimes confirms that the morphological patterning in *L. striata* occurs, despite the presence of intense gene flow. Whether this also means that shell variation in *L. striata* is ecophenotypic, needs further study. However, if selection would be the underlying mechanism responsible for the shell variation in *L. striata*, as is the case for the sculptural shell variation in the planktonic littorinid *L. picta*, it will have to be strong, in order to overcome the homogenising effects of intense gene flow.

Acknowledgements

This work was supported by the MAST III programme of the European Commission under contract number MAS3-CT95-0042 (AMBIOS) and PRAXIS XXI 2/2.1/BIA/169/94 (Portugal). H. D. W. holds an IWT scholarship.

References

Backeljau, T., K. Breugelmans, H. Leirs, T. Rodriguez, D. Sherbakov, T. Sitnikova, J.-M. Timmermans, J. L. Van Goethem & E. Verheyen, 1994. Application of isoelectric focussing in Molluscan systematics. The Nautilus, Suppl. 2: 156–167.

Backeljau, T., K. Breugelmans, C. Brito, L. De Bruyn, H. De Wolf & J.-M. Timmermans, 1995. Macrogeographic genetic homogeneity in *Littorina striata* from the Azores (Mollusca, Prosobranchia). Açoreana, Suppl. 1995: 159–171.

Berger, E. M., 1977. Gene-enzyme variation in three sympatric species of *Littorina*. II. the Roscoff populations, with a note on the origin of North American *L. littorea*. Biol. Bull. 153: 255–264.

Britton, J. C., 1995. The relationship between position on shore and shell ornamentation in two size-dependent morphotypes of *Littorina striata*, with an estimate of evaporative water loss in these morphotypes and in *Melarhaphe neritoides*. Hydrobiologia 309: 129–142.

De Wolf, H., S. Van Dongen, C. Brito & T. Backeljau, 1994. Population genetics of *Littorina striata* at a microgeographical scale. Cah. Biol. mar. 35: 243–244.

De Wolf, H., T. Backeljau, R. Medeiros & R. Verhagen, 1997. Microgeographical shell variation in *Littorina striata*, a planktonic developing periwinkle. Mar. Biol. 129: 331–342.

De Wolf, H., T. Backeljau, S. Van Dongen & R. Verhagen, 1998a. Large scale patterns of shell variation in *Littorina striata*, a planktonic developing periwinkle from Macaronesia (Mollusca: Prosobranchia). Mar. Biol. 131: 309–317.

De Wolf, H., T. Backeljau & R. Verhagen, 1998b. Spatio-temporal genetic structure and gene flow between two distinct shell morphs of the planktonic developing periwinkle *Littorina striata* (Mollusca: Prosobranchia). Mar. Ecol. Prog. Ser. 163: 155–163.

Dytham, C., J. Grahame & P. J. Mill, 1992. Esterase variation in *Littorina arcana* and *L. saxatilis* at Robin Hood's bay, Yorkshire. In Grahame, J., P. J. Mill & D. G. Reid (eds), Proceedings of the Third International Symposium on Littorinid Biology. The Malacological Society of London, London: 39–44.

Gaines, M. S., J. Caldwell & A. M. Vivas, 1974. Genetic variation in the mangrove periwinkle *Littorina angulifera*. Mar. Biol. 27: 327–332.

Medeiros, R., L. Serpa, C. Brito, H. De Wolf, K. Jordaens, B. Winnepenninckx & T. Backeljau, (1998). Radular myoglobin and protein variation within and among some littorinid species (Mollusca: Gastropoda). Hydrobiologia 378: 43–51.

Mill, P. J. & J. Grahame, 1988. Esterase variability in the gastropods *Littorina saxatilis* (Olivi) and *L. arcana* Ellis. J. moll. Stud. 54: 347–353.

Mill, P. J. & J. Grahame, 1992. Clinal changes in esterase variability *in Littorina saxatilis* (Olivi) and *L. arcana* Hannaford Ellis in southern Britain. In Grahame, J., P. J. Mill & D. G. Reid (eds), Proceedings of the Third International Symposium on Littorinid Biology. The Malacological Society of London, London: 31–38.

Morton, B., 1990. The intertidal ecology of Ilheu de Vila Franca – A drowned volcanic crater in the Azores. Açoreana, Suppl. 1990: 3–20.

Newkirk, G. F. & R. W. Doyle, 1979. Clinal variation at an esterase locus in *Littorina saxatilis* and *L. obtusata*. Can. J. Genet. Cytol. 21: 505–513.

Olabarria, C., J.-M. Timmermans & T. Backeljau, 1998. Electrophoretic heterogeneity within and between flat periwinkles (Mollusca: Gastropoda) along an intertidal transect at Ria Ferrol, northwest Spain. Hydrobiologia 378: 11–19.

Olsson, I., U.-B. Axio-Fredriksson, M. Degerman & B. Olsson, 1988. Fast horizontal electrophoresis. I. Isoelectric focusing and polyacrylamide gel electrophoresis using Phastsystem. Electrophoresis 9: 16–22.

Reid, D. G., 1996. Systematics and evolution of *Littorina*. The Ray Society, London. 463 pp.

Struhsaker, J. W., 1968. Selection mechanisms associated with intraspecific shell variation in *Littorina picta* (Prosobranchia: Mesogastropoda). Evolution 22: 459–480.

Vuilleumier, F. & M. B. Matteo, 1972. Esterase polymorphisms in European and American populations of the periwinkle *Littorina littorea* (Gastropoda). Experientia 28: 1241–1242.

Wium-Andersen, G., 1970. Haemoglobin and protein variation in three species of *Littorina*. Ophelia 8: 267–273.

Hydrobiologia **378:** 33–42, 1998.
R. M. O'Riordan, G. M. Burnell, M. S. Davies & N. F. Ramsay (eds), Aspects of Littorinid Biology.

A comparison of different protocols for RAPD analysis of *Littorina*

N. Mikhailova[1,2] & K. Johannesson[2,*]
[1] *Institute of Cytology RAS, St Petersburg 194064, Russia*
[2] *Tjärnö Marine Biological Laboratory, S-452 96 Strömstad, Sweden*
(author for correspondence)*

Key words: protocols, RAPD, *Littorina saxatilis, L. obtusata, L. fabalis, L. littorea*, extracting DNA, amplifying DNA

Abstract

Randomly amplified polymorphic DNA (RAPD) is a fast and useful method of genome marking that is useful for studies of, for example, parentage, mating patterns, taxonomy of sibling species and intra-specific population genetic structures. Here we compare three different procedures for extracting high molecular weight genomic DNA; phenol-chloroform, hexadecyltrimethyl ammonium-bromide (CTAB) and Chelex 100. Double phenol-chloroform and CTAB extractions both generated high amounts of high quality DNA while Chelex 100 failed to do so. We also compared PCR-amplification with different concentrations of template DNA and found that 1–2 ng per 25 μl of amplification cocktail gave the best results. Amplifying DNA prepared by the three extraction methods revealed that DNA extracted with double phenol-chloroform gave the clearest bands. The double phenol-chloroform extraction seems thus the most suitable extraction method for RAPD in *Littorina*, however Chelex may be the only method useful for extracting DNA from very small individuals, for example, pre-hatching stages.

Introduction

Allozyme markers have successfully been used in studies of littorinid gastropods for more than 15 years, particularly in the fields of taxonomy (e.g. Ward & Warwick, 1980; Mastro et al., 1982; Janson, 1985; Johannesson & Johannesson, 1990) and evolutionary ecology (Janson, 1987; Johannesson et al., 1995; Rolán-Alvarez et al., 1995). Recently, new molecular techniques have become available and there is already an increasing emphasis on the use of DNA characteristics as markers in taxonomic studies (Crossland et al., 1993; Rumbak et al., 1994; Crossland et al., 1996; Reid et al., 1996). Furthermore, it may be expected that studies of gene flow, hybridisation, parentage and mating patterns in littorinid species may soon profit from the use of DNA-markers as in a number of other species (Avise, 1994; Schierwater et al., 1994). Rumbak et al. (1994) and Reid et al. (1996) successfully used mitochondrial cytochrome-b, 12S and 16S ribosomal RNA sequences to reconstruct the phylogeny of the genus *Littorina*. Likewise mtDNA has been used for identification of sibling species of blue mussels, *Mytilus* (Geller & Powers, 1994; Heath et al., 1996). In these cases PCR (polymerase chain reaction) and widely applicable universal primers were used to amplify regions of the mtDNA genes. For example, fragments of the 12S and 16S ribosomal RNA genes and of the cytochrome-b gene, were amplified and sequenced in the phylogenetic analysis of the species of *Littorina* (Rumbak et al., 1994; Reid et al., 1996).

An alternative to sequence amplified genes and a much simpler method is RAPD (random amplified polymorphic DNA) (Welch & McClelland, 1990; Williams et al., 1990). This technique also relies on PCR to amplify arbitrary sequences of DNA using short synthetic oligonucleotide primers. The amplified products are, however, directly used as anonymous genetic markers. RAPD has successfully been applied in a number of taxonomic and population genetic studies, for example, in marine mussels (Adamkewicz & Harasewych, 1994; Patwary et al., 1994; Tyler-Walters &

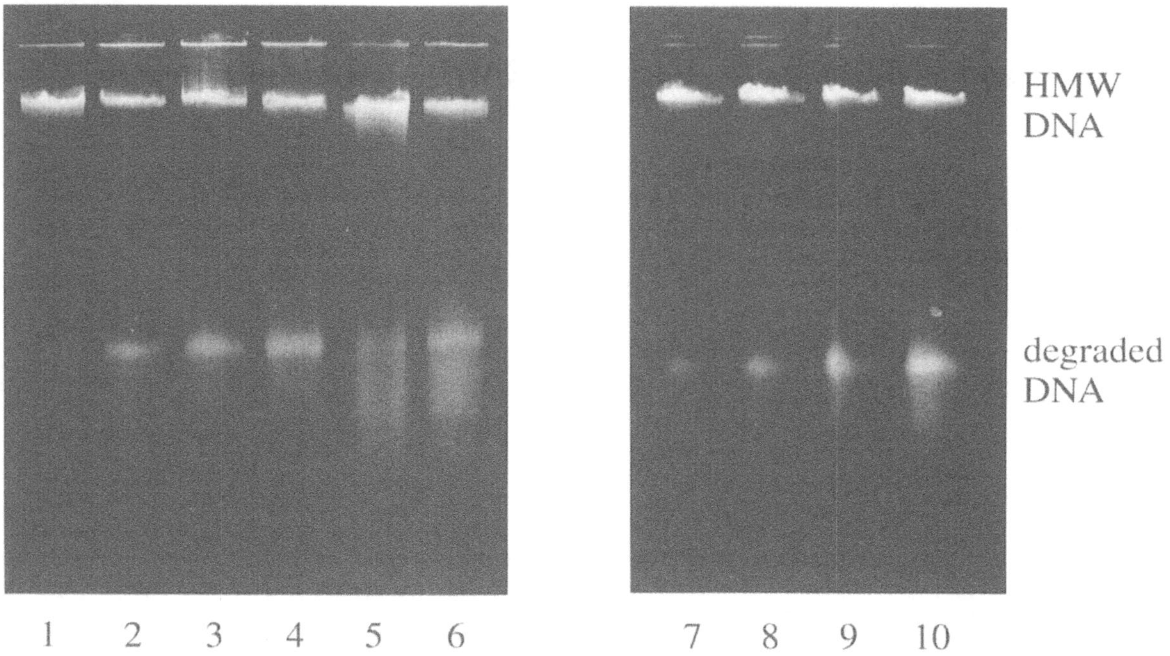

Figure 1. Electrophoresis of the extracted nucleic material from double phenol-chloroform extractions. Gels such as these were used to verify the presence of high molecular weight (HMW) DNA. Partly degraded DNA is seen as faint bands on the mid part of the gel. 250 ng of DNA per line was used and the gels were of 0.8% agarose in ×1 TBE buffer. Individual snails are Swedish *L. saxatilis* (1–2), *L. fabalis* (3–4), and *L. littorea* (5–6), and Icelandic *L. saxatilis* (7–8) and *L. obtusata* (9–10).

Hawkins, 1995) and in land snails (Pfenninger et al., 1995).

Despite the extent of taxonomic studies devoted to the systematics and evolution of littorinid species (thoroughly reviewed by Reid, 1996), there are yet few studies of littorinids in which RAPD has been used (Crossland et al., 1993, 1996).

Possibly the shortage of RAPD studies in littorinids partly reflects the technical problems associated with this method. Despite this method being superficially simple there are problems to overcome especially in the extraction of DNA. One problem is that the mucopolysaccharides present in mollusc tissues accompany the DNA and suppress Taq polymerase activity during amplification (Winnepenninckx et al., 1993; Rumpho et al., 1994; Palumbi, 1996). Here we present a comparison of three methods of extracting DNA and details about the PCR amplification of the target DNA using RAPD.

Materials and methods

Samples of Littorina

We sampled four species of *Littorina*, (*L. saxatilis*, *L. obtusata*, *L. fabalis* and *L. littorea*) to assess the usefulness of this method to find species-diagnostic DNA markers. Individuals of all species were collected from two islands on the Swedish west coast (Saltö and Lökholmen) about 1 km apart, and in one species (*L. fabalis*) populations at different sites from one of the islands were used. In addition to the Swedish populations we used two samples from Iceland, one of *L. saxatilis* and one of *L. obtusata*, both from the Westman Islands. We used only males of *L. obtusata* and *L. fabalis*, as these are easily recognised by the morphology of their penis, while females may only be distinguished by dissection (Reid, 1996). In *L. saxatilis* and *L. littorea*, however, individuals of both sexes were used for DNA preparation. We used either both muscular and liver tissues, or only muscular tissue for DNA preparation. We found no differences between these two treatments, probably because the gonads and digestive glands of all specimens were carefully inspected under the microscope and all trematode infected specimens were discarded. Thus we probably eliminated the risk of parasite contamination.

Preparation of tissue and DNA extraction

We used three methods to extract total high molecular weight (HMW) DNA from individual snails. One method was the phenol-chloroform extraction described by Sambrook et al. (1989) which we modified slightly (see below). The second method we used was a CTAB (hexadecyltrimethyl ammonium-bromide) extraction buffer as described by Levitan & Grosberg (1993). The third method was Chelex 100 extraction according to the description by Walsh et al. (1991) and Sundberg & Andersson (1995). It turned out that these three methods gave rather different results, and we will therefore describe in detail how we proceeded with the different protocols of extraction.

The phenol-chloroform extraction

Fresh tissue of a snail was washed in extraction buffer without proteinase K (see below), cut in small pieces and mixed with extraction buffer containing 20 mM Tris-HCl (pH = 8.0), 200 mM EDTA (pH = 8.0), 1% SDS (sodium dodecyl sulfate) and 0.1 mg ml^{-1} proteinase K (Mikhailova et al., 1995). The homogenate was incubated in a dry bath at 56 °C during 2.5–3 hours. The lysate was thereafter extracted with an equal volume of phenol. The mixture was gently but persistently shaken during 5–6 minutes and centrifuged at 6000 rpm for 5 minutes. The supernatant was transferred to a new eppendorf tube and an equal volume of a mixture of chloroform and isoamyl alcohol (24:1) was added. The new mixture was shaken and spun at the same conditions as before. The phenol-chloroform procedure was repeated to thoroughly extract the DNA. Following the extraction we used two volumes of cold 95% ethanol to precipitate the DNA and thereafter it was spun down at 6000 rpm for 5 minutes. After centrifugation the DNA pellet was washed twice with 70° ethanol to remove excessive salts and then it was left overnight to air-dry. The next day we resuspended the DNA pellet in autoclaved double-distilled water and quantified the approximate DNA concentration in each sample with a DyNA Quant™ 200 fluorometer (Hoefer, Pharmacia Biotech. Inc.). We used # 33258 Hoechst dye and assay solution A and calibrated the instrument with known concentrations of calf thymus DNA (Clontech Laboratories Inc., USA, concentration 100 ng μl^{-1}).

The phenol-chloroform extraction yielded DNA at concentrations of 30–130 ng μl^{-1} from 27–87 mg of original tissue. The DNA solution was diluted to a concentration of 1–2 ng μl^{-1} before the amplification by adding ×1 TE buffer (10 mM Tris-HCl, pH = 8.0, 1 mM EDTA, pH = 8.0) or sterile double-distilled water.

The CTAB extraction

We used the method described by Levitan & Grosberg (1993) without modifications. The CTAB extraction yielded DNA at adequate concentrations (31–43 ng μl^{-1}), which was diluted as described above before amplification.

The Chelex 100 extraction

This extraction, as described by Sundberg & Andersson (1995), has earlier been proved to be useful for very small sample sizes (e.g. Walsh et al., 1991; André et al., unpublished). We successfully scored individual embryos of *Littorina saxatilis,* as described below. When analysing adult snails we extracted DNA from about 2 mg muscle tissue taken from the head and mantle. To the tissue we added 50 μl 5% Chelex 100 (Sigma Chemical Co.) and proteinase K to a final concentration of 0.1 mg ml^{-1}. When analysing individual embryos we added 10 μl 5% Chelex 100 solution to one embryo taken directly from the brood pouch of a female, and in this case we used the exact protocol of Sundberg & Andersson (1995). The concentration of extracted DNA from one embryo of *L. saxatilis* was about 7–8 ng μl^{-1}, and this was obviously more than enough for the amplification procedure.

After DNA extraction we measured the DNA concentration with the fluorometer as described above. We also assessed the quality of the DNA in 0.8% agarose gels run in ×1 TBE buffer (0.089M Tris-borate and 0.089M boric acid, Sambrook et al., 1989). In this we applied 250 ng of DNA solution per line, and stained with ethidium bromide (0.5 ng ml^{-1}).

Random amplified polymorphic DNA (RAPD)

We used the RAPD protocol of Williams et al. (1990) with the modifications described by Sundberg & Andersson (1995) and André et al. (unpublished). The amplifications were carried out in 25 μl cocktails of ×1 Stoffel Buffer (Perkin-Elmer), 4 mM $MgCl_2$ (Perkin-Elmer), 200 μM each of dATP, dCTP, dGTP and dTTP (Gene Amp®, Perkin-Elmer), and 1.2 nM primer. We used 1 μl of extracted snail DNA (concentration 1–2 ng μl^{-1}) and 2U of Ampli Taq® DNA Polymerase, Stoffel fragment. Finally we topped the cocktail with one drop of mineral oil to prevent desiccation.

We amplified the DNA in a PTC-100^{TM} thermal controller (MJ Research Inc.). The PCR used 40 cycles with an initial denaturing stage at 94 °C for 2 minutes. Each cycle contained the following steps: 94 °C for 1 minute, 36 °C for 1 minute and 72 °C for 2 minutes. The amplification products were separated by electrophoresis at 3.3 V cm^{-1} in 1.5% agarose gels using the same ×1 TBE buffer as before. The DNA fragments were detected by staining with ethidium bromide (0.5 ng ml^{-1}) visible in ultraviolet light. Individual fragments were sized using a 100 bp DNA-ladder (Gibco BRL) (0.5 μg $line^{-1}$) run at two or more positions on the gel. A Nicon F-601 camera with a 100 mm macro lens and a yellow filter (Wratten #2 and #22) loaded with Kodak T-MAX 400 film was used to document the gel-mapping. The negatives were digitised into a computer using a SprintScan 35 film scanner (Polaroid Corporation).

We both amplified with 1 ng of template DNA and with larger quantities of DNA (up to 30 ng per reaction) to evaluate the effects of varying amounts of template DNA.

Twenty-seven ten-base oligonucleotide RAPD™ primers (Operon Technologies, Du Pont, USA) were tried in this study, seven from kit G and twenty from kit H. These primers have arbitrary sequences and the G+C content varies from 60 to 70%. Twenty of the primers yielded either too much variation, that is, each individual had unique bands, or they yielded too little variation with all or nearly all individuals of the different species being monomorphic for the same bands. These twenty primers were run only once and not further considered. The remaining seven primers, on the other hand, showed banding characters that revealed different genetic structures in the material. These primers were run 2–3 times each. In all cases we found ample variation for band presence or absence and did not score bands' brightness as a character.

Results

Amounts and quality of extracted DNA

The three methods of DNA extraction yielded very different concentrations of non-degraded DNA useful for the PCR amplification.

The phenol-chloroform extractions resulted in DNA concentrations from 30 to 129 ng μl^{-1} per individual ($n = 6$). In addition to single phenol-chloroform treatments we also used double treatments (e.g. Barral et al., 1993) with the DNA from the same six snails. In the double treatment, however, large parts of the earlier amounts of DNA were lost and the remaining concentrations varied from 11 to 45 ng μl^{-1}. The DNA quality, on the other hand, was better than in the single treatment. That is, the content of high molecular weight (non-degraded) DNA was higher, and comparable to that of others (e.g. compare Hoelzel, 1992: 230 with Figure 1). Thus the double extraction seemed more appropriate than the single extraction for DNA amplification. There was no evidence of RNA in these samples that, if present, would have been seen in the lower part of Figure 1 (see also Figure 2). However, partly degraded DNA was obvious as faint fast-moving bands slightly below the mid part of Figure 1. Thus the phenol-chloroform extraction, and particularly the double extraction, successfully extracted a large quantity of high molecular weight DNA without heavy contamination of degraded DNA.

The extraction using the CTAB buffer yielded concentrations of DNA between 30 and 40 ng per individual. The dominating part of which seemed to be high molecular weight DNA (Figure 2b). These extractions yielded some RNA in the nucleic acid solution, which could be seen as smears in the low molecular weight area at the bottom of the gel (Figure 2b).

Extraction of DNA using Chelex 100 is faster than both the phenol-chloroform and the CTAB extraction, because of the fewer steps of the extraction. However, when extracted DNA (250 ng of DNA per lane) was separated using agarose gel electrophoresis no high molecular weight bands were present in the Chelex extracts (Figure 2a). All DNA, except that remaining at the top of the gel (at the start position), was degraded. The low molecular weight material present as a smear was probably RNA. Thus, the Chelex extracted DNA appeared to be of low quality.

However, the attempts to amplify whatever DNA was present in the nucleic acid solution revealed, to our surprise, quite satisfactory results (see below). A promising property of the Chelex method is that it is possible to prepare DNA from a very small quantity of tissue. We used the Chelex method to extract DNA from single individual embryos of *L. saxatilis* (eggs, veligers and metamorphosed stages) as well as from their mothers. The extracted concentrations of DNA

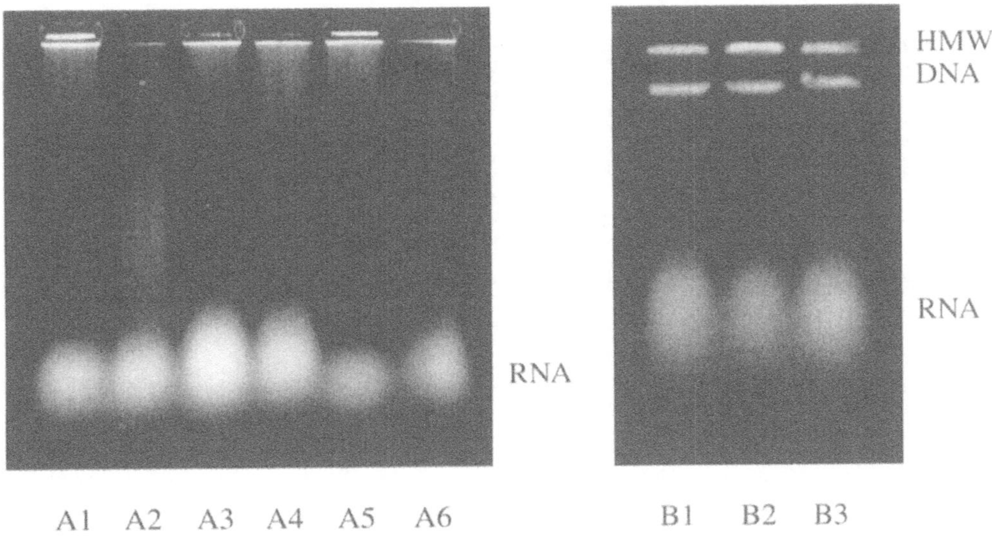

Figure 2. Electrophoresis of the extracted nucleic material from Chelex 100 (A) and CTAB (B) extractions. High molecular weight (HMW) DNA is only present in the CTAB extraction, while smears of low molecular weight material (likely RNA) are present on both gels. Gels were prepared as in Figure 1. A1–2 and B1–3 are individuals of *L. saxatilis*, A3–4 are *L. fabalis* and A5–6 are *L. littorea*.

from the embryos were 2–8 ng μl^{-1} and enough for amplification of DNA.

Results of amplification

PCR amplification of varying quantities of template DNA (from 1 ng to 30 ng per 25 μl of cocktail) gave different results. Concentrations higher than 4 ng per 25 μl of amplification cocktail gave, strangely enough, no amplification at all. The best results were obtained when the concentration of template DNA was diluted to 1–2 ng μl^{-1} and we took only 1 μl per reaction.

We tried to amplify DNA prepared by the three methods. The best results were achieved with DNA extracted by the double phenol-chloroform method (Figure 3, lanes 7–14). In contrast, single phenol-chloroform extraction gave a poor amplification (Figure 3, lanes 1–6; 1, 2, 5 and 6 use the same template DNA as in 7–10). The CTAB extraction gave no bands at all (lanes 15–17), while the Chelex extraction gave amplification in one of two lanes (lanes 18 and 19). Thus the double phenol-chloroform extraction seems the most suitable one of the different extraction methods we tested for RAPD in species of *Littorina*. Chelex 100 extraction is, however, the only possible method, for extracting DNA from embryo *Littorina*.

Mother-offspring comparisons

Two short ten-base arbitrary primers (OPG-3 and OPG-8) were used for comparing female and offspring DNA in *L. saxatilis*. Most bands are common to both mother and offspring, but additional paternal bands and losses of maternal bands are seen in most offspring pairs (Figure 4), as expected from recombination during sexual reproduction. Possibly, a small fraction of the bands unique to the offspring may be non-parental bands introduced by mutation (Scott & Williams, 1994).

Preliminary accounts of variation among and within species

Eight ten-base primers were used in the amplification of DNA from the four species of *Littorina* (OPG-3, OPG-4, OPG-5, OPG-8, OPG-9, OPG-10, OPH-15 and OPH-19). Only one, OPH-15, revealed differences between all these congeneric species (Figure 5). In this primer, the amplification products of *L. obtusata* were very different from these of the other three species. There was one major band of about 420 bp (basepair) and two very dim bands, of about 210 bp and 150 bp (Figure 5, lanes 7 and 8). Minor differences were evident among the three other species (*L. saxatilis*, *L. littorea* and *L. fabalis*). However, *L. littorea* was easily distinguished on bands representing fragments of approximately 250 bp size, furthermore

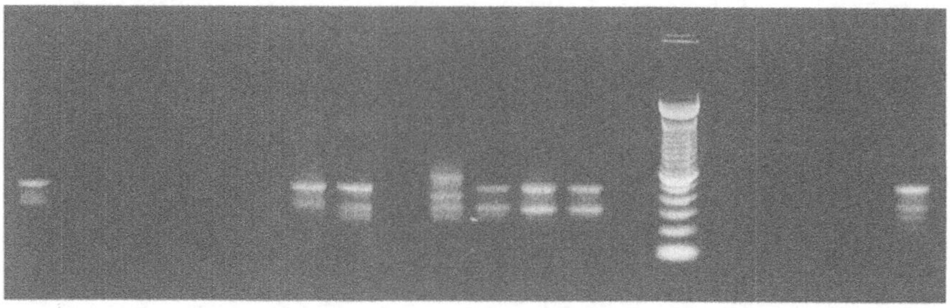

Figure 3. Amplification products of primer OPG-17 with different methods of DNA extraction; single phenol-chloroform (1–6), double phenol-chloroform (7–14), CTAB (15–17) and Chelex (18–19). 'M' indicates molecular weights at marking intervals of 100 bp (see Figure 5). Species of *Littorina* are: *L. fabalis* (3, 4), *L. littorea* (5, 6, 9, 10), *L. obtusata* (13, 14) and *L. saxatilis* (remaining lanes).

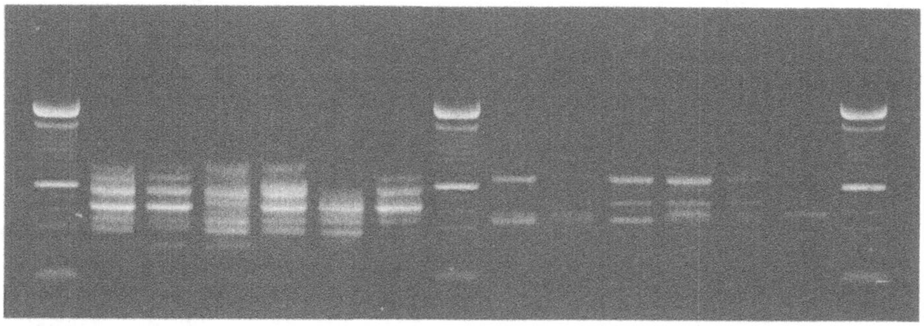

Figure 4. Amplification products of primers OPG-3 (1–6) and OPG-8 (7–12) in females of *L. saxatilis* and their offsprings. The amplification used DNA extracted with Chelex 100. Lanes 1, 3, 5, 7, 9 and 11 are different females and lanes 2, 4, 6, 8, 10 and 12 are their offspring, respectively. 'M' indicates molecular weights at marking intervals of 100 bp (see Figure 5).

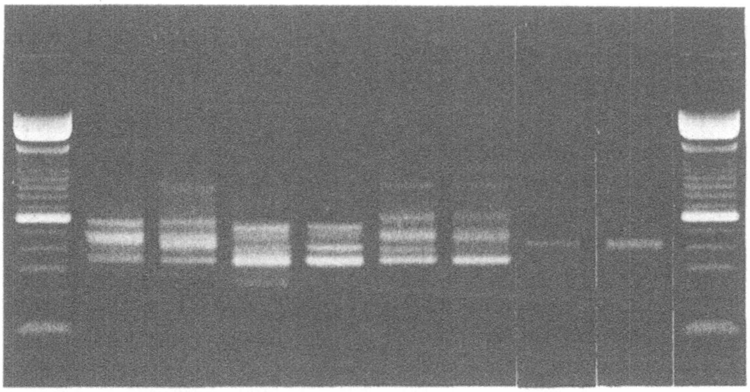

Figure 5. Amplification products of primer OPH-15 in four species of *Littorina*; *L. saxatilis* (1, 2), *L. littorea* (3, 4), *L. fabalis* (5, 6) and *L. obtusata* (7, 8), all individuals were from Sweden. 'M' indicates molecular weights at marking intervals of 100 bp.

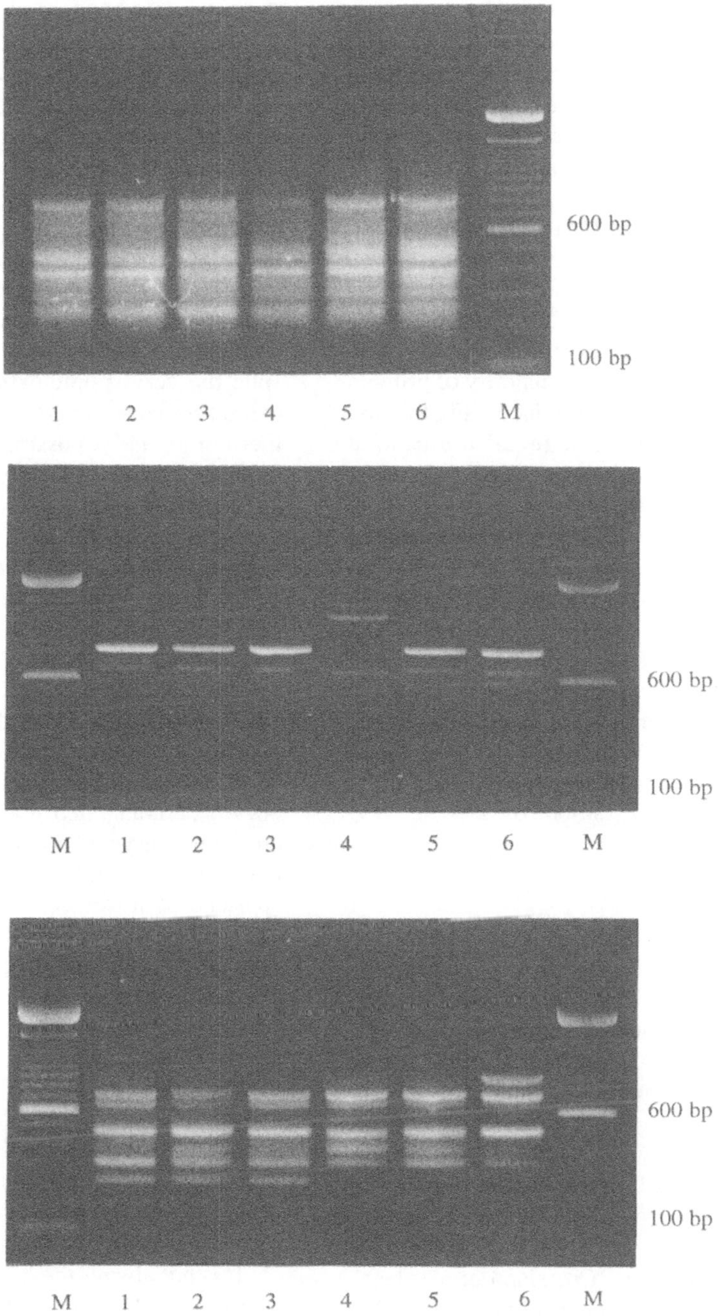

Figure 6. Amplification products of three primers in three individuals of *L. fabalis* from a wave exposed part of a Swedish boulder shore (1–3) and in three individuals from a sheltered part of the same shore (4–6). Primer OPH-12 (top) exhibits a monomorphic pattern, OPH-20 (mid) show polymorphic fragments of about 850 and 1300 bp size, and OPH-11 (bottom) reveals what may be a site-specific fragment of about 240 bp size in the exposed individuals. 'M' indicates molecular weights at marking intervals of 100 bp.

bands at approximately 550 bp were unique to this species. The two remaining species, *L. fabalis* and *L. saxatilis* were distinct on bands at 580 bp (*L. saxatilis*) and 620 bp (*L. fabalis*). As we used only two individuals of each species these results should be considered as very preliminary. However, it seems as if the primer OPH-15 might be a good candidate for a diagnostic marker for the four species of *Littorina* screened in this study.

To assess the amount of intraspecific variation we screened 19 primers from kit H (Operon Technol. Inc.) in six individuals of one of the species, *L. fabalis*. Four of the investigated primers (OPH-2, OPH-6, OPH-12 and OPH-16) yielded monomorphic amplification products (e.g. Figure 6a) while the majority of primers (15 of 19) gave polymorphic products (Figure 6b–c). One of the primers, OPH-11, revealed a band at approximately 240 bp in all three individuals of *L. fabalis* from an exposed part of a shore while none of the three conspecific snails from a sheltered part of the same shore about 150 m away had this band (Figure 6c). The same result was obtained by re-running the same primer with the same six individuals, and repeating the amplification using ten new snails, five from exposed and five from the sheltered part of the shore. This primer may be a useful marker of specific populations of this species, although we must stress the uncertainty of this interpretation with only eight individuals analysed per population.

Discussion

RAPD has successfully been applied to both analysis of parentage and taxonomic problems in a broad range of, for example, marine organisms (hydroids, Levitan & Grosberg, 1993; gorgonians, Coffroth & Mulawka III, 1995; nemertines, Sundberg & Andersson, 1995; and bivalves, Patwary et al., 1994; Adamkewicz & Harashewych, 1994; Tyler-Walters & Hawkins, 1995). RAPD has previously shown to be successful in separating sibling species of the *Littorina saxatilis* species-group, namely *L. arcana*, *L. nigrolineata* and *L. saxatilis* (Crossland et al., 1993, 1996). Furthermore, RAPD analysis of Crossland et al. (1996) supports earlier suggestions of a gene flow between *L. neglecta* and *L. saxatilis*.

Most earlier studies used various modifications of the phenol-chloroform DNA extraction (e.g. Crossland et al., 1993, 1996; Adamkewicz & Harashewych, 1994; Patwary et al., 1994; Tyler-Walters & Hawkins, 1995), while the CTAB extraction (Levitan & Grosberg, 1993) and the Chelex 100 extraction (Sundberg & Andersson, 1995) are less often used.

As we show in the present study the choice of extraction procedure may be critical to the capturing of high-molecular weight DNA suitable for the molecular analysis. For example, the double phenol-chloroform extraction yielded the most promising amplification results in our samples. We found that it gave DNA of a better quality for amplification by PCR than did the single extraction. In contrast, the CTAB preparation gave DNA with much RNA but no amplification products, still others have used this extraction method successfully (e.g. Rumbak et al., 1994). Despite the very simple extraction protocol, the Chelex extraction gave amplified products. This method furthermore made it possible to extract minute amounts of DNA from snail embryos that could be used in 7–8 amplifications. Chelex extraction is thus a very useful approach for analysis of paternity and mating patterns in species of *Littorina*. It seems also useful for taxonomic identification of small juvenile individuals in sibling species of *Littorina* (see also Coffroth & Mulawka III, 1995).

The concentration of template DNA may be critical in RAPD. Earlier studies have yielded successful amplifications inside certain ranges of template DNA concentrations. For example, Welsh & McClelland (1990) demonstrated that the amplification products are very consistent for template DNA concentrations between 7.5 ng and 30 pg per 10 μl of cocktail but only sporadic at lower concentrations in *Staphylococcus* strains. In a study of sea scallop DNA the number and amount of amplified products were best in reactions containing over 40 ng of DNA per 25 μl of cocktail (Patwary et al., 1994). Crossland et al. (1993) used about 100 ng of template DNA per 100 μl of cocktail in species of *Littorina*. We found, however, that DNA at very low concentrations (1–2 ng per 25 μl of reactionary mixture) gave the best results in the present study. The reason for this is not clear, but possibly a diluted template concentration prevented interference by mucopolysaccharides (see below).

It is not always the case that the extraction method that yields a high amount of DNA will give better results after amplification than a method that gives less DNA. In our study the CTAB extraction gave DNA at a high concentration compared to the Chelex method but still the latter gave a better result after amplification. Furthermore, the minute sizes of the embryo prevented the use of the CTAB and phenol-chloroform methods, while the Chelex extraction was possible.

Mucus is a serious problem in DNA preparation because the mucopolysaccharides will co-precipitate with the DNA during the extraction and this may impede the amplification (Palumbi, 1996). Marine slugs, for example, secrete large amounts of mucus that is found attached to most, if not all, of the cellular components including nucleic acids (Rumpho et al., 1994). By applying an initial precipitation of the slugs' tissue in ice-cold absolute ethanol and thereafter centrifugation of the crude DNA on a CsCl gradient the mucous problem may be overcome (Rumpho et al., 1994). Another suggestion is to use the CTAB extraction buffer as this buffer contains 2% hexadecyltrimethyl ammonium-bromide, 0.2% ß-mercaptoethanol and 1.4 m NaCl that bind proteins and polysaccharides (Levitan & Grosberg, 1993; Winnepenninckx et al., 1993). Diluting the concentration of the template DNA may prevent negative effects of the accompanying mucus, and our results suggest that mucus inhibited the PCR reaction at moderate concentrations of template. However, diluting the template DNA to 1–2 ng μl^{-1} and using only 1 μl per 25 μl of reaction cocktail gave good results.

RAPD is very sensitive to the reaction conditions and the reproducibility has to be improved by standardising every step of the analysis (e.g. Hadrys & Siva-Jothy, 1994). We repeated the amplification with the same template DNA 2–3 times and we used only those results that showed consistent patterns among repeated amplifications of the same individual.

The usefulness of RAPD in littorinid taxonomy has already been shown (Crossland et al., 1993, 1996). Our results suggest that species-specific RAPD-characters may also exist for the four species *L. saxatilis*, *L. obtusata*, *L. fabalis* and *L. littorea*. However, for RAPD to be useful in ecological studies (e.g. studies of parentage, sperm competition and mating patterns), intraspecific markers distinguishing populations or even family groups will be needed. From low sample sizes ($n = 16$) we have results that give preliminary support to this.

Acknowledgements

We are grateful to Susanne Andersson, Maria Hellsten, Per Jonsson, Mats Lindegarth, Per Nilsson and Per Sundberg for advice about and discussion of methods and results. The work was financially supported by a grant to N. M. from the Swedish Institute, and with funds from the Swedish Natural Science Research Council to K. J.

References

Adamkewicz, L. S. & M. G. Harasewych, 1994. Use of random amplified polymorphic DNA (RAPD) markers to assess relationships among beach clams of the genus *Donax*. The Nautilus Suppl. 2: 51–60.

Avise, J. C., 1994. Molecular markers, natural history and evolution. Chapman and Hall, London, 511 pp.

Barral, V., P. This, D. Imbert-Establet, C. Combes & M. Delseny, 1993. Genetic variability and evolution of the *Schistosoma* genome analysed by using random amplified polymorphic DNA markers. Mol. Biochem. Parasitol. 59: 211–222.

Coffroth, M. A. & J. M. Mulawka III, 1995. Identification of marine invertebrate larvae by means of PCR-RAPD species-specific markers. Limnol. Oceanogr. 40: 181–189.

Crossland, S., J. Coates, J. Grahame & P. J. Mill, 1993. Use of random amplified polymorphic DNAs (RAPDs) in separating two sibling species of *Littorina*. Mar. Ecol. Prog. Ser. 96: 301–305.

Crossland, S., J. Coates, J. Grahame & P. J. Mill, 1996. The *Littorina saxatilis* species complex – interpretation using random amplified polymorphic DNAs. In J. Taylor (ed.), Origin and Evolutionary Radiation of the Mollusca. Oxford University Press, New York: 205–209.

Geller, J. B. & D. A. Powers, 1994. Site directed mutagenesis with the polymerase chain reaction for identification of sibling species of *Mytilus*. The Nautilus Suppl. 2: 141–144.

Hadrys, H. & M. T. Siva-Jothy, 1994. Unravelling the components that underlie insect reproductive traits using a simple molecular approach. In B. Schierwater, B. Streit, G. P. Wagner & R. DeSalle (eds), Molecular Ecology and Evolution: Approaches and Applications. Birkhäuser, Basel: 75–90.

Heath, D. D., D. R. Hatcher & T. J. Hilbish, 1996. Ecological interaction between sympatric *Mytilus* species on the west coast of Canada investigated using PCR markers. Mol. Ecol. 5: 443–447.

Hoelzel, A. R., 1992. Molecular genetic analysis of populations. Oxford Univ. Press, New York, 315 pp.

Janson, K., 1985. Genetic variation in three species of Caribbean periwinkles, *Littorina angustior*, *L. lineolata*, and *L. ziczac* (Gastropoda: Prosobranchia). Bull. mar. Sci. 37: 871–879.

Janson, K., 1987. Genetic drift in small and recently founded populations of the marine snail *Littorina saxatilis*. Heredity 58: 31–37.

Johannesson, K. & B. Johannesson, 1990. Genetic variation within *Littorina saxatilis* (Olivi) and *Littorina neglecta* Bean: Is *L. neglecta* a good species? Hydrobiologia 193: 89–97.

Johannesson, K., B. Johannesson & U. Lundgren, 1995. Strong natural selection causes microscale allozyme variation in a marine snail. Proc. natn. Acad. Sci. USA 92: 2602–2606.

Levitan, D. R. & R. K. Grosberg, 1993. The analysis of paternity and maternity in the marine hydrozoan *Hydractinia symbiolongicarpus* using randomly amplified polymorphic DNA (RAPD) markers. Mol. Ecol. 2: 315–326.

Mastro, E., V. Chow & D. Hedgecock, 1982. *Littorina scutulata* and *Littorina plena*: Sibling species status of two prosobranch gastropod species confirmed by electrophoresis. Veliger 24: 239–246.

Mikhailova, N. A., I. B. Lobov, O. I. Podgornaya & V. J. Birstein, 1995. Restriction analysis of the nuclear DNA of three species of *Acipenser*. The Sturg. Quarterly 3, 8–9.

Palumbi, S. R., 1996. Nucleic Acids II: The polymerase chain reaction. In D. M. Hillis, C. Moritz & B. K. Mable (eds). Molecular Systematics, 2nd edition. Sinauer, Sunderland MA.: 205–247.

Patwary, M. U., E. L. Kenchington, C. J. Bird & E. Zouros, 1994. The use of random polymorphic DNA markers in genetic studies of the sea scallop *Placopecten magellanicus* (Gmelin, 1791). J. Shellfish Res. 13: 547–553.

Pfenninger, M., M. Frye, A. Bahl & B. Streit, 1995. Discrimination of three conchologically similar *Helicellinae* (*Helicella*, Gastropoda) species using RAPD-fingerprinting. Mol. Ecol. 4: 521–522.

Reid, D. G., 1996. Systematics and evolution of *Littorina*. Ray Society, London. 463 pp.

Reid, D. G., E. Rumbak & R. H. Thomas, 1996. DNA, morphology and fossils: phylogeny and evolutionary rates of the gastropod genus *Littorina*. Phil. Trans. r. Soc., Lond. B 351: 877–895.

Rolán-Alvarez, E., C. Zapata & G. Alvarez, 1995. Multilocus heterozygosity and sexual selection in a natural population of the marine snail *Littorina mariae* (Gastropoda: Prosobranchia). Heredity 75: 17–25.

Rumbak, E., D. G. Reid & R. H. Thomas, 1994. Reconstruction of phylogeny of 11 species of *Littorina* (Gastropoda: Littorinidae) using mitochondrial DNA sequence data. The Nautilus Suppl. 2: 91–97.

Rumpho, M. E., C. V. Mujer, D. L. Andrews, J. R. Manhart & S. K. Pierce, 1994. Extraction of DNA from mucilaginous tissues of a sea slug (*Elysia chlorotica*). BioTechniques 17: 1097–1102.

Sambrook, J., E. F. Fritsch & T. Maniatis, 1989. Molecular cloning: a laboratory manual. Cold Spring Harbor Laboratory Press, New York, 1659 pp.

Schierwater, B., B. Streit, G. P. Wagner & R. DeSalle, 1994. Molecular Ecology and Evolution: Approaches and Applications. Birkhäuser, Basel, 622 pp.

Scott, M. P. & S. M. Williams, 1994. Measuring reproductive success in insects. In B. Schierwater, B. Streit, G. P. Wagner & R. DeSalle (eds), Molecular Ecology and Evolution: Approaches and Applications. Birkhäuser, Basel: 61–74.

Sundberg, P. & S. Andersson, 1995. Random amplified polymorphic DNA (RAPD) and intraspecific variation in *Oerstedia dorsalis* (Hoplonemertea, Nemertea). J. mar. biol. Ass. U.K. 75: 483–490.

Tyler-Walters, H. & A. R. Hawkins, 1995. The application of RAPD markers to the study of the bivalve mollusc *Lasea rubra*. J. mar. biol. Ass. U.K. 75: 563–569.

Walsh, P. S., D. A. Metzger & R. Higuchi, 1991. Chelex® 100 as a medium for simple extraction of DNA for PCR-based typing from forensic material. BioTechniques 10: 506–513.

Ward, R. D. & T. Warwick, 1980. Genetic differentiation in the molluscan species *Littorina rudis* and *Littorina arcana* (Prosobranchia: Littorinidae). Biol. J. linn. Soc. 14: 417–428.

Welsh, J. & M. McClelland, 1990. Fingerprinting genomes using PCR with arbitrary primers. Nucleic Acids Res. 18: 7213–7218.

Williams, J. G. K., A. R. Kubelik, K. J. Livak, J. A. Rafalski & S. V. Tingey, 1990. DNA polymorphisms amplified by arbitrary primers are useful as genetic markers. Nucleic Acids Res. 18: 6531–6535.

Winnepenninckx, B., T. Backeljau & R. De Wachter, 1993. Extraction of high molecular weight DNA from molluscs. Trends in Genetics 9: 407.

Hydrobiologia **378:** 43–51, 1998.
R. M. O'Riordan, G. M. Burnell, M. S. Davies & N. F. Ramsay (eds), Aspects of Littorinid Biology.

Radular myoglobin and protein variation within and among some littorinid species (Mollusca: Gastropoda)

R. Medeiros[1], L. Serpa[1], C. Brito[1], H. De Wolf[3], K. Jordaens[2], B. Winnepenninckx[3] & T. Backeljau[3,*]
[1] *Dept. Biology, University of the Azores, Rua da Mãe de Deus 58, Apartado 1422, P-9502 Ponta Delgada, Azores, Portugal*
[2] *Dept. Biology, University of Antwerp, Groenenborgerlaan 171, B-2020 Antwerp, Belgium*
[3] *Royal Belgian Institute of Natural Sciences, Vautierstraat 29, B-1000 Brussels, Belgium*
(author for correspondence)*

Key words: electrophoresis, soluble proteins, Littorinidae, myoglobin, systematics

Abstract

The radular muscles of several littorinid species, including *Littorina littorea*, *L. saxatilis*, *L. obtusata*, *L. striata* and *Melarhaphe neritoides*, contain myoglobin (Mb). Here we report on the presence of radular Mb in eight other littorinids: *L. compressa*, *L. arcana*, *L. fabalis*, *Nodilittorina punctata*, *N. trochoides*, *N. radiata*, *Littoraria undulata* and *Littoraria cingulifera*. Using native polyacrylamide gel electrophoresis (PAGE) and isoelectric focusing (IEF) we compared the Mb and soluble protein (SP) profiles of these species. This suggested that: (1) *L. saxatilis* and *L. arcana* may have specific Mb/SP profiles, (2) *Littoraria* spp., *Nodilittorina* spp. and *L. striata* share similar Mb patterns, (3) Mb is remarkably diverse in the genus *Littorina*, (4) *L. littorea* shows intraspecific Mb/SP variation, (5) *L. saxatilis* does not show geographic Mb/SP differences, and (6) IEF uncovers substantial hidden Mb/SP heterogeneity not shown by PAGE (particularly for *Melarhaphe neritoides*). Hence, littorinid Mb/SP may be a useful taxonomic marker whose ecophysiological significance deserves further study, even if its genetic basis remains unclear.

Introduction

Although several prosobranch gastropods are known to contain myoglobin (Mb) in their radular muscles (e.g. Read, 1966; Bonaventura & Bonaventura, 1983), data on the presence of radular Mb still need to be collected for the majority of species (Bonaventura & Bonaventura, 1983). Yet, within the Littorinidae (periwinkles), the occurrence of radular Mb has been demonstrated in at least 17 species, including *Littorina littorea*, *L. saxatilis* (formerly referred to as *L. rudis*), *L. obtusata*, *L. striata*, *Melarhaphe neritoides* and several species of *Nodilittorina* and *Littoraria* (Read, 1966, 1968; Bannister et al., 1968; Wium-Andersen, 1970; Jones, 1972; Alyakrinskaya, 1986, 1989, 1994). Despite the relative wealth of data on littorinid Mb only two authors, viz. Wium-Andersen (1970) and Jones (1972), have attempted to use Mb variation in littorinid population genetics and systematics.

Wium-Andersen (1970) applied native polyacrylamide gel electrophoresis (PAGE) to compare radular Mb profiles of *L. striata*, *L. littorea* and *L. saxatilis*. The results of this work suggested that *L. striata* and *L. littorea* have monomorphic Mb profiles with *L. striata* being fixed for the two slow migrating bands and *L. littorea* for the two fast migrating bands. Furthermore, it appeared that *L. saxatilis* shows Mendelian variation of the two Mb bands observed in *L. littorea*, such that (1) genotype frequencies in Danish populations show a nice fit to those expected under Hardy-Weinberg equilibrium conditions and (2) Mb gene frequencies follow a cline along the Danish west coast. Finally, Wium-Andersen (1970) reported that radular proteins other than Mb, are fixed (but different) in *L. striata* and *L. littorea*, whereas in *L. saxatilis* they are highly variable, suggesting a strong (species-specific?) divergence of *L. groenlandica*, but no clear separation of *L. tenebrosa* [currently the latter two taxa are considered as varieties or ecotypes of *L. saxatilis* (Reid, 1996)].

Jones (1972) also used native PAGE of radular Mb in a systematic study of 11 Panamanian '*Littorina*' species (currently all these species are assigned to *Nodilittorina* and *Littoraria*; David Reid in litt., 1996). The aim of this work was to identify so-called Pacific-Caribbean 'geminate' (twin or sister) species on both sides of the Panamanian Isthmus. However, although Jones (1972) observed species-specific differences in both the mobility and the number of Mb bands, she could not define 'geminate' species pairs. Moreover, the phylogenetic inferences based on Jones' (1972) data were inconsistent (Reid, 1986).

Nevertheless, the appealing Mb results of Wium-Andersen (1970), combined with the fact that haemoglobin/myoglobin polymorphisms have been used in the systematics and population genetics of bivalves (e.g. O'Gower & Nicol, 1968), basommatophoran gastropods (e.g. Bailey et al., 1986; Mulvey et al., 1988; Woodruff & Mulvey, 1997) and other animal groups (e.g. Basaglia & Callegarini, 1987; Lavrenchenko et al., 1992; Alexandrino et al., 1996; Das & Handique, 1996; Macaranas et al., 1996), prompted us to initiate an electrophoretic survey of littorinid Mbs. Therefore we aimed at: (1) extending Wium-Andersen's (1970) observations, (2) screening a number of species for which the occurrence of Mb was not yet documented, (3) assessing the degree of intraspecific variation of littorinid Mbs, (4) testing the utility of Mb (and other radular proteins) as markers to distinguish closely related species, (5) evaluating the utility of Mb for supraspecific littorinid systematics, and (6) comparing the power of different electrophoretic methods to detect Mb variation.

Throughout this paper we will follow the taxonomy and nomenclature proposed by Reid (1986, 1992, 1996). Furthermore, we will maintain the abbreviation '*L.*' for *Littorina*, while the name *Littoraria* will be written in full.

Materials and methods

Radular muscle Mb and soluble proteins (SP) were investigated in 482 specimens from 19 populations involving 11 littorinid species. Two additional species were represented by pooled samples of about 20 individuals each. The following material was included: *Melarhaphe neritoides* (Linnaeus, 1758): São Miguel (Azores) (No. = 10); *Littoraria cingulifera* (Dunker, 1845): Gorée (Senegal) (No. = 10); *Littoraria undulata* (Gray, 1839), Laing Island (Papua New Guinea) (No. = 10); *Nodilittorina (N.) trochoides* (Gray, 1839) [formerly *N. pyramidalis* (Quoy & Gaimard, 1833)], Cape d'Aguilar (Hong Kong) (No. = 20 pooled); *N. (N.) radiata* (Eydoux & Souleyet, 1852) [formerly *N. exigua* (Dunker, 1860)], Cape d'Aguilar (Hong Kong) (No. = 20 pooled); *N. (Echinolittorina) punctata* (Gmelin, 1791), Brucoli (Sicily) (No. = 10); São Vicente (Cape Verde Islands) (No. = 10), *Littorina (Liralittorina) striata* King & Broderip, 1832, São Miguel (Azores) (No. = 10); *L. (Neritrema) obtusata* (Linnaeus, 1758), Roscoff (France) (No. = 10); *L. (Neritrema) fabalis* (Turton, 1825) [formerly *L. mariae* Sacchi & Rastelli, 1966], Zeebrugge (Belgium) (No. = 10); *L. (Neritrema) compressa* Jeffreys, 1865 [formerly *L. nigrolineata* Gray, 1839], Roscoff (France) (No. = 10); *L. (Neritrema) arcana* Hannaford Ellis, 1978, Dale Fort 'Great Castle Head' (U.K.) (No. = 40); *L. (Neritrema) saxatilis* (Olivi, 1792), Trondheim (Norway) (No. = 40), Ria de Arosa (Spain) (No. = 40); *L. (Littorina) littorea* (Linnaeus, 1758), Brouwersdam (The Netherlands) (No. = 40), Oostende (Belgium) (No. = 38), Wimereux 'Dique de la Crèche' (France) (No. = 40), Ile de Ré (France) (No. = 40), Ria Ferrol (Spain) (No. = 40), Ria de Arosa (Spain) (No. = 40), Milford Haven (U.K.) (No. = 34).

After field collection, all material was transported alive or frozen in liquid nitrogen to the laboratory where it was stored at –80 °C until prepared for electrophoresis. Prior to sample preparation all specimens were sexed. In *L. littorea* and *L. saxatilis* five shell measurements were taken with vernier callipers: height, width, height of the aperture, width of the aperture and height of the shell top. In both species the colour pattern and total weight were determined. Individual tissue homogenates were prepared by thawing frozen snails, crushing their shells and dissecting the radular muscles in cold distilled water. The muscles were blotted on filter paper and, depending on the tissue volume, transferred into 40, 25 or 15 μl of a 20% (w/v) aqueous sucrose solution, in which the muscles were homogenised. Crude homogenates were centrifuged for 45 min at 27000 × g and 4 °C. The resulting supernates were stored at –80 °C until used for electrophoresis.

Native vertical PAGE was performed in 80 × 80 × 0.5 mm gels ('Mighty Small SE 250' apparatus

Table 1. Programmed conditions for IEF separations of Mb/SP in polyacrylamide gels (pH gradient: 3–9) using PhastSystem. The conditions for 4–6.5 pH gradients were the same except that SEP 1.1 and SEP 1.2 were set at 2.0 mA, while SEP 1.3 was set at 5.0 mA

SAMPLE APPLICATOR DOWN AT				1.2	0 Vh
SAMPLE APPLICATOR UP AT				1.3	0 Vh
EXTRA ALARM TO SOUND AT				1.1	73 Vh
SEP 1.1	2000 V	2.5 mA	3.5 W	5 °C	75 Vh
SEP 1.2	200 V	2.5 mA	3.5 W	5 °C	15 Vh
SEP 1.3	2000 V	2.5 mA	3.5 W	5 °C	410 Vh

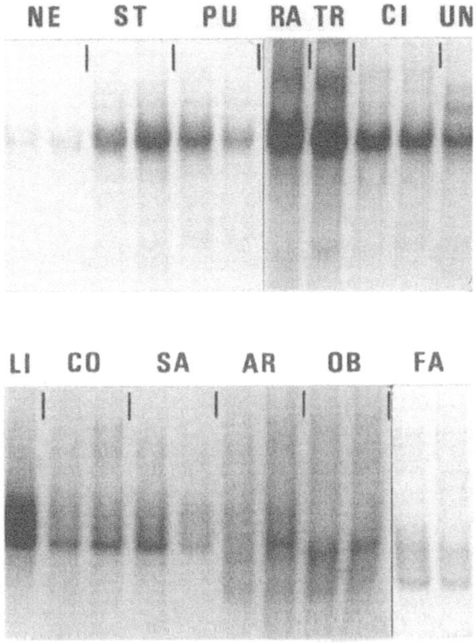

Figure 1. Native vertical PAGE profiles (top = cathodal gel side) of Mb in 13 littorinid species. Abbreviations: AR = *L. arcana*; CI = *Littoraria cingulifera*; CO = *L. compressa*; FA = *L. fabalis*; LI = *L. littorea*; NE = *M. neritoides*; OB = *L. obtusata*; PU = *N. punctata*; RA = *N. radiata*; SA = *L. saxatilis*; ST = *L. striata*; TR = *N. trochoides*; UN = *Littoraria undulata*.

of Hoefer Scientific Instruments), with a gel concentration of 7% and using a discontinuous buffer system consisting of Tris-HCl (pH 9.0) in the gel and Tris-Glycine (pH 9.0) in the tray (Backeljau, 1989). Electrophoreses were started at 25 V for 15 min and continued at 150 V until the migrating front reached the anodal (lower) gel side. During the experiments, temperatures were kept below 15 °C. Depending on the concentration of the homogenates (roughly estimated by the intensity of the red colour), 1–7 μl of sample was applied per run.

Horizontal IEF was performed in precast mini polyacrylamide gels of 50 x 43 x 0.35 mm containing a 3–9 or a 4–6.5 pH gradient. Gels were run on LKB-Pharmacia's PhastSystem (Olsson et al., 1988; Backeljau et al., 1994), according to the programmed conditions outlined in Table 1 and with the sample applicator positioned in the cathodal region of the gels. IEF runs took 30–45 min.

After IEF, soluble proteins were stained with Coomassie Brilliant Blue R-250. Therefore, gels were first placed for 15 min in an aqueous solution containing 5% (w/v) 5-sulfosalicylic acid and 10% (w/v) trichloroacetic acid. Subsequently, gels were rinsed for 2 x 10 min in an aqueous stock solution containing 35% (v/v) ethanol and 10% (v/v) glacial acetic acid. Next, gels were stained for 5 min in stock solution in which 0.2% (w/v) Coomassie Brilliant Blue R-250 was dissolved. Finally, gels were destained for 2 x 10 min in pure stock solution. Coomassie staining of PAGE gels was done in the same way, but with each step taking more time.

Mb bands were identified by a specific and sensitive benzidine staining (Broyles et al., 1979). To this end gels were placed for 10 min in a benzidine solution (0.4 g benzidine dissolved in 200 ml of a 1:1:1 mixture of acetic acid, ethanol and water), after which they were transferred into a 2% (v/v) aqueous hydrogen peroxide solution. As benzidine is a highly potent carcinogen (Broyles et al., 1979; Riggs, 1981), this staining procedure was only used for the initial detection of Mb in the profiles and for the evaluation of new patterns.

Given the preliminary nature of the present analyses, we neither quantified protein concentrations, nor tried to infer exact molecular weights or isoelectric points. Yet, such data are not a prerequisite for systematic comparisons (e.g. Macaranas et al., 1996;

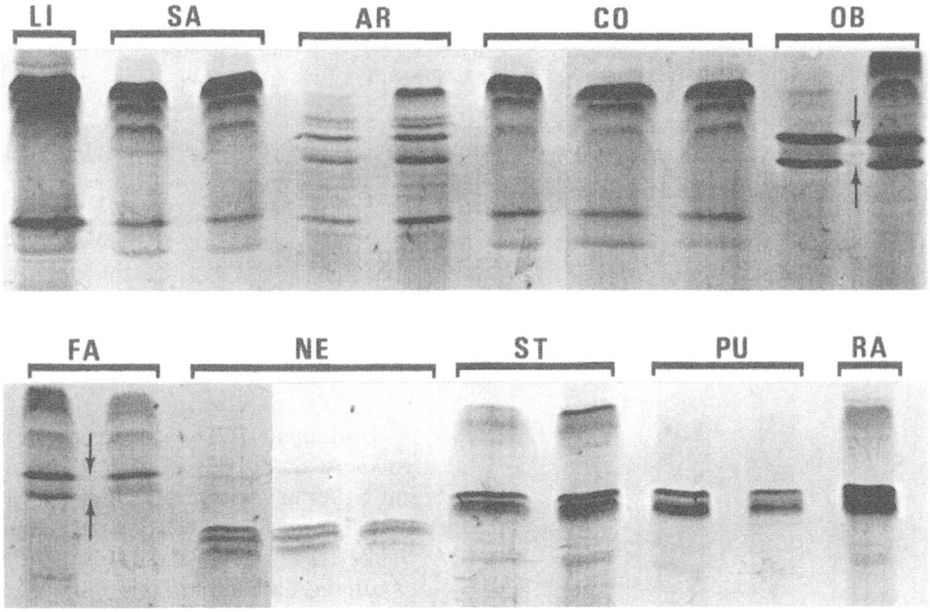

Figure 2. IEF patterns (top = anodal gel side) in a 4–6.5 pH gradient of Mb/SP in nine littorinid species. Abbreviations as in Figure 1. The bands marked by an arrow serve as references.

Mangum, 1996; Mangum & Greaves, 1996; Mangum & McKenney, 1996).

Results

All species studied had conspicuously red radular muscles suggesting the presence of Mb. This red colour remained visible during, but disappeared soon after, electrophoresis. Yet, in all cases the presence of Mb was further supported by a strong benzidine staining reaction (Figures 1, 3A).

PAGE was unable consistently to resolve protein bands other than Mb. Yet, it did reveal four different, more or less well-resolved Mb patterns (Figure 1): (1) a single, monomorphic, slow migrating (cathodal) band shared by *M. neritoides*, *Littoraria* spp., *Nodilittorina* spp. and *L. striata*, (2) a nearly monomorphic two-banded, faster migrating pattern shared by *L. littorea*, *L. compressa* and *L. saxatilis*, (3) a complex pattern with a partially more anodal position in *L. obtusata* and *L. fabalis*, and (4) a unique, distinct three- to four-banded pattern in *L. arcana*.

Although in two-banded PAGE patterns, the more cathodal band was sometimes poorly resolved (Figure 1), we found little or no Mb variation within or among the seven *L. littorea* populations or the two *L. saxatilis* populations. In *L. arcana* we detected one individual with four bands instead of the usual three, while another specimen showed a Mb profile typical of the group *L. saxatilis/L. compressa/L. littorea*. It is possible that this individual was misidentified.

In contrast to PAGE, IEF yielded very well-resolved, but complex Mb/SP patterns, and this with both the 3–9 and 4–6.5 pH gradients. In all species, the Mbs had rather low isoelectric points (estimated $pI < 5.0$). The IEF profiles of Mb confirmed the four species groups defined by PAGE, except for *M. neritoides*, which revealed a unique fifth Mb pattern that differed from all other species, including those of the *Littoraria/Nodilittorina/L. striata* group (Figure 2). In this latter group, IEF demonstrated SP differences between the two *Littoraria* species, as well as between *L. striata*, *N. punctata* and *N. radiata*. However, this latter species could not be distinguished from *N. trochoides*. Similarly, *L. obtusata* and *L. fabalis* could not be separated by their Mb patterns (see Olabarria et al., this volume), but both species were strongly differentiated from any other littorinid included in this work (Figure 2).

IEF also uncovered hidden heterogeneity with respect to intraspecific Mb/SP variation in *L. littorea*. In this species the benzidine staining of the IEF profiles consistently revealed two heavy 'Mb' bands (Figure 3A). Yet, the Coomassie staining showed that both

Table 2. Intraspecific Mb/SP variation in *L. littorea*, showing per population the numbers of individuals with a given overall Mb/SP profile (A–F) as resolved with IEF in a 4–6.5 pH gradient. *N* = total number of specimens investigated

Population (*N*)	A	B	C	D	E	F
Brouwersdam (40)	40	–	–	–	–	–
Oostende (38)	38	–	–	–	–	–
Wimereux (40)	3	–	–	1	36	–
Ile de Ré (40)	36	3	1	–	–	–
Ria Ferrol (40)	40	–	–	–	–	–
Ria de Arosa (40)	39	–	–	–	–	1
Milford Haven (34)	34	–	–	–	–	–

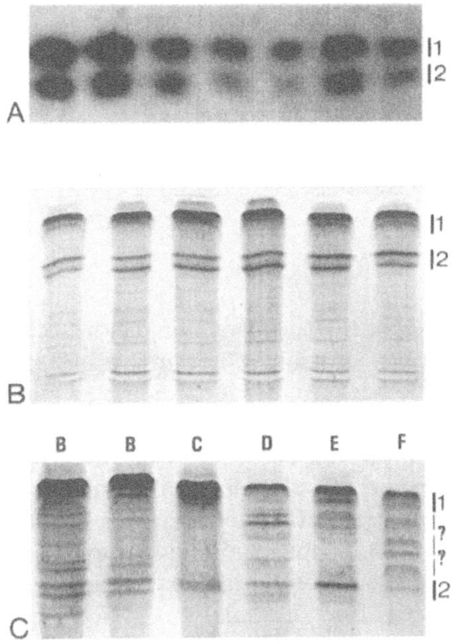

Figure 3. IEF patterns (top = anodal gel side) of Mb/SP variation in *L. littorea*. A. Mb profiles in a 3–9 pH gradient stained by benzidine; B. Most common IEF profile of Mb/SP (pattern A) (3–9 pH gradient) in the Oostende population; C. Variant IEF profiles of Mb/SP resolved in a 4–6.5 gradient. The corresponding zones in the three gels are numbered (1–2), but the bands between the zones cannot be assigned unambiguously

bands consist of several protein fractions with similar pI values. Although some of these fractions must involve Mb, it is likely that others represent SP. Anyway, six different IEF patterns could be distinguished (A–F) with the Mb/SP bands (Figures 3B–C). Pattern A was clearly the common one since four populations (Brouwersdam, Oostende, Milford Haven and Ria Ferrol) were fixed for it (Figure 3B), while in two other populations (Ile de Ré and Ria de Arosa) it was present in more than 90% of the snails (Table 2). In the Wimereux population, however, pattern A was found in only 7.5% of the animals, while 90% of the specimens showed pattern E (one individual had profile D) (Table 2). None of these IEF patterns showed an appreciable Mb/SP variation with PAGE, except for the F profile, which was also quite distinct (one individual) with PAGE. In any case, the observed Mb/SP variation in *L. littorea* could not be correlated with sex, size or colour characteristics of the shell.

In contrast, *L. saxatilis* did not reveal hidden Mb/SP variation with IEF, so that the lack of differentiation between the Norwegian and Spanish populations, as suggested by PAGE, was confirmed by IEF (Figure 4C). Yet, the Mb differentiation between *L. saxatilis/L. compressa/L. littorea* and *L. arcana* was well-resolved and consistent with IEF (Figures 4A–B). The SP profiles reinforced this differentiation.

Finally, it appeared that long term storage at − 80 °C did not affect Mb/SP patterns as is shown by the comparison of the profiles of *L. littorea* (1992–1995). Because Wittenberg & Wittenberg (1981) suggested that Mb in 0.05 M Tris-HCl buffer at pH 8.0 remains stable even after repeated freezing and thawing, we compared the stability of *L. littorea* Mb in this buffer (with 1 mM EDTA) and in our 20% (w/v) aqueous sucrose solution. After repeated freezing and thawing, both Mb preparations yielded exactly the same IEF profiles. Similarly, the EDTA experiment suggested that oxidation had little or no effect on our results. Moreover, even if oxidation affected our Mb profiles, we still think that this would hardly influence the systematic implications of our analyses, because all samples were treated in the same way.

Discussion

This study is the first report on the occurrence of radular Mb in *Littoraria undulata*, *Littoraria cingulifera*, *N. punctata*, *N. trochoides*, *N. radiata*, *L. compressa*, *L. arcana* and *L. fabalis*. Yet, the red colour of the radular tissue combined with a strong benzidine reaction is still not a foolproof identification of Mb, so that additional evidence from spectrophotometric absorption spectra would be desirable.

Our PAGE results confirmed only two observations of Wium-Andersen (1970), viz. (1) *L. striata* has a monomorphic Mb electromorph which moves slower than that of *L. littorea* or that of *L. saxatilis*, and (2)

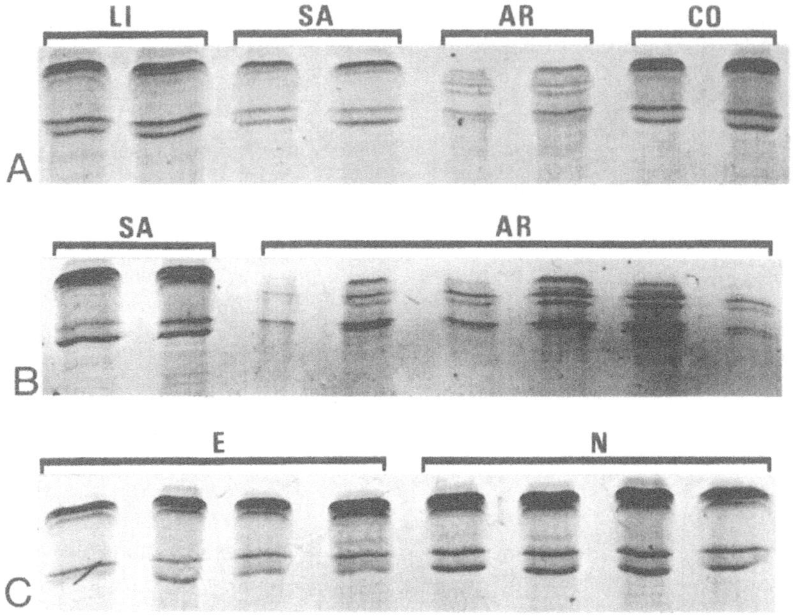

Figure 4. IEF patterns (top = anodal gel side) of Mb/SP in *L. littorea*, *L. saxatilis*, *L. arcana* and *L. compressa* in a 3–9 pH gradient. A–B. Differentiation and consistency of the *L. arcana* pattern; C. Lack of differentiation between *L. saxatilis* from Spain (E) and Norway (N). Abbreviations as in Figure 1.

L. littorea and *L. saxatilis* have a similar two-banded Mb pattern. However, our PAGE results differed from those of Wium-Andersen (1970) by (1) the single Mb band in *L. striata*, (2) the presence of a rare Mb variant in *L. littorea*, and (3) the absence of (Mendelian) variation in the Mb of *L. saxatilis*. Further analyses are required to determine whether these discrepancies are due to technical artefacts (e.g. the single or two banded Mb of *L. striata*), or biases in sample size and geographic origin of the populations studied (e.g. the variation or lack thereof in *L. littorea* and *L. saxatilis*). Anyway, we currently doubt that Wium-Andersen's (1970) Mendelian interpretation of the Mb variation in *L. saxatilis* can be maintained. This interpretation is indeed difficult to reconcile with the fixed two-banded Mb patterns the same author reported for *L. striata* and *L. littorea*, unless (1) both species show fixed heterozygosity at the Mb locus, (2) they have a fundamentally different genetic basis of Mb expression, and/or (3) the quaternary structure of Mb in *L. littorea* and *L. striata* is different from that of rough periwinkles (see Olabarria et al., this volume). In this context it is worth mentioning that the Mb of *L. littorea* is reported to be a dimer, that at low concentrations mainly exists as a dissociated monomer (Read, 1968; Terwilliger & Read, 1969).

Although IEF yielded a higher resolution than PAGE, it was difficult, if not impossible, to interpret the genetic background of the complex banding patterns (see Olabarria et al., this volume). The increased number of Mb bands observed with IEF may be due to several factors. Firstly, electrophoresis or IEF of homogeneous preparations of ferric Mb often yield many fractions with identical amino acid compositions, ligand kinetics and conformations, but different charges (e.g. Rumen, 1959; Edmundson & Hirs, 1962; Atassi, 1964; Parkhurst & LaGow, 1975; Wittenberg & Wittenberg, 1981). Yet, these fractions are not artefacts of the isolation procedure or proteolysis in the muscle homogenates (Wittenberg & Wittenberg, 1981). Multiple Mb bands can also be caused by *in vitro* auto-oxidation (Righetti, 1983), because fully oxidised, partially oxidised and reduced Mb may have different charges. Thirdly, complex Mb band profiles may also result if the dissociation products of the dimeric littorinid Mb have different isoelectric points. Since we diluted our samples it is possible that (part of) the Mb was dissociated. Anyway, under physiological conditions some haemoglobins are known to dissociate readily and reversibly (Righetti, 1983) and even human myoglobin, which is definitely a monomer, may produce multiple fractions with different pI values (e.g. Giraldino et al., 1995). Hence, in the absence of further data, the interpretation of the *L. littorea* Mb polymorphism remains unclear. However, we suspect that at least part of the observed intraspecific Mb variation may have an ecophysiological significance reflecting feeding regimes, intertidal zonation patterns, pollution, etc. Alyakrinskaya (1986, 1989, 1994), for example, related differential Mb concentrations in prosobranch radular muscles with feeding and activity patterns, while Bonaventura et al. (1975) more generally emphasised the ecophysiological significance of multiple haemoglobins and haemoglobin polymorphisms in vertebrates. Anyway, the *L. littorea* Mb/SP polymorphism strongly contrasts with some allozyme data that suggest a high genetic homogeneity among European *L. littorea* populations (e.g. Johannesson, 1992). Nevertheless, Johannesson (1992) remarked (1) that there was a handful of genotypically 'abnormal' specimens along the Belgian coast, and (2) that the population from Roscoff (France) studied by Berger (1977) may be an 'odd' representative of European *L. littorea*. This led Johannesson (1992) to the very speculative suggestion that two sibling species may be involved. Without expanding on this issue, it is noteworthy that in our study it was the two French *L. littorea* populations that showed the most conspicuous Mb/SP heterogeneity.

Another promising result of our Mb/SP data was the suggestive diagnostic differentiation between *L. arcana* and *L. saxatilis/L. compressa*. This result was unexpected because, hitherto, no diagnostic protein markers were known for *L. arcana* (review by Reid, 1996). Yet, given the limited number of populations studied and the relatively small sample sizes, it is evident that the value of Mb/SP as a diagnostic marker must be verified on the basis of a more extensive survey of *L. arcana*, *L. saxatilis* and *L. compressa*.

In general, the combination of Mb with SP increased the value of radular protein patterns as taxonomic markers. This was not only evident for the differentiation of *L. arcana* and *L. saxatilis*, but also of *L. littorea* and *L. saxatilis*, which had the same Mb profile, but differed consistently by other radular SP. A similar observation applied to the two *Littoraria* species. However, neither *N. trochoides* and *N. radiata* nor *L. saxatilis* and *L. compressa* could be differentiated during this study (for the *L. obtusata/fabalis* complex see Olabarria et al., (1998)).

The last surprising result of our Mb work was the relative intra- and interspecific Mb diversity in *Littorina* spp. compared to the electrophoretic (PAGE and IEF) Mb monomorphism within and among the *Littoraria/Nodilittorina* species. Obviously, the lack of Mb variation in these latter species needs further confirmation with larger sample sizes, particularly since Jones (1972) found a relatively high degree of 'species-specific' Mb differentiation among Panamanian *Nodilittorina* and *Littoraria* species. The fact that our analyses suggest that the Mb of *L. striata* belongs to the *Littoraria/Nodilittorina* group and thus differs from any other *Littorina* Mb so far studied, emphasises the still somewhat problematic generic assignment and phylogenetic position of this species, which has been variously associated with *Melarhaphe*, *Nodilittorina* or (currently) *Littorina* (for recent discussions see Reid, 1989, 1990, 1996; Backeljau & Warmoes, 1992; Reid et al., 1996). Hence, if the placement of *L. striata* in the genus *Littorina* is correct, then the Mb similarity between *L. striata* and *Littoraria/Nodilittorina* may be a plesiomorphic condition. However, the Mb data of *L. striata* should be interpreted with much caution, because (1) the possibility of hidden heterogeneity cannot yet be ruled out, and (2) they may reflect ecological or functional, rather than phylogenetic, correlates. Reid (1996), for example, remarked that the ecology and radula morphology of *L. striata* resembles that of many *Nodilittorina* species. It is tempting to hypothesise that 'ecological or functional correlations' also account for the strong Mb differentiation of *L. obtusata/L. fabalis* (the only obligatory macrophyte-associated littorinids included here; cf. Reid, 1996) and for the Mb characteristics of *L. littorea*. Yet, it must be emphasised that these issues are highly speculative and thus require more detailed investigations.

In conclusion, although the high expectations raised by Wium-Andersen's (1970) paper were not satisfied, and despite the difficulties with the genetic interpretation of littorinid Mb patterns, our results suggest that the electrophoretic analysis of radular Mb/SP may provide a convenient taxonomic marker that can be used in a similar way as haemocyanins (e.g. Symondson & Walton, 1994; Mangum, 1996; Mangum & Greaves, 1996; Mangum & McKenney, 1996). Moreover the ecological and functional significance of littorinid Mb clearly needs to be investigated further as we suspect that such data may yield valuable insights as to how periwinkles interact with their environment.

Acknowledgements

We are indebted to Philippe Bouchet, Patrick Grootaert, Brian Morton, Jon Arne Sneli, Anders Warén and Thierry Warmoes for their help in obtaining some of the material. We gratefully acknowledge the constructive comments of Peter Mill, David Reid and two anonymous referees. This research was supported by the MAST 3 programme of the European Commission under contract number MAS3-CT95-0042 (AMBIOS). Additional funding was received from PRAXIS XXI 2/2.1/BIA/169/94 (Portugal) and F. J. B. R. grant 2.0023.94. R. M. was funded by ERASMUS and H. D. W. holds an IWT scholarship.

References

Alexandrino, P. J., N. Ferrand & J. Rocha, 1996. Genetic polymorphism of a haemoglobin chain and adenosine deaminase in European shads: evidence for the existence of two distinct genetic entities with natural hybridization. J. Fish Biol. 48: 447–456.

Alyakrinskaya, I. O., 1986. Hemoglobin content in subradular cartilages of the gastropod mollusc *Littorina littorea*. Zh. Evol. Biokhim. Fisiol. 22: 575–577. [in Russian].

Alyakrinskaya, I. O., 1989. Biochemical adaptations of some littoral molluscs to feeding. Zh. Evol. Biokhim. Fisiol. 25: 112–115. [in Russian].

Alyakrinskaya, I. O., 1994. Haemoglobin concentration and its functional meaning in White Sea *Littorina*. Gidrobiol. Zh. 30: 69–78. [in Russian].

Atassi, M. Z., 1964. Properties of components of myoglobin of the sperm whale. Nature 202: 496–498.

Backeljau, T., 1989. Electrophoresis of albumen gland proteins as a tool to elucidate taxonomic problems in the genus *Arion* (Gastropoda, Pulmonata). J. med. appl. malacol. 1: 29–41.

Backeljau, T. & T. Warmoes, 1992. The phylogenetic relationships of ten Atlantic littorinids assessed by allozyme electrophoresis. In J. Grahame, P. J. Mill & D. G. Reid (eds), Proceedings of the Third International Symposium on Littorinid Biology. The Malacological Society of London, London: 9–24.

Backeljau, T., K. Breugelmans, H. Leirs, T. Rodriguez, D. Sherbakov, T. Sitnikova, J.-M. Timmermans, J. L. Van Goethem & E. Verheyen, 1994. Application of isoelectric focusing in molluscan systematics.The Nautilus, Suppl. 2: 156–167.

Bailey, J. B., E. H. Michelson & W. L. Paraense, 1986. Differentiation of the sibling species *Biomphalaria occidentalis* and *Biomphalaria tenagophila* by the electrophoretic patterns of their hemoglobin. Mem. Inst. Oswaldo Cruz 81: 319–322.

Bannister, W. H., J. V. Bannister & H. Micallef, 1968. Occurrence of haemoglobin in the radular muscles of some prosobranch molluscs. Comp. Biochem. Physiol. 24: 1061–1063.

Basaglia, F. & C. Callegarini, 1987. Electrophoretic and isoelectrophoretic characteristics of hemoglobins of Italian ictalurids. Comp. Biochem. Physiol. 86B: 269–271.

Berger, E., 1977. Gene-enzyme variation in three sympatric species of *Littorina*. II. The Roscoff population, with a note on the origin of North American *L. littorea*. Biol. Bull. 153: 255–264.

Bonaventura, C. & J. Bonaventura, 1983. Respiratory pigments: structure and function. In K. M. Wilbur & P. W. Hochachka, (eds), The Mollusca, Volume 2, Environmental Biochemistry and Physiology. Academic Press, New York: 1–50.

Bonaventura, J., C. Bonaventura & B. Sullivan, 1975. Hemoglobins and hemocyanins: comparative aspects of structure and function. J. exp. Zool. 194: 155–174.

Broyles, R. H., B. M. Pack, S. Berger & A. R. Dorn, 1979. Quantification of small amounts of hemoglobin in polyacrylamide gels with benzidine. Analyt. Biochem. 94: 211–219.

Das, R. & R. Handique, 1996. Hemoglobin in *Chironomus ramosus* (Insecta, Diptera): an electrophoretic study of polymorphism, developmental sequence and interspecific relationship. Hydrobiologia 318: 43–50.

Edmundson, A. B. & C. H. W. Hirs, 1962. On the structure of sperm whale myoglobin I. The amino acid composition and terminal groups of the chromatographically purified protein. J. Mol. Biol. 5: 663–682.

Giraldino, I., M. C. Pico & R. Alegre, 1995. Mioglobina humana: aislamiento y purificacion. Rev. Biol. 9: 23–29.

Johannesson, K., 1992. Genetic variability and large scale differentiation in two species of littorinid gastropods with planktotrophic development, *Littorina littorea* (L.) and *Melarhaphe (Littorina) neritoides* (L.) (Prosobranchia: Littorinacea), with notes on a mass occurrence of *M. neritoides* in Sweden. Biol. J. linn. Soc. 47: 285–299.

Jones, M. L., 1972. Comparisons of electrophoretic patterns of littorine snails of Panama: an attempt to define geminate species. XVIIe Congrès International de Zoologie, Monte Carlo, 25–30 Septembre 1972, Thème 3: 1–10.

Lavrenchenko, L. A., O. P. Kihnova & V. N. Orlov, 1992. Hemoglobin patterns: a possible implication in systematics of multimammate rats *Mastomys* (Muridae, Rodentia). Zool. Zh. 71: 85–93. [in Russian].

Macaranas, J. M., L. Q. Agustin & A. E. Eknath, 1996. Multiple haemoglobins in three tilapiine species of the genus *Oreochromis* and in eight strains of *O. niloticus* (L.). Aquacult. Res. 27: 597–601.

Mangum, C. P., 1996. Subunit composition of polymorphic hemocyanins in the decapod crustaceans: differences between sibling species. Physiol. Zool. 69: 568–585.

Mangum, C. P. & J. Greaves, 1996. Hemocyanins of the genus *Uca*: structural polymorphisms and native oligomers. J. exp. mar. Biol. Ecol. 199: 1–15.

Mangum, C. P. & A. L. McKenney, 1996. Subunit composition of the crustacean hemocyanins: divergence in incipient speciation. Biol. Bull. 191: 33–41.

Mulvey, M., M. C. Newman & D. S. Woodruff, 1988. Genetic differentiation among west Indian populations of the schistosome-transmitting snail *Biomphalaria glabrata*. Malacologia 29: 309–317.

O'Gower, A. K. & P. I. Nicol, 1968. A latitudinal cline of haemoglobins in a bivalve mollusc. Heredity 23: 485–491.

Olabarria, C., J.-M. Timmermans & T. Backeljau, 1998. Electrophoretic heterogeneity within and between flat periwinkles (Mollusca: Gastropoda) along an intertidal transect at Ria Ferrol, northwest Spain. Hydrobiologia 378: 11–19.

Olsson, I., U.-B. Axio-Fredriksson, M. Degerman, & B. Olsson, 1988. Fast horizontal electrophoresis. I. Isoelectric focusing and polyacrylamide gel electrophoresis using Phastsystem. Electrophoresis 9: 16–22.

Parkhurst, L. J. & J. LaGow, 1975. Kinetic and equilibrium studies of the ligand binding reactions of eight electrophoretic components of sperm whale ferrimyoglobin. Biochemistry 14: 1200–1205.

Read, K. R. H., 1966. Molluscan hemoglobin and myoglobin. In K. M. Wilbur & C. M. Yonge (eds), Physiology of Mollusca, Volume II. Academic Press, New York: 209–232.

Read, K. R. H., 1968. The myoglobins of the gastropod molluscs *Busycon contrarium* Conrad, *Lunatia heros* Say, *Littorina littorea* L. and *Siphonaria gigas* Sowerby. Comp. Biochem. Physiol. 25: 81–94.

Reid, D. G., 1986. The littorinid molluscs of mangrove forests in the Indo-Pacific region. British Museum (Natural History), London. 228 pp.

Reid, D. G., 1989. The comparative morphology, phylogeny and evolution of the gastropod family Littorinidae. Phil. Trans. r. Soc. Lond. B 324: 1–110.

Reid, D. G., 1990. A cladistic phylogeny of the genus *Littorina* (Gastropoda): implications for evolution of reproductive strategies and for classification. Hydrobiologia 193: 1–19.

Reid, D. G., 1992. The gastropod family Littorinidae in Hong Kong. In B. Morton (ed.), The Marine Flora and Fauna of Hong Kong and Southern China III. Hong Kong University Press, Hong Kong: 187–210.

Reid, D. G., 1996. Systematics and evolution of *Littorina*. The Ray Society, London. 463 pp.

Reid, D. G., E. Rumbak & R. H. Thomas, 1996. DNA, morphology and fossils: phylogeny and evolutionary rates of the gastropod genus *Littorina*. Phil. Trans. r. Soc. Lond. B 351: 877–895.

Riggs, A., 1981. Preparation of the blood hemoglobins of vertebrates. Meth. Enzym. 76: 5–29.

Righetti, P. G., 1983. Isoelectric focusing: theory, methodology and applications. Elsevier, Amsterdam. 386 pp.

Rumen, N. M., 1959. Isolation of five myoglobins from seal (*Phoca vitulina*). Acta Chem. Scand. 13: 1542–1546.

Symondson, W. O. C. & M. P. Walton, 1994. Electrophoretic separation of pulmonate haemocyanins: a simple taxonomic tool. J. moll. Stud. 60: 351–354.

Terwilliger, R. C. & R. H. Read, 1969. The radular muscle myoglobins of the amphineuran mollusc, *Acanthopleura granulata* Gmelin. Comp. Biochem. Physiol. 29: 551–560.

Wittenberg, J. B. & B. A. Wittenberg, 1981. Preparation of myoglobins. Meth. Enzymol. 76: 29–42.

Wium-Andersen, G., 1970. Haemoglobin and protein variation in three species of *Littorina*. Ophelia 8: 267–273.

Woodruff, D. S. & M. Mulvey, 1997. Neotropical schistosomiasis: African affinities of the host snail *Biomphalaria glabrata* (Gastropoda: Planorbidae). Biol. J. linn. Soc. 60: 505–516.

Hydrobiologia **378:** 53–57, 1998.
R. M. O'Riordan, G. M. Burnell, M. S. Davies & N. F. Ramsay (eds), Aspects of Littorinid Biology.

Allozyme variation and behavioural dimorphism among populations of *Littorina brevicula* (Philippi) from Japan

N. I. Zaslavskaya[1] & Y. Takada[2]
[1] *Institute of Marine Biology, Vladivostok 690041, Russia*
[2] *Seikai National Fisheries Research Institute, Fukai-ota 148–446, Ishigaki 907–0451, Japan*

Key words: Littorina brevicula, migration behaviour, genetic differentiation

Abstract

In Amakusa (Japan), an apparently homogeneous population of *Littorina brevicula*, divides into two groups during the reproductive season (winter). Some snails migrate to the lower intertidal zone, while the rest remain in the upper zone. This migration behaviour might be controlled genetically and the behavioural dimorphism may be maintained by assortative mating. To evaluate this hypothesis, we studied genetic variation at four polymorphic enzyme loci (Pgm, Aldh, Ipp, Gpt) in both groups, as well as in seven other samples from the coast of Japan and Vostok Bay (Russian coast of the Sea of Japan). There were no significant allele frequency differences between the two behavioural groups of snails with differing migration patterns ($\chi^2 = 8.411$, df = 12, $p > 0.5$). The samples from Japan belong to a panmictic population. Yet, a significant heterogeneity was found between the sample from Vostok Bay and the samples from Japan, indicating a reduced gene flow between these regions. Generally, observed genotype frequencies showed good fits to Hardy-Weinberg equilibrium expectations at three of the four loci investigated.

Introduction

Littorina brevicula (Philippi) is one of the most common snails found in the intertidal zone of Japan. Seasonal migration behaviour is characteristic of this species (Kojima, 1959; Luckens, 1970). In summer the snails occur in the upper intertidal zone and in winter they migrate downward to the lower zone. Kojima (1959) suggested that this downward winter migration is related to reproduction. However, Takada (1992, 1995) found that the snails in Amakusa, southern Japan, show two alternative migratory patterns during the reproductive season (i.e. winter): one group of snails migrates to lower zones, while the rest remain in the upper zone. Experiments with marked specimens showed that individual migration behaviour was consistent over two successive winters, i.e. the same group of snails migrated downward every winter and the same group of snails remained in the upper zone. Moreover, transplanted snails moved toward the original zones where they were caught. Hence the snails actively selected their habitat in winter. Most snails copulated within each group. This led to a hypothesis that the migratory behaviour of *L. brevicula* might be genetically controlled and that this behavioural dimorphism may be maintained by partial reproductive isolation between the two groups (Takada, 1995).

In order to test the hypothesis of genetic differentiation between the two migration groups, we used allozyme electrophoresis to assess the degree of genetic differences between the both groups. These data were then compared with the degree of intraspecific macrogeographic genetic differentiation of *L. brevicula* in the Sea of Japan.

Materials and methods

Two samples from Amakusa, subsequently referred to as 'upper' and 'lower', were taken near the town of Tomioka (Figure 2).

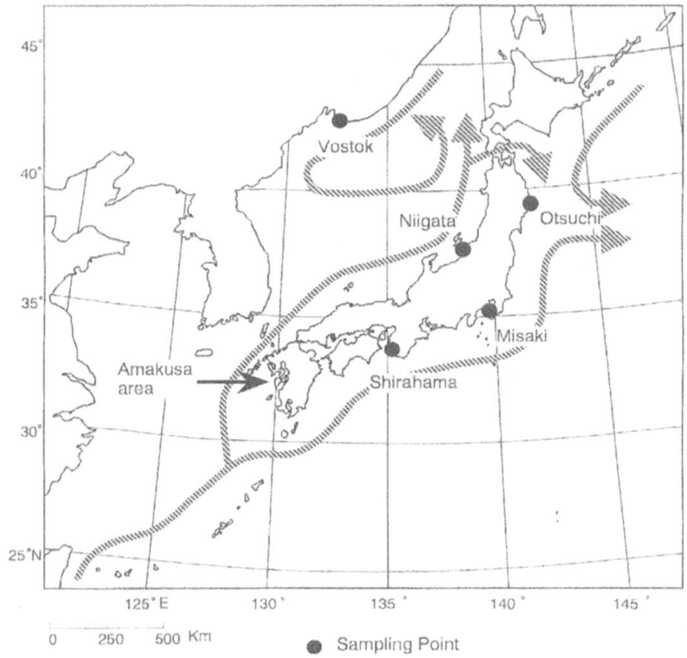

Figure 1. Map to show the sample sites (the arrows show a scheme of the sea currents by Yurasov & Yarichin, 1991).

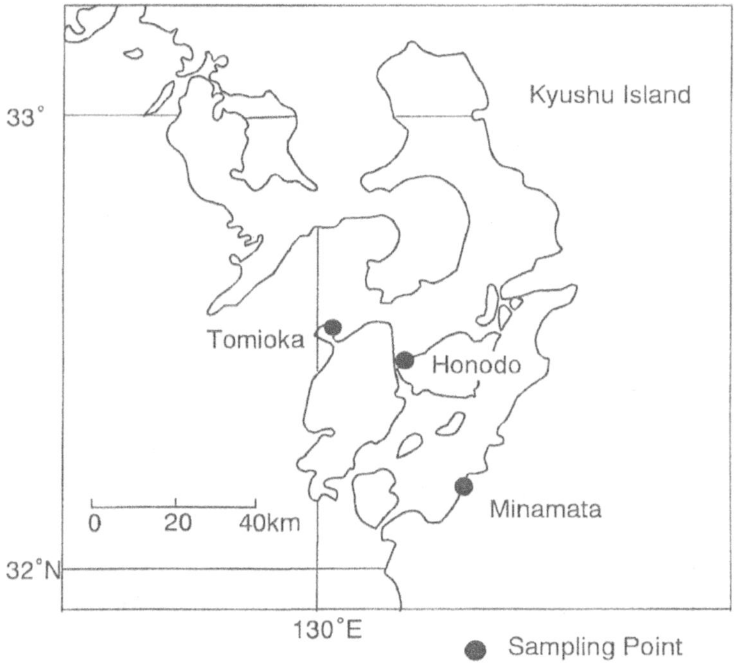

Figure 2. Sample sites around the Amakusa area.

Table 1. Allele frequencies in 9 studied localities of *Littorina brevicula*

Locus	Localities								
	Mina-mata	Shira-chama	Misaki	Vostok	Otsuc-hi	Niig-ata	Honodo	Tomioka	
								lower	*upper*
Pgm									
(*N*)*	94	94	90	84	80	78	78	80	80
A	0.005	0.000	0.000	0.000	0.006	0.026	0.013	0.000	0.000
B	0.027	0.011	0.039	0.006	0.038	0.026	0.026	0.019	0.025
C	0.899	0.931	0.900	0.982	0.881	0.897	0.949	0.913	0.919
D	0.064	0.053	0.061	0.012	0.050	0.045	0.013	0.069	0.500
E	0.005	0.005	0.000	0.000	0.025	0.006	0.000	0.000	0.006
Aldh									
(*N*)	94	93	90	79	79	78	78	80	80
A	0.000	0.005	0.006	0.006	0.000	0.000	0.000	0.000	0.000
B	0.064	0.059	0.134	0.070	0.146	0.135	0.084	0.044	0.087
C	0.840	0.801	0.728	0.835	0.715	0.705	0.788	0.793	0.788
D	0.016	0.022	0.017	0.057	0.013	0.026	0.019	0.019	0.025
E	0.080	0.113	0.117	0.032	0.127	0.134	0.109	0.144	0.100
Ipp									
(*N*)	93	93	90	84	80	78	78	80	80
A	0.065	0.043	0.067	0.077	0.056	0.038	0.071	0.063	0.069
B	0.005	0.005	0.022	0.030	0.006	0.013	0.038	0.006	0.025
C	0.925	0.941	0.911	0.827	0.906	0.936	0.891	0.925	0.900
D	0.005	0.011	0.000	0.066	0.032	0.013	0.000	0.006	0.006
Gpt									
(*N*)	89	89	85	81	80	78	78	80	80
A	0.000	0.006	0.000	0.000	0.006	0.006	0.000	0.000	0.000
B	0.028	0.022	0.035	0.006	0.044	0.058	0.013	0.019	0.019
C	0.949	0.922	0.935	0.957	0.925	0.923	0.968	0.962	0.950
D	0.017	0.028	0.018	0.037	0.006	0.013	0.013	0.019	0.025
E	0.006	0.022	0.012	0.000	0.019	0.000	0.006	0.000	0.006

* *N* – number of snails studied

Snails were marked in February when the population was still separated and sampled in May when they mixed together. In addition, six samples of *L. brevicula* were collected from other localities off the coast of Japan (Niigata, Otsuchi, Misaki, Shirahama, Minamata, Honodo) (Figures 1, 2). Snails were brought alive to Russia to the Vostok Biological Station of the Institute of Marine Biology located on the coast of Vostok Bay (Peter the Great Bay, northern part of the Sea of Japan). A sample of the snails from Vostok Bay was also collected (Figure 1).

Horizontal starch gel electrophoresis was carried out by standard methods as described by Zaslavskaya, (1989). Four enzymes coded by four polymorphic loci were studied: phosphoglucomutase (Pgm; E.C.2.7.5.1), alanopine dehydrogenase (Aldh; E.C.1.5.1.17), inorganic pyrophosphatase (Ipp; E.C.3.6.1.1) and glutamate pyruvate transaminase (Gpt; E.C.2.6.1.2).

Table 2. Contingency χ^2-test of allele frequencies between 'upper' and 'lower' groups of snails

Locus	χ^2	df	*p*
Pgm	1.616	3	0.656
Aldh	3.737	3	0.291
Ipp	1.902	3	0.593
Gpt	1.156	3	0.764
Overall loci	8.412	12	0.752

The data were analysed with the BIOSYS package (Swofford & Selander, 1981). Variation in allelic

Table 3. χ^2 heterogeneity values for 9 and 8 populations of *L. brevicula*

9 populations				8 populations (without Vostok's population)		
Locus	χ^2	df	p	χ^2	df	p
Pgm	54.200	32	0.0084	39.443	28	0.0741
Aldh	53.807	32	0.0093	33.503	28	0.2177
Ipp	56.582	24	0.0002	27.529	21	0.1540
Gpt	35.575	32	0.3037	25.811	28	0.5835
Overall loci	200.172	120	0.00001	126.285	105	0.0771

frequencies between localities was analysed using χ^2 contingency tests. χ^2 goodness of fit tests were used to study the agreement with Hardy-Weinberg expectations. Pseudoprobability contingency χ^2 tests were employed using the program CHIRXC (Zaykin & Pudovkin, 1993). Shidak's multiplicative inequality for calculation of critical values of χ^2-distribution was used when several independent tests were calculated (Zaykin & Pudovkin, 1991).

Allele frequencies at the four loci for the nine samples are given in Table 1. At all loci observed genotype frequencies showed a good fit to those expected under Hardy-Weinberg equilibrium except at Ipp in the 'lower' group ($\chi^2 = 26.229$; df$=6$, $k=9$, $p = 0.013$). This fact can be explained by an extremely rare event: the heterozygote AB was met (the expected value is 0.063).

No significant allele frequency differences were observed between the samples from the two behavioural groups (Table 2).

The following results were obtained for the other localities. When we examine all populations a high degree of heterogeneity was observed (Table 3): $p < 0.00001$. We calculated the χ^2-heterogeneity values for pair-wise comparisons and it was found that the heterogeneity is mainly due to the sample from Vostok Bay. When this sample was eliminated from the analysis the heterogeneity among the rest of the localities was not significant (Table 3).

Results

Discussion

Takada (1995) came to the conclusion, that in Amakusa the dimorphism in migration behaviour of *L. brevicula* during the reproductive season results in partial reproductive isolation between the two subpopulations and this reproductive isolation promotes the dimorphic behaviour. He suggested the next explanation of this dimorphism. Two groups are descendants of geographically different populations which evolved under different selection pressures. Amakusa is a hybridising zone for two populations. In the northern part of the sea of Japan all the adults migrate downwards to avoid low temperature and establishment of ice. The 'lower' group in Amakusa shows a similar migration behaviour to the northern population. On the other hand, the 'upper' group may be a descendant of a southern population which does not migrate downwards because of high predation pressure on the low shore. Most likely such behaviour is genetically controlled. Our results did not show the genetic differentiation among these groups. Yet, if these groups are recent it might happen they have not yet acquired substantial differentiation. Furthermore, as copulation and downward migration start simultaneously in November (Takada, 1995) and copulation continues until the end of upward migration in March some snails from the 'upper' and 'lower' subpopulations may copulate with each other. In all probability the small gene exchange slows down the process of an accumulation of the genetic differences.

Generally, species with a planktonic stage are believed to have low genetic diversity between local populations (Johannesson, 1988; Behrens Yamada, 1989). *L. brevicula* shows little morphological variation (Reid, 1996) and this can be explained by the homogenising effect of presumably substantial gene flow in a widely dispersing species. Studies of allozyme variation support the expectation of relatively low genetic differentiation, although some loci have been found to show significant differences in allele frequencies even between samples from the same shore, probably as a result of selection (Tatarenkov, 1994).

Our data support low interpopulation differentiation. It seems plausible that the samples from Japan belong to a single population. Overall life-span in plankton of eggs and larvae is more than 4 weeks (Golikov, 1976). The system of currents (Yurasov & Yarichin, 1991; Figure 1) makes it possible for propagation of the larvae along the coast line of Japan. On the other hand this system of currents as well as a long expanse of sea water between Vostok Bay and Japan prevent the exchange of the larvae between them.

Acknowledgements

We thank Dr A. I. Pudovkin of the Laboratory of Genetics, Institute of Marine Biology for comments on the manuscript and T. P. Priima for the assistance in the allozyme analysis of the snails.

References

Behrens Yamada, S., 1989. Are direct developers more locally adapted than planktonic developers? Mar. Biol. 103: 403–411.

Golikov, A. N., 1976. Gastropoda. The Animals and Plants of Peter the Great Bay. Nauka, Leningrad: 79–92.

Johannesson, K., 1988. The paradox of Rockall: why is a brooding gastropod (*Littorina saxatilis*) more widespread than one having a planktonic larval dispersal stage (*L. littorea*)? Mar. Biol. 99: 507–513.

Kojima, K., 1959. The relation between seasonal migration and spawning of a periwinkle, *Littorina brevicula* (Philippi). Bull. biol. Stn. Asamushi 9: 183–186.

Luckens, P. A., 1970. Seasonal distributional variation within a limited shore area at Asamushi. Sci. Rep. Tohoku Univ. Ser. IV (Biol.) 35: 161–170.

Reid, D. G., 1996. Systematics and evolution of *Littorina*. The Ray Society, London. 463 pp.

Swofford, D. L. & R. B. Selander, 1981. BIOSYS-1: a Fortran program for the comprehensive analysis of electrophoretic data in population genetics and systematics. J. Hered. 72: 281–283.

Takada, Y., 1992. The migration and growth of *Littorina brevicula* on a boulder shore in Amakusa, Japan. In J. Grahame, P. J. Mill & D. G. Reid (eds), Proceedings of the Third International Symposium on Littorinid Biology, The Malacological Society of London, London: 277–279.

Takada, Y., 1995. Seasonal migration promoting assortative mating in *Littorina brevicula* on a boulder shore in Japan. Hydrobiologia 309: 151–159.

Tatarenkov, A. N., 1994. Genetic heterogeneity in populations of *Littorina brevicula* (Philippi) (Mollusca: Gastropoda) in the northern part of Peter the Great Bay (Sea of Japan). Cah. Biol. mar. 35: 264.

Yurasov G. I. & V. G. Yarichin, 1991. The Water Currents in the Sea of Japan. Far East Branch of the Academy Science of the USSR, Vladivostok, 173 pp.

Zaslavskaya, N. I., 1989. Genetic variability in four Pacific species of periwinkles (Mollusca: Gastropoda). Genetika 25: 1636–1644.

Zaykin, D. V. & A. I. Pudovkin, 1991. The program MULTEST: calculation of indices of statistical significance for multiple tests. Genetika 27: 2034–2038.

Zaykin, D. V. & A. I. Pudovkin, 1993. Two programs to estimate significance of χ^2 values using pseudoprobability test. J. Hered. 84: 152.

Hydrobiologia **378:** 59–69, 1998.
R. M. O'Riordan, G. M. Burnell, M. S. Davies & N. F. Ramsay (eds), Aspects of Littorinid Biology.

Sexual selection and assortative mating by size and their roles in the maintenance of a polymorphism in Swedish *Littorina saxatilis* populations

J. Erlandsson[1] & E. Rolán-Alvarez[2,*]
[1] *Tjärnö Marine Biological Laboratory, 452 96 Strömstad, Sweden*
[2] *Departamento de Biologia Fundamental (Genetica), Universidad de Vigo, 36200 Vigo, Spain*
(author for correspondence)*

Key words: assortative mating, *Littorina saxatilis*, mating pattern, sexual isolation, sexual selection, size polymorphism

Abstract

Two independent components of mating behaviour, sexual selection and assortative mating, were studied in two allopatric morphs, one sheltered boulder shore form (S-morph) and one exposed cliff shore form (E-morph), of *Littorina saxatilis* from the west coast of Sweden. Sexual selection was studied by comparing the sizes of copulating and non-copulating snails in the field. Size assortative mating was studied by collecting copulating pairs in the field, while assortative mating between morphs was investigated by bringing the pure morphs together in intermediary habitats and then noting the matings.

The S-morph mated randomly in relation to size in two of the studied populations and exhibited a trend towards size assortative mating in a third, while the E-morph showed size assortative mating in both studied populations. The microdistribution of sizes of snails on the shores could not explain all the size assortative mating found, and instead it is argued that a size-based mate rejection behaviour also contributes to the assortative mating in at least some of these populations. There was sexual selection on size in both males and females in the S-morph, with large individuals being favoured as mates. In contrast, copulating snails of the E-morph were smaller than non-copulating ones. The significantly different sexual selection intensities between the two morphs may help to explain the size differences between them. There was random mating between the E- and the S-morphs of *L. saxatilis*, which suggests no incipient reproductive isolation between morphs on Swedish rocky shores. This is in agreement with earlier studies of Swedish populations, but is in contrast to the situation found in other geographical areas.

Introduction

Mate recognition, that is the ability to discriminate between different species and sexes, is of great evolutionary significance for an animal. Mismatings may produce less viable or sterile offspring, or none at all, and reduce the time available for activities that increase individual fitness. Moreover, mating behaviour may be of fundamental importance in the recognition of conspecifics, and thus be responsible for prezygotic reproductive isolation and speciation (Coyne, 1992). However, few species allow detailed studies on mating behaviour in nature, partially owing to the difficulty in finding and studying mating pairs *in situ*.

Mating behaviour in the laboratory has been partitioned into two statistically independent components (Merrel, 1950; Spieth & Ringo, 1983): sexual selection (mating success) and sexual isolation (mating pattern or deviation from 'random mating'). Thus, sexual selection is estimated by comparing the frequencies of mating versus non-mating individuals, while sexual isolation is exclusively estimated from the mating pattern among copulating individuals. The same approach can be extended further to study mating behaviour in the wild (Anderson & McGuire, 1978; Spieth & Ringo, 1983; Santos et al., 1986; Rolán-Alvarez et al., 1995a & b). There are two principal advantages for this statistical partitioning: firstly, each mating component can have different causes in the wild (see Rolán-Alvarez et al., 1995a and Johannesson et al., 1995); and secondly and more importantly, they show different evolutionary consequences. Thus, sexual selection, as a partial component of natural selection, is responsible for the change of gene frequencies in populations and thus directly involved in microevolution. On the other hand, sexual isolation does not change gene frequencies in populations, but may be the factor mainly responsible for isolation between different subpopulations and thus is important for speciation.

Size is one of the most important traits which can affect the life history of a particular species. It is, for example, common that males prefer to mate with large, often more fecund females (Ridley, 1983; Crespi, 1989). Female fecundity has been shown to increase with female size in littorinid snails, being positively correlated with shell height (Hughes & Answer 1982; Janson, 1985). This is a phenomenon that can generate both sexual selection and assortative mating. Sexual selection on female size has been shown in *Littorina littorea*, as large females were found in copulating pairs more often than expected by chance in the field, and also large females copulated more often and for longer than did small females in laboratory experiments (Erlandsson & Johannesson, 1994). Sexual selection on size has also been detected in male *L. fabalis* (Rolán-Alvarez et al., 1995a). In *L. saxatilis*, size assortative mating (Saur, 1990; Johannesson et al., 1995) as well as sexual selection on size (Johannesson et al., 1995) have been observed in a number of natural populations. However, the biological or ethological mechanisms responsible for the assortative mating and the sexual selection have not been fully determined.

Littorina saxatilis has separate sexes and internal fertilisation, and shows promiscuity (Reid, 1996). It lacks pelagic larval dispersal, i.e. it is ovoviviparous, and its reproductive activity is almost constant during the whole year. The growth rate of females of *L. saxatilis* is higher than that of males, resulting in sexual size dimorphism (see Reid, 1996). Females of *L. saxatilis* can store sperm over more than one year since females isolated for more than a year still release viable embryos (K. Johannesson, unpubl. data). There is often striking polymorphism in *L. saxatilis* perhaps because of the lack of larval dispersal, and low adult dispersal leading to the potential for low gene flow (Janson, 1983). On the Swedish west coast different morphs or ecotypes of *L. saxatilis* occupy different rocky shore habitats continuously replacing each other along the coast (Janson, 1982). Allozyme studies suggest that potential genetic exchange between populations of different morphs is of the same magnitude as the exchange between populations of the same morph at similar distances from each other and thus the distinct morphs are considered to be conspecifics (Janson & Ward, 1984; Johannesson & Tatarenkov, 1997). These morphs differ in a number of characters, such as size, shell morphology and migratory behaviour (Janson, 1982, 1983; Janson & Sundberg, 1983), as well as in egg number and embryo size (Janson, 1985). Small thin-shelled snails with a comparatively large foot inhabit exposed rocky cliffs (E-morph), while large thick-shelled snails with a small foot occur on sheltered boulder shores (S-morph). In transitionary zones between exposed cliff and sheltered boulder shores there are several intermediary forms.

This work aims to determine if sexual selection on size and assortative mating by size and between distinct morphs commonly occur in Swedish populations of *L. saxatilis*, and to investigate some of the possible causes and evolutionary consequences.

Materials and methods

Sampling procedure in natural populations

We sampled copulating pairs of the S- and the E-morphs in natural populations on sheltered boulder shores and exposed cliff shores, respectively, during spring (April–May) 1994. Five populations, 3 S- (Saltö West-facing, Saltö South-facing, Saltö Ängklåvebukten) and 2 E-sites (Saltö 1, Saltö 2) were sampled at the island of Saltö (58°2′ N, 11°7′ E), close to Tjärnö Marine Biological Laboratory on the Swedish west coast (see Figure 1 for exact localities). We found 22, 23 and 22 male–female pairs respectively of the S-morph in the three populations, and 18 and 20 male-female pairs of the E-morph in the populations sampled. About 35% of the pairs were maladaptive (between males or between a male and a juvenile). The snails were sampled for a few hours over 1–2 days for each population. The Swedish west coast has a maximum tidal amplitude of about 30 cm, and the water level fluctuations are irregular and caused mainly by changes of such factors as the atmospheric pressure and the strength of wind, waves and currents.

Each pair was gently picked up, separated and the mates were put in separate containers and brought to the laboratory. All individuals of *L. saxatilis* found within an approximately 5 cm radius of a copulating

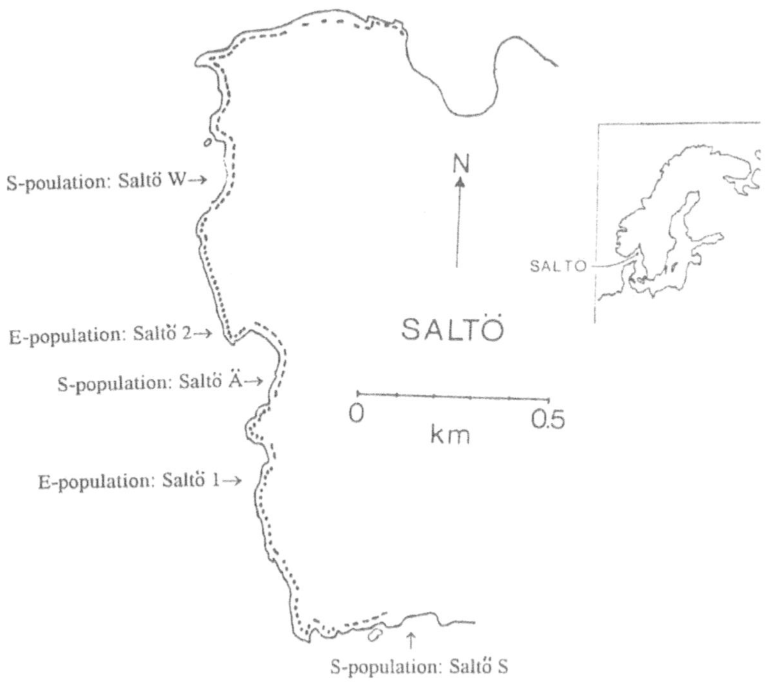

Figure 1. The map shows the different populations of Exposed (2 sites) and Sheltered (3 sites) *Littorina saxatilis* ecotypes sampled on the Swedish west coast. (Saltö W = Saltö West, Saltö S = Saltö South, Saltö Ä = Saltö Ängklåvebukten).

pair were collected to represent non-copulating potential mates. In the laboratory both copulating and non-copulating individuals were measured for shell height and dissected to determine the sex, sexual status (mature or juvenile) and parasitic infection. Females were defined as sexually mature if any eggs or embryos were present in the embryo-pouch, while males with a fully developed penis were considered sexually mature.

Sexual selection on size was studied by comparing the sizes of copulating and non-copulating individuals in each sex and population, to see if any size of mate was favoured over any other. We used the copulating pairs collected to test the correlation between male and female sizes in each population in order to examine if they mated at random or size assortatively.

Experiments in rock-pools

To investigate the mating pattern between the two ecotypes, S and E, we collected individuals for one experimental set-up during one day in early April 1995 and for another set-up during one day in mid May 1995. For each experimental set-up we collected 100 E- and 100 S-individuals and the snails were kept for one day in an aquarium with running sea water before being sorted, marked and used in the experiment. The sex ratio of this species is usually considered to be 1:1 (see Saur, 1990), and therefore it may be argued that there were approximately equal numbers of males and females in the experiments. The different morphs were collected at Saltö Ängklåvebukten (S-morph) and Saltö 2 (E-morph) for experimental set-up 1, and at Saltö West (S-morph) and Saltö 1 (E-morph) for set-up 2. For experimental set-up 1 a rock-pool in an intermediary habitat adjacent to Saltö South was used, and for set-up 2 the experiment was carried out within a rock-pool in a north-facing intermediary habitat.

Rock-pool 1 had an approximate area of 80 × 50 cm and rock-pool 2 an area of 80 × 20 cm. All other snails occupying the pool before the experiment were removed, and the experimental snails were put in the centre of the rock-pool (within an area of 25 × 25 cm) with the different morphs mixed together. For rock-pool 1 the size distribution of experimental snails was: E-morph individuals, eighty-five between 6–9 mm, fourteen between 9–12 mm, one over 12mm; S-morph individuals, thirty-seven between 6–9 mm, fifty-six between 9–12 mm, seven over 12 mm. For rock-pool 2 the similar size distribution was: E-morph individuals, forty snails 6–9 mm, sixty snails 9–12 mm; S-morph individuals, thirty-one snails 6–9 mm, nine snails of 9 mm, sixty snails 9–12 mm.

Both experiments were conducted in the afternoon over 3–3.5 h. Every 30 min we monitored the experimental snails to determine if there were any copulating pairs in the rock-pool. In rock-pool 1 there were 27 copulations altogether (17 male-female pairs, 10 maladaptive pairs), while in rock-pool 2 there were 22 copulations (8 male-female pairs, 14 maladaptive pairs).

Statistics

Sexual selection on size was measured by the sexual selection intensity index (S.S.I.), which is the mean size of mating snails minus the mean size of all snails (mating plus non-mating), divided by the standard deviation of all snails. This index can be used to compare the sexual selection effect on size among populations with different average sizes (Falconer, 1981). The statistical significance of sexual selection was evaluated by comparing mating and non-mating sizes in each sex separately by a one-way randomisation ANOVA, using a compiled BASIC program. This program randomised data of sizes among mating and non-mating individuals within every population and sex, and calculated the re-sampling distribution of the F estimate. The probability that the observed F (calculated directly from the samples) is represented in the re-sampling F distribution gives us an analogous statistical inference to the classical oneway ANOVA, but without the typical assumptions and limitations of classical parametric tests (see Manly, 1991).

Assortative mating by size was investigated by collecting copulating pairs in natural populations and estimating the Pearson correlation coefficient (r) for size in mating pairs. Statistical significances of correlations were evaluated using classical parametric tests (Sokal & Rohlf, 1995). A computer simulation was used to check the null hypothesis that the observed r coefficient in the samples was a consequence of random mating but with a non-random microdistribution of snail sizes. Following this, a compiled BASIC program was used to simulate a re-sampling of mating pairs from the N microareas studied in every population. Thus, a mating pair was randomly chosen among the adult females and males present within every microarea (both mating and non-mating individuals). The Pearson correlation coefficient was calculated for the N mating pairs re-sampled. This process was repeated 5000 times, and thus a re-sampling distribution for the r coefficient was obtained in such circumstances. The two tailed probability that the observed r coefficient (calculated directly from the sample) was present in such a re-sampling distribution, gives us the statistical confidence of the above null hypothesis (Efron, 1982; Manly, 1991).

Assortative mating between ecotypes (sheltered and exposed) was evaluated by the Yule's V index (Gilbert & Starmer, 1985). This index is estimated according to the formula,

$$V = (\mathrm{EE}^*\mathrm{SS}\text{-}\mathrm{ES}^*\mathrm{SE})/(\mathrm{ME}^*\mathrm{MS}^*\mathrm{FE}^*\mathrm{FS}), \quad (1)$$

where EE is the number of homotypic (within ecotype) exposed pairs, SS is the homotypic pairs of the sheltered ecotype, and ES and SE are the two types of heterotypic (between ecotypes) pairs. Furthermore, ME, MS, FE, and FS are the total number of E-males, S-males, E-females and S-females, respectively. The statistical significance of the assortative mating between ecotypes was evaluated using pseudoprobability contingency χ^2 tests (Zaykin & Pudovkin, 1993).

Results

All mating and non-mating individuals taken together were used to check the known sex dimorphism occurring in *L. saxatilis* with respect to size and to check the size differences between ecotypes. In most populations of both sheltered and exposed morphs, females showed significantly larger sizes than males: Saltö West ($F = 1.96$, $p = 0.164$), Saltö South ($F = 9.78$, $p = 0.002$), Saltö Ängklåvebukten ($F = 8.78$, $p = 0.0037$), Saltö 1 ($F = 7.61$, $p = 0.007$), and Saltö 2 ($F = 16.65$, $p = 0.000$). This difference was quantified by a standardised differentiation index (mean of female minus mean of male, divided by the standard deviation of both male and female sizes), giving a mean of 0.46 (SD = 0.176) for the sheltered populations and of 0.6 (SD = 0.057) for the exposed ones. That is, the size differences between sexes were larger in the E-morph than in the S-morph.

A total of 60 mating pairs of the Exposed ecotype, and 103 pairs of the Sheltered ecotype were sampled. However, more than one third of the mating pairs were maladaptive, i.e. male–male or male–juvenile pairs. The proportions of male–male and male–juvenile mating pairs in the E-morph were, respectively, 5/29 and 6/29 at Saltö 1, and 8/31 and 3/31 at Saltö 2. The same proportions in the S-morph were 9/33 and 2/33 at Saltö

Table 1. Sexual selection on male size in the different populations of the two ecotypes of *Littorina saxatilis*. Mean sizes, standard deviations and sample sizes for mating and non-mating individuals. The *F*-estimate for a one-factor ANOVA is presented, and its significance (*p*) estimated after 5000 randomisations. S.S.I. is the Sexual Selection Intensity (mean size of mating snails-mean size of all snails/standard deviation of all snails). * $p < 0.05$ ** $p < 0.01$ *** $p = 0.001$

		Copulating snails	Non-copulating snails	S.S.I.	*F*	*p*
S-morph:						
	Saltö West					
				0.34	4.68	0.033*
	Mean size	11.53	10.42			
	S.D.	1.21	2.17			
	N	22	30			
	Saltö South					
				0.45	10.16	0.003**
	Mean size	11.17	9.33			
	SD	1.39	2.47			
	N	23	28			
	Saltö Ängklåvebukten					
				0.57	12.54	0.001***
	Mean size	10.07	8.75			
	SD	1.04	1.58			
	N	22	45			
E-morph:						
	Saltö 1					
				–0.20	1.18	0.284
	Mean size	5.51	6.04			
	SD	1.44	1.65			
	N	18	25			
	Saltö 2					
				–0.33	3.01	0.086
	Mean size	4.72	5.08			
	SD	0.91	0.77			
	N	20	58			

West, 9/35 and 4/35 at Saltö South, and 6/35 and 6/35 at Saltö Ängklåvebukten.

The incidence of parasitic infection was less than 3% among the sampled S-morph individuals and less than 2% among the E-morph individuals, and therefore we argue that the effect of parasitism on the present results is negligible although severe parasitism is known to sterilise *L. littorea* (Hughes & Answer, 1982).

We used exclusively the male-female mating pairs to study the two independent components of mating behaviour. The sexual selection component was studied by comparing the mean sizes of copulating and non-copulating males and females from the different populations. The mean sizes of copulating and non-copulating males and females of the E- and the S-morphs, their standard deviations and their sexual selection intensities are shown in Tables 1 and 2. Copulating males and females of the S-morph were significantly larger than non-copulating ones in all populations, except for females from Saltö West, where the tendency was not significant. Moreover,

Table 2. Sexual selection on female size in the different populations of the two ecotypes of *Littorina saxatilis*. Mean sizes, standard deviations and sample sizes for mating and non-mating individuals. The F-estimate for a one-factor ANOVA is presented, and its significance (p) estimated after 5000 randomisations. S.S.I. is the Sexual Selection Intensity (mean size of mating snails-mean size of all snails/standard deviation of all snails). * $p < 0.05$ ** $p < 0.01$

		Copulating snails	Non-copulating snails	S.S.I.	F	p
S-morph:						
	Saltö West					
				0.29	2.97	0.090
	Mean size	12.00	11.10			
	SD	1.17	2.31			
	N	22	44			
	Saltö South					
				0.35	5.72	0.020*
	Mean size	12.09	10.87			
	SD	1.77	1.75			
	N	23	25			
	Saltö Ängklåvebukten					
				0.45	7.94	0.006**
	Mean size	10.69	9.59			
	SD	1.01	1.65			
	N	22	38			
E-morph:						
	Saltö 1					
				–0.38	4.54	0.035*
	Mean size	6.16	7.15			
	SD	1.06	1.76			
	N	18		26		
	Saltö 2					
				–0.06	0.10	0.751
	Mean size	5.58	5.68			
	SD	1.05	1.19			
	N	20	58			

copulating males and females of the E-morph were of smaller sizes than the non-copulating ones in all populations, which was significant for females from Saltö 1 (Tables 1 and 2). Thus, the S-morph always showed positive sexual selection intensities (copulating individuals larger than non-copulating ones) while the E-morph always showed negative sexual selection intensities (non-copulating individuals larger than copulating ones). These differences between morphs in sexual selection intensities (S.S.I.) was demonstrated by a t-test comparing the average estimates between morphs in males ($T=7.3$, df$=3$, $p < 0.01$) and females ($T=6.6$, df$=3$, $p < 0.01$) separately. In addition, the great size difference between these two ecotypes is evident from the Tables 1 and 2.

The sexual isolation component for size was estimated by the Pearson correlation coefficients (see r in Table 3). The S-morph of *L. saxatilis* mated randomly in relation to size in two populations, while a third population (Saltö West) showed significant correlation between mate sizes using Monte Carlo test (see below), but it was non-significant using a clas-

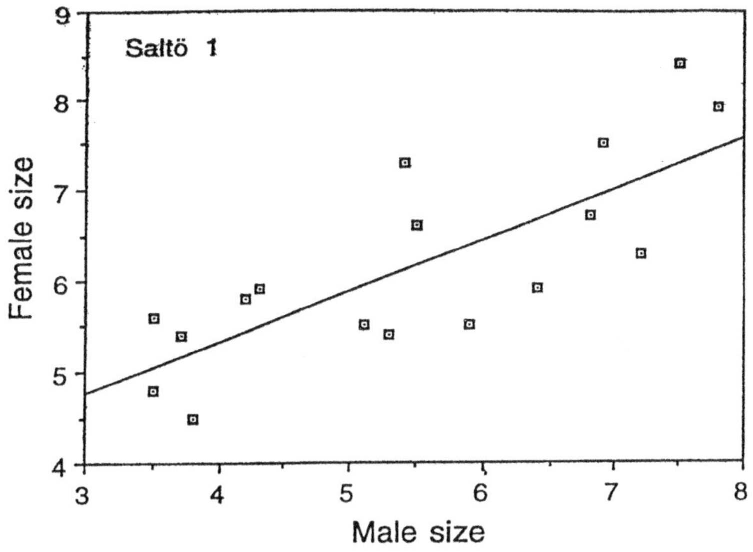

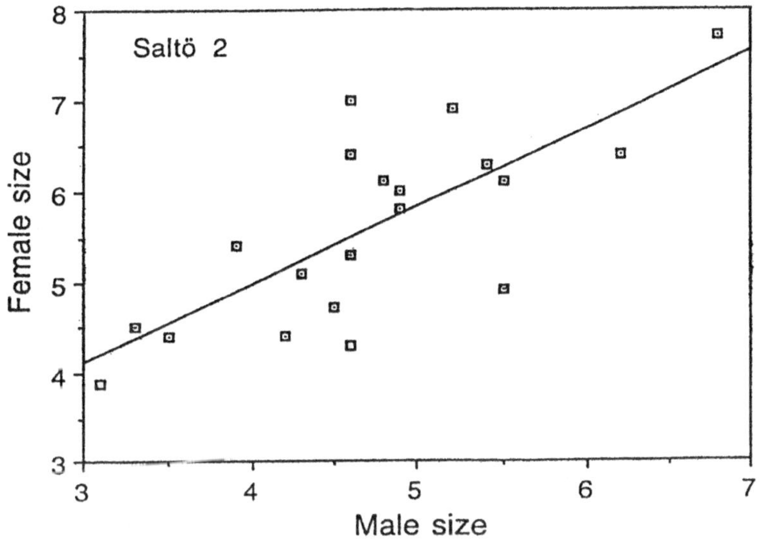

Figure 2. Correlations of sizes (mm) of male and female mates of the E-morph in two different cliff shore populations. Pearson's correlation analysis: Saltö 1, $r=0.764$, df$=17$, $p = 0.0002$; Saltö 2, $r=0.745$, df$=19$, $p = 0.0002$.

sical parametric test ($0.05 < p < 0.10$: see Table 3). The E-morph showed strong size assortative mating in both studied populations (Table 3; Figure 2). However, these significant values could be partially explained by a non-random microdistribution with respect to size (the average sizes of males and females within each microarea were positively correlated in the E-morph populations: Saltö 1 ($r=0.48$, $n=15$, $p = 0.07$) and Saltö 2 ($r=0.44$, $n=26$, $p = 0.02$). For this reason, we used a computer program to test the null hypothesis that the assortative mating was exclusively caused by the snail microdistribution. This test gave similar results, all significant r coefficients could not be totally explained by a non-random microdistribution of snails (Table 3). Also in the sheltered population (Saltö West), the size assortative mating, which was less distinct compared to the two E-populations, could not be explained by non-random microdistributions. The

Table 3. Size assortative mating and the influence of microdistribution of sizes of snails in the different populations of the two ecotypes of *Littorina saxatilis*. The Pearson correlation coefficient (*r*) of sizes of male and female mates, and the average Pearson correlation coefficient (rm) of sizes of mating pairs resampled 5000 times from the *N* studied micro-areas. Every resampling supposes the random choice of the size of a male and a female (a mating pair) from the different microareas (see text). Prob. is the probability (not the significance level) that the observed assortative mating may be explained by a non-random microdistribution of sizes among microareas

		r	rm	*N*	prob.
S-morph:	Saltö W	0.410	–0.015	22	0.020
	Saltö S	0.241	0.237	23	0.992
	Saltö Ä	0.168	–0.015	22	0.408
E-morph:	Saltö 1	0.764	0.344	18	0.006
	Saltö 2	0.745	0.155	20	0.000

Monte Carlo approach seems safer than the parametric test, as it takes care of the particular microdistribution of the sample and does not depend on as many statistical assumptions. Moreover, the effect of the microdistribution could only explain on average 32.9% (SD = 41.1) of the observed size assortative mating.

In the study of assortative mating between the two parental morphs, E and S, Yule's *V* index (Gilbert & Starmer, 1985) showed that the two morphs mated randomly with each other in both experimental set-ups and for both male-female pairs and maladaptive pairs (Table 4).

Table 4. Mating pattern between the two distinct ecotypes (E and S) of *Littorina saxatilis* in two rock pools on the Swedish west coast. Number of copulations between males and females, and between either 'active' males and 'passive' males or between males and juveniles (maladaptive pairs) are presented. The Yule's *V* index is a sexual isolation index between the morphs (see e.g. Gilbert & Starmer, 1985) and ranges from –1 to 1 (1, 0 and –1 are complete assortative, random and disassortative mating respectively). ns = not significant

	Male/female pairs	maladaptive pairs
Rock pool 1:		
S/S pairs	3	6
S/E pairs	6	1
E/S pairs	3	2
E/E pairs	5	1
Yule's *V*	–0.044 ns	0.218 ns
Rock pool 2:		
S/S pairs	2	6
S/E pairs	4	4
E/S pairs	0	3
E/E pairs	2	1
Yule's *V*	0.333 ns	–0.141 ns

Discussion

As in former studies (e.g. Janson, 1982, 1983; Janson & Sundberg, 1983; Johannesson & Johannesson, 1996), we have found sexual size dimorphism and ecotype size differences in Swedish populations of *L. saxatilis*. Janson (1982, 1983) and Janson & Sundberg (1983) showed that the natural distribution of these two ecotypes could be successfully explained owing to the low dispersal capability of this species, and to the creative modelling force of natural selection in the different habitats (sheltered, intermediate and exposed). Furthermore, Johannesson & Johannesson (1996) have shown that the size differences between females and males are determined by hereditary differences in growth rates. Here, we focus on the mating behaviour with respect to size and among morphs, although as we comment below, a more general picture on this genetic polymorphism may emerge from the present data. However, it is possible that differences in mating behaviour between ecotypes have an environmental explanation (different shore types) and not only an adaptive genetic explanation, even if many morphological and behavioural characters to a great extent are genetically governed (Johannesson & Johannesson, 1996).

In this study between 33 and 38% of the copulations were maladaptive pairs. Similar proportions have been found in other studies of this species (Saur, 1990; Johannesson et al., 1995). In *L. littorea* only about 5% of the copulating pairs found on a shore are male-male copulations (Saur, 1990; Erlandsson & Johannesson, 1994), which is explained by the ability of males to discriminate between male and female mucus trails prior to copulation (Erlandsson & Kostylev, 1995) and the much shorter copulation duration for male-male pairs than for male-female pairs (Saur, 1990). Males of Swedish *L. saxatilis* cannot, however, discriminate between male and female mucus trails (which may explain the larger number of maladaptive pairs), but they seem to increase the encounter rate of mates by following any mucus trail (irrespective of the sex of the depositor of the trail) for longer distances than females follow any trails (Erlandsson, Kostylev & Rolán-Alvarez, unpubl. data).

There were clear differences in sexual selection intensities on size between the two morphs of *L. saxatilis*. Large males and females of the S-morph were favoured as mates, while both copulating males and females of the E-morph were smaller than non-copulating ones. The higher mating success may be a consequence of either female choice, male choice or male competition. However, these hypotheses are unlikely to explain the contrasting trends for sexual selection on size in the two ecotypes. Female choice cannot easily explain why S-females would choose larger males, while E-females would prefer smaller ones. Similarly, since large females of this species have a higher fecundity (Janson, 1985), it is not obvious why E-males would not choose large female mates as S-males would do. The male competition mechanism seems *'a priori'* even more difficult to explain these different trends. How can small E-males succeed in competition against large E-males? However, if we think about competition, not as a direct struggle for mates, but rather as a general capability to spend time and effort in reproduction, then smaller E-individuals (both males and females) would be at an advantage in exposed habitats (greater resistance against wave action), while larger S-individuals (both males and females) would show greater success in sheltered habitats (greater resistance against crab predation). This possibility is in agreement with the viabilities that these two morphs (with distinct size differences) showed when transplanted to the different habitats (Janson, 1983). It means that there may not be a choice involved at all producing the sexual selection patterns observed. In exposed habitats larger snails may have to spend most of their time searching for crevices and protection against waves, while in sheltered habitats smaller snails have to do likewise to escape predation by crabs. We must emphasise that, irrespective of the mechanism involved, the positive sexual selection intensities in the S-morph compared to the negative selection intensities in the E-morph may be evolutionarily important since it suggests that sexual selection contributes to the size differences between the two distinct morphs of *L. saxatilis* on the Swedish west coast.

Among the S-morphs of *L. saxatilis*, random mating by size occurred in two populations and weak size assortative mating in a third. In another study, this morph showed size assortative mating at a sheltered boulder shore site (Saur, 1990). Thus, there seems to be some variation in the mating patterns of different populations of the S-morph. The assortative mating by size found in the E-morph, however, was very clear. Aggregations of similar sized individuals on the shore have been suggested to partially explain the size assortative mating observed in Galician populations of *L. saxatilis* on exposed rocky shores (Johannesson et al., 1995). However, in our samples, the microdistribution of sizes of snails could not fully explain the non-random mating pattern with respect to size. Indeed, the size microdistributions only explained approximately 30% of the observed assortative mating. In Sweden, some kind of size-based mate choice behaviour, with males, females, or both males and females choosing mates of a particular size, may be one explanation in some populations. A non-random mating pattern can be suggested from males and females preferring a mate size dependent on their own size. Alternatively, a size-dependent copulation duration mechanism would be, perhaps, the simplest explanation, as large-sized males and females would show, respectively, larger ejaculates and larger seminal receptacles to be filled. However, this mechanism cannot produce strong assortative mating, except perhaps in some particular situations. Ridley (1983) suggested an alternative theory, that size assortative mating in different species is generally found in populations with a positive correlation between female size and fecundity, a great investment in time for a pair staying in sexual association or pair formation, and a higher mating success for larger males. This mechanism seems, however, unlikely in *L. saxatilis*, as large males were favoured as mates only in S-morph populations, and as *L. saxatilis* does not seem to invest very much time in each mating and in staying in pair formation (Saur, 1990).

Size assortative mating would also be produced if matings were difficult or impossible (mechanical incompatibility) between extremely different sized mates. Preliminary computer modelling using Monte Carlo simulations on this mechanism indicate that this is a possibility in the sampled populations (Rolán-Alvarez, unpubl. data). Mating pairs were resampled from every microarea following different 'mating algorithms' depending on the behavioural mechanism tested. The mechanism that could explain most of the natural mating pattern was where a mating is interrupted (either the female or the male rejects a mate) if snails differ from a particular threshold size difference between mates. Thus, there seems to be a certain size threshold where snails reject different sized mates as the genitals make contact with each other. This size-based mate rejection mechanism is more likely than a size-based mate choice mechanism according to the different 'mating algorithms' (Rolán-Alvarez, unpubl. data), although we cannot exclude other mechanisms contributing to size assortative mating in Swedish populations of *L. saxatilis*.

Although the power of our rock-pool experiments may have been low, there were no indications of any deviation from random mating between the two distinct allopatric morphs of *L. saxatilis* (in Galician populations significant deviations from random mating can often be revealed with similar sample sizes). This supports the view of Johannesson & Sundberg (1992) that this species is not likely to be subject of incipient speciation on Swedish rocky shores in the present conditions. Each morph, including intermediates, has the highest survival rate in its native habitat (Janson, 1983), and gene flow between the pure morphs is possible through the intermediary subpopulation. In Galician populations, however, there is a pronounced assortative mating between two parental morphs of *L. saxatilis* in a narrow midshore zone of overlap (Johannesson et al., 1995). Hull et al., (1996), also report a likely case of incipient reproductive isolation between two forms of *L. saxatilis* on east English rocky shores and a laboratory experiment also showed that the two forms mated assortatively (Hull, 1998). In both of these two examples, the ecotypes vary in their vertical distribution at different shore levels (Johannesson et al., 1993; Hull et al., 1996), while Swedish ecotypes of *L. saxatilis*, that were found to mate randomly, replace each other horizontally on the shore. This may suggest that different morphs of *L. saxatilis* have evolved different levels of incipient speciation in different situations. At least on Galician shores, morphological incompatibilities between morphs during matings, and sexual competition seem the most plausible mechanisms maintaining the natural polymorphisms (Rolán-Alvarez et al., unpubl. data).

In contrast to assortative mating, sexual selection is usually less affected by other biological mechanisms than choice and competition. Following these complex relationships which may appear in natural populations with respect to sexual isolation, sexual selection and the different mechanisms involved, we must emphasise the importance of the statistical partitioning made in this and similar studies. This seems the only safe approach to focus on the biological mechanisms and evolutionary consequences of these processes.

Acknowledgements

We thank Drs K. Johannesson and M. S. Davies for important comments on an earlier version of this paper. We are also grateful to Dr D. Reid for ideas and discussion, and to two anonymous referees for improving the manuscript. This work was partially supported by grants from The Royal Swedish Academy of Sciences to J. E. and from the University of Vigo to E. R.-A.

References

Anderson, W. W. & P. R. McGuire, 1978. Mating pattern and mating success of *Drosophila pseudoobscura* karyotypes in large experimental populations. Evolution 32: 416–423.

Coyne, J. A., 1992. Genetics and speciation. Nature 355: 511–515.

Crespi, B. J., 1989. Causes of assortative mating in arthropods. Anim. Behav. 38: 980–1000.

Efron, B., 1982. The Jacknife, the bootstrap and other resampling plans. Society for industrial and applied mathematics, Philadelphia, 92 pp.

Erlandsson, J. & K. Johannesson, 1994. Sexual selection on female size in a marine snail, *Littorina littorea* (L.). J. exp. mar. Biol. Ecol. 181: 145–157.

Erlandsson, J. & V. Kostylev, 1995. Trail following, speed and fractal dimension of movement in a marine prosobranch, *Littorina littorea*, during a mating and a non-mating season. Mar. Biol. 122: 87–94.

Falconer, D. S., 1981. Introduction to quantitative genetics. Longman, London, 383 pp.

Gilbert, D. G. & W. T. Starmer, 1985. Statistics of sexual isolation. Evolution 39: 1380–1383.

Hughes, R. N. & P. Answer, 1982. Growth, spawning and trematode infection of *Littorina littorea* (L.) from an exposed shore in north Wales. J. moll. Stud. 48: 321–330.

Hull, S. L., J. Grahame & P. J. Mill, 1996. Morphological divergence and evidence for reproductive isolation in *Littorina saxatilis* (Olivi) in northeast England. J. moll. Stud. 62: 89–99.

Hull, S. L., 1998. Assortative mating between two distinct microallopatric populations of *Littorina saxatilis* (Olivi) on the northeast coast of England. Hydrobiologia 378: 79–88.

Janson, K., 1982. Phenotypic differentiation in *Littorina saxatilis* (Olivi) (Mollusca, Prosobranchia) in a small area on the Swedish west coast. J. moll. Stud. 48: 167–173.

Janson, K., 1983. Selection and migration in two distinct phenotypes of *Littorina saxatilis* in Sweden. Oecologia 59: 58–61.

Janson, K., 1985. Variation in the occurrence of abnormal embryos in females of the intertidal gastropod species *Littorina saxatilis* (Olivi). J. moll. Stud. 51: 64–68.

Janson, K. & P. Sundberg, 1983. Multivariate morphometric analysis of two varieties of *Littorina saxatilis* from the Swedish west coast. Mar. Biol. 74: 49–53.

Janson, K. & R. D. Ward, 1984. Microgeographic variation in allozyme and shell characters in *Littorina saxatilis* (Olivi) (Prosobranchia: Littorinidae). Biol. J. linn. Soc. 22: 289–307.

Johannesson, B. & K. Johannesson, 1996. Population differences in behaviour and morphology in the snail *Littorina saxatilis*: phenotypic plasticity or genetic differentiation? J. zool. Lond. 240: 475–493.

Johannesson, K. & P. Sundberg, 1992. Speciation in *Littorina saxatilis*? A one-dimensional selection-migration model. In 1990 J. Grahame, P. J. Mill, D. G. Reid (eds), Proceedings of the Third International Symposium on Littorinid Biology. The Malacological Society of London: 1–8.

Johannesson, K. & A. Tatarenkov, 1997. Allozyme variation in a snail (*Littorina saxatilis*) – deconfounding the effects of microhabitat and gene flow. Evolution 51: 402–409.

Johannesson, K., B. Johannesson & E. Rolan-Alvarez, 1993. Morphological differentiation and genetic cohesiveness over a microenvironmental gradient in the marine snail *Littorina saxatilis*. Evolution 47: 1770–1787.

Johannesson, K., E. Rolan-Alvarez & A. Ekendahl, 1995. Incipient reproductive isolation between two sympatric morphs of the intertidal snail *Littorina saxatilis*. Evolution 49: 1180–1190.

Manly. B. F. J., 1991. Randomization and Monte Carlo Methods in Biology. Chapman & Hall. London, 281 pp.

Merrel, D. J., 1950. Measurement of sexual isolation and selective mating. Evolution 4: 326–331.

Reid, D. G., 1996. Systematics and evolution of *Littorina*. Ray Society, London, 463 pp.

Ridley, M., 1983. The Explanation of Organic Diversity. The Comparative Method and Adaptations for Mating. Clarendon Press, Oxford, UK, 719 pp.

Rolán-Alvarez, E., C. Zapata & G. Alvarez, 1995a. Multilocus heterozygosity and sexual selection in a natural population of the marine snail *Littorina mariae* (Gastropoda: Prosobranchia). Heredity 75: 17–25.

Rolán-Alvarez, E., K. Johannesson, & A. Ekendahl, 1995b. Frequency- and density-dependent sexual selection in natural populations of Galician *Littorina saxatilis* (Olivi). Hydrobiologia 309: 167–172.

Santos, M., R. Tarrio, C. Zapata & G. Alvarez, 1986. Sexual selection on chromosomal polymorphism in *Drosophila subobscura*. Heredity 57: 161–169.

Saur, M., 1990. Mate discrimination in *Littorina littorea* (L.) and *L. saxatilis* (Olivi) (Mollusca: Prosobranchia). Hydrobiologia 193: 261–270.

Sokal, R. R. & F. J. Rohlf, 1995. Biometry. Freeman & Co., New York, 887 pp.

Spieth, H. T. & J. M. Ringo, 1983. Mating behavior and sexual isolation in *Drosophila*. In M. Ashburner, H. L. Carson & N. J. Thompson (eds), The Genetics and Biology of *Drosophila*, Vol 3c. Academic Press, London: 224–270.

Zaykin, D. V. & A. Pudovkin, A. 1993. Two programs to estimate significance of χ^2 values using pseudo-probability tests. J. Hered. 84: 152.

Hydrobiologia **378:** 71–78, 1998.
R. M. O'Riordan, G. M. Burnell, M. S. Davies & N. F. Ramsay (eds), Aspects of Littorinid Biology.

Rough periwinkle polymorphism on the east coast of Yorkshire: comparison of RAPD-DNA data with morphotype

C. S. Wilding, J. Grahame & P. J. Mill
Department of Biology, The University of Leeds, Leeds LS2 9JT, U.K.

Key words: RAPD, *Littorina saxatilis*, *L. neglecta*, phylogeny

Abstract

A genetic analysis of 'morphotypes' of *Littorina saxatilis* from two locations on the north-east coast of England (Filey and Ravenscar), using randomly amplified DNA polymorphisms (RAPD) generated with a single primer, revealed quite different patterns of variation. Thin shelled, wide-apertured (H-form) animals from Ravenscar tended to cluster separately from thick shelled (M) forms, indicating genetic differentiation of these morphs. Animals of similar morphology (H and M) from Filey (about 30 km distant) did not display such an obvious pattern, and although there was still evidence of differentiation from discriminant analysis of RAPD data, levels of correct classification were reduced at Filey. This suggests that the utility of a single RAPD primer for separation of such forms varies over a relatively small distance. *L. arcana* from Ravenscar, included as an outgroup, were generally well differentiated from *L. saxatilis* and were noted to exhibit less variation, a phenomenon that has been noted previously in some allozyme and RAPD analyses. A similar RAPD analysis undertaken on small, barnacle dwelling, brooding forms from Peak Steel, Ravenscar revealed that animals appeared to have as great a tendency to cluster together on a microgeographic scale (by collection patch) as by 'species' (*L. neglecta* or *L. saxatilis* b) although predominance of certain species in individual patches largely explains this. Discriminant analysis of RAPD presence/absence data did correctly place over 90% of barnacle dwelling animals to their respective species, and we consider this as evidence of separate gene pools. RAPD is taken to be a useful tool for screening genetic variation in this complex of animals on a local scale when either a pre-selected informative primer is utilised or a battery of primers is used, but its efficacy may be reduced when a single primer is employed for screening animals from different shores.

Introduction

Currently three species are recognised in the *Littorina saxatilis* group of periwinkles, *L. saxatilis*, *L. arcana* and *L. compressa* (Reid, 1996). The status of a fourth taxon *L. neglecta* is somewhat contentious (Reid, 1993; Caley et al., 1995). Allozyme studies aimed at studying systematics and phylogeny have tended to confirm the status of *L. arcana*, *L. saxatilis* and *L. compressa* but studies involving *L. neglecta* remain inconclusive, Johannesson & Johannesson (1990) arguing that *L. neglecta* is actually an ecotype of *L. saxatilis*. However as a result of high variability and overlapping allele frequencies at most allozyme loci, potential selection, and because allele frequencies can alter significantly over relatively short distances (Ward, 1990), the utility of the allozyme technique for examining the status of these forms is reduced and further detailed investigation will likely require the use of alternative technologies. Mitochondrial DNA methodologies have been applied to the 19 extant *Littorina* species, demonstrating that at the loci studied, a phylogeny compatible with fossil and morphological data is produced (Rumbak et al., 1994; Reid et al., 1996). However for the recently diverged *L. saxatilis* species complex the phylogeny is still not fully resolved, and no data were presented on *L. neglecta*. The inability of the mtDNA gene sequences studied by Reid et al. (1996) to resolve the phylogeny within this complex is perhaps unsurprising given the likely timescale available for differentiation in these species (3.5–4 my maximum Reid et al., 1996) and the relatively conserved nature of the studied genes (mitochondrial 12S rRNA, 16S rRNA, cytochrome *b*). In order to gather suitable data for reconstructing the phylogeny of these species, a more rapidly evolving locus or set of loci will be required for which sufficient differentiation will have accumulated in the possibly separate lineages.

Although the polymerase chain reaction (PCR) is often used to amplify suitable loci from individuals prior to restriction enzyme analysis or sequencing, little sequence information is available for *Littorina*, causing difficulties for the design of primers for use in the PCR. One method to surmount this problem involves the use of random amplified polymorphic DNA (RAPD) PCR in which only a single, short (typically 10 base pairs) primer is utilised in the amplification (Hadrys et al., 1992; Grosberg et al., 1996). Because of the high likelihood that primers will anneal in inverted orientation within close proximity, amplification typically yields a number of products which can then be used to assess variability and genetic differentiation from presence/absence data of bands. Since the majority of genomic DNA is non-coding, most RAPD products are believed to be functionless and therefore neutral and potentially rapidly evolving (Haymer, 1994). They are consequently considered useful for systematic studies where such criteria are valuable, if not necessary. Use of RAPDs in phylogenetic studies is problematical at higher taxonomic levels, due to non-homology of co-migrating RAPD bands, but is considered a valid technique for examining closely related species (Hadrys et al., 1992). Crossland et al. (1993, 1996) have applied RAPD-PCR to littorinids and concluded that the applicability of the technique varies between primers such that recognised species are differentiable with some primers but not with others. This implies that screening with a number of primers may be necessary to uncover a primer useful for a particular study.

Recently, Hull et al. (1996) have provided convincing evidence that *L. saxatilis* from the north-east coast of England are composed of two morphologically distinct forms with a possible reproductive barrier. These two forms, a thin-shelled patulous animal (*L. saxatilis* 'H') and a thick shelled form (*L. saxatilis* 'M') are probably equivalent to the *L. saxatilis* and *L. rudis* described by Smith (1981). The nature of gene flow between these forms has yet to be examined in detail and they therefore require study in order to place their level of genetic differentiation in context with that of accepted species. RAPD technologies promise to be a useful technique for a preliminary survey of these forms and here we apply a single primer, previously identified to be potentially useful for separating *L. saxatilis* H and M forms (Grahame et al., 1997) to the analysis of the phylogeny of *L. arcana*, *L. saxatilis* 'H' and 'M', and small barnacle-dwelling forms of *L. saxatilis*.

Materials and methods

Collection

Animals were collected from two sites on the north-east coast of England (Figure 1), Old Peak, Ravenscar (British National Grid Reference NZ984021) and Filey Brigg (TA128816), at both of which *L. saxatilis* H and M are found (Hull et al., 1996). *L. saxatilis* H and *L. arcana* were collected from high shore massive boulders, whilst *L. saxatilis* M were taken from on and under mid shore small boulders. Animals were identified on the basis of shell characters (see Figure 2 of Hull et al., 1996). Occasionally rare intermediate forms are found (see Figure 2 of Hull et al., 1996) with shell characters mid-way between *L. saxatilis* H and M, and these are termed *L. saxatilis* I. On the high shore at Old Peak, five discrete collections were taken from crevices or patches less than 1 m^2 in size of massive boulders. Distance between collection patches ranged from 10 cm to 200 m and thus over the latter scale would likely preclude the sampling of single families. Mid shore (M form) animals were collected from 3 areas approximately 100 m distant from the collection of the H animals and *L. arcana* and spaced between 10 and 20 m apart. Inspection of the data for all animals revealed no clear sign of spatial structuring which probably would not be evident from samples of this size. A more rigorous evaluation of RAPD variation involving fine scale sampling is the subject of future developments. In total 76 animals (15 H, 20 M, 1 I and 40 *L. arcana*) were collected from boulders and bedrock at Ravenscar. 'H' and 'M' forms of *L. saxatilis* were collected also from Filey Brigg where H forms were collected from crevices in boulders (< 50 cm^2) ranging from 3–25 m apart on the north side of the Brigg, whilst the M forms were taken from on and under boulders on the south side from patches of 1 m^2 spaced 50–100 m apart.

Small barnacle dwelling morphs, classified as *L. neglecta* (smooth, globose) or *L. saxatilis* b (the

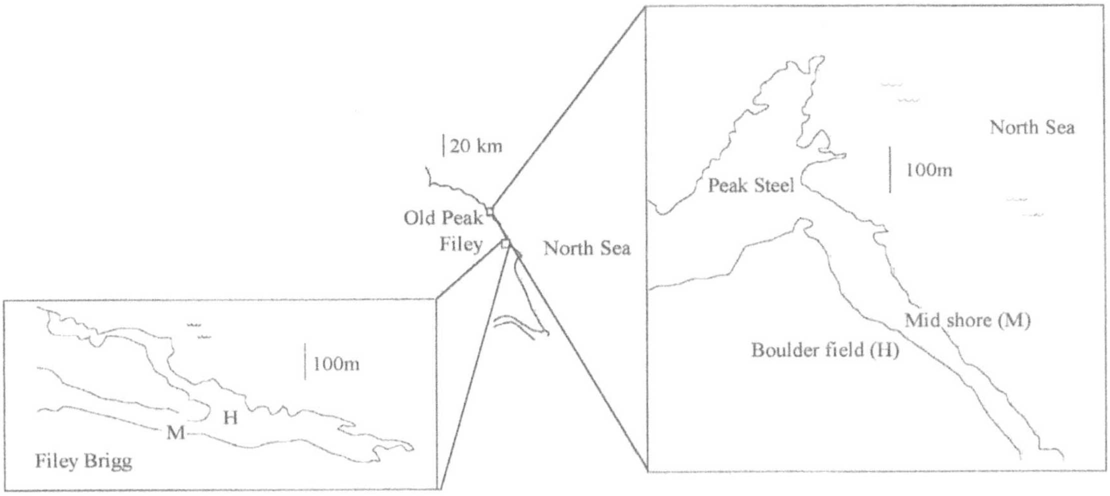

Figure 1. Collection sites for animals used in this study. Scale bars point North-South (up-down). H and M refer to collection sites for *L. saxatilis* H (and *L. arcana*) and *L. saxatilis* M.

small, reproductively mature littorinid with sculptured shell) on the basis of shell characters (Grahame et al., 1995) were collected from Peak Steel, Ravenscar (NZ979026) from discrete patches of rock of approximately 2 m^2. Three areas were sampled in duplicate (2 m between paired samples). The three collection sites formed a triangle with right-angled sides of dimensions 14 m and 40 m. All animals were held live at 4 °C until utilised for DNA extraction.

DNA extraction

Genomic DNA was extracted, essentially according to the protocol of Ashburner (1989) from reproductively mature female individuals (male *L. saxatilis* and *L. arcana* can be easily confused). The digestive gland area can harbour parasites that may cause contamination of DNA, and although parasitised animals were not used for extraction, the head-foot region only was used for DNA extraction. Pallial oviduct and operculum were also discarded. Tissue was homogenised in 200 μl homogenisation buffer (10 mm Tris-HCl pH 7.5, 10 mm EDTA, 5% [w/v] sucrose, 0.15 mm spermine, 0.15 mm spermidine) with a plastic pestle in a microfuge tube. Following maceration, 200 μl lysis buffer (300 mm Tris-HCl pH 9.0, 100 mm EDTA, 0.625% [w/v] SDS, 5% [w/v] sucrose) was added and the tube(s) incubated at 70 °C for 15 mins. On completion of the incubation step, tubes were removed and allowed to cool to room temperature, then 60 μl 5 m potassium acetate was added and mixed. Proteins were then allowed to precipitate on ice for 30 min. Two phenol-chloroform-isoamyl alcohol (24:24:1 pH 8.0) and 1 chloroform extraction were performed and then DNA was precipitated from the supernatant with 2 vols ice-cold ethanol for 10 min on ice. DNA was pelleted at 13 000 r.p.m. for 10 min and the resultant pellet washed with 70% ethanol then re-spun for 5 min at 13 000 r.p.m. Pellets were dried in a vacuum centrifuge then resuspended in 20 μl 0.1 × TE buffer. DNA concentrations were measured by fluorometry and adjusted to 10 ng μl^{-1} since DNA concentrations must be standardised for RAPD amplification (Muralidharan & Wakeland, 1993).

DNA amplification

RAPD-PCR was undertaken in 50 μl total volume using thin-walled reaction vessels in a Perkin-Elmer 480 thermal cycler. Reaction conditions of 200 μM each dNTP, 100 pmol primer, 10 ng template DNA and 1U *Taq* polymerase (Supertaq, HT Biotechnologies) were used. The thermal cycler profile was optimised to 1 min denaturation at 94 °C, 1 min annealing at 35 °C and 2 min extension at 72 °C with a 3 min ramp time from 35 °C to 72 °C. A single primer, RAPD-H (5'-GCCGTGGTTA-3') previously noted to generate informative polymorphic RAPD profiles (Grahame et al., 1997) was applied. Following amplification, 10 μl of amplified product was electrophoresed in a 3% agarose gel (1% standard agarose, 2% nu-seive) and stained with ethidium bromide. Bands were visu-

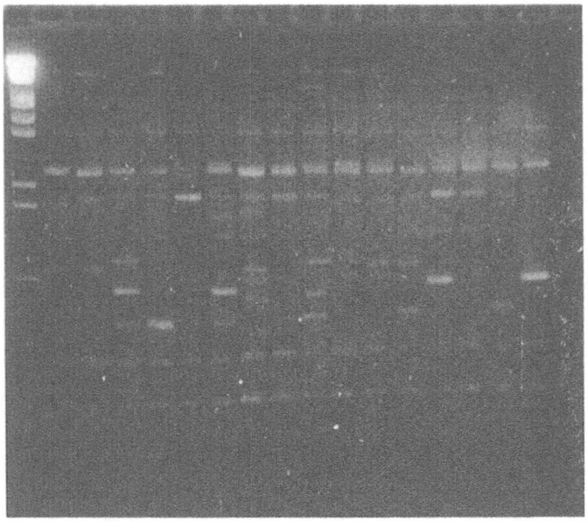

Figure 2. Photograph of a typical ethidium bromide stained, 3% agarose gel used to separate RAPD products. Lane 1 marker lane (*Eco*RI-*Hind*III cutλ DNA). Lanes 2–8 barnacle dwelling *L. saxatilis*. Lanes 9–16 *L. neglecta*. Lane 17 *L. saxatilis*. Lane 18 no template added.

alised under UV light and Polaroid photographs taken as permanent records. In order to standardise conditions between reaction vessels, a PCR master mix was aliquoted into reaction tubes prior to addition of template. Preliminary trials demonstrated no effect of template quantity ranging from 1–50 ng and since concentrations were accurately determined through fluorometrical readings, all reaction conditions are as constant as possible. In our laboratory where 5–15% of all reactions are repeated, we do not see variation in RAPD profiles from repeat reactions.

Analysis of results

For each individual, bands were scored as present (1) or absent (0). Data were analysed using RAPDPLOT (Kambhampati et al., 1992) to produce a matrix of similarity values calculated from

$$M = \frac{N_{\mathrm{AB}}}{N_{\mathrm{T}}} \tag{1}$$

where N_{AB} is the number of matches between individuals A and B, that is the band is either present or absent in both individuals, and N_{T} is the total number of bands scored. Similarity measures were clustered using the NEIGHBOR option of PHYLIP (Felsenstein, 1993) and trees drawn using DRAWGRAM from this package. The dataset was also manipulated by detecting those bands that were polymorphic and aided most in the separation of the forms by utilising the F_{ST} option of RAPDPLOT and removing bands that produced a $N_{\mathrm{e}}m$ estimate > 1. $N_{\mathrm{e}}m$ levels (the effective number of migrants) below 1 are typically accepted as indicative of the level at which gene flow is insufficient for maintenance of all alleles in a population (Kimura & Ohta, 1971), and RAPD bands (loci) that produce $N_{\mathrm{e}}m$ values below 1 should therefore be the most useful for differentiating populations. Discriminant analysis was also applied to the dataset to quantify the usefulness of the technique at each collection site for discriminating between forms. Discriminant analysis with resubstitution was performed using MINITAB (v9.2). This analysis also enabled identification of animals which were incorrectly assigned to the designated groupings (species or morphotypes).

Table 1. Variability in *L. saxatilis L. neglecta* and *L. arcana* at Filey, Old Peak and Peak Steel. H and M refer to *L. saxatilis* H and M morphs. I refers to intermediate forms

Locality	Species	No.	Mean no. bands	Range
Filey	*L. saxatilis* 'H'	29	7.21	2–10
	L. saxatilis 'I'	1	8.00	8
	L. saxatilis 'M'	35	7.20	2–10
Old Peak	*L. saxatilis* 'H'	16	8.38	5–12
	L. saxatilis 'I'	1	5.00	5
	L. saxatilis 'M'	19	9.11	5–13
	L. arcana	40	3.58	1–8
Peak Steel	*L. arcana*	8	3.00	1–8
	L. saxatilis b	18	8.94	3–13
	L. neglecta	14	7.21	3–13

Results

Amplification using the primer RAPD-H typically yielded numerous bands after electrophoretic separation and staining. In all 25 discrete bands were recognised. A typical stained gel displaying a RAPD-H profile is depicted in Figure 2.

L. saxatilis exhibited the highest variability with a mean of 8.66 bands per individual at Old Peak and *L. arcana* the lowest of 3.75 bands per individual at Old Peak (Table 1). *L. neglecta* (7.21) and '*saxatilis*-b' (8.94) were similar in value to typical *L. saxatilis*. Table 1 describes the variability displayed by the Old Peak, Peak Steel and Filey animals.

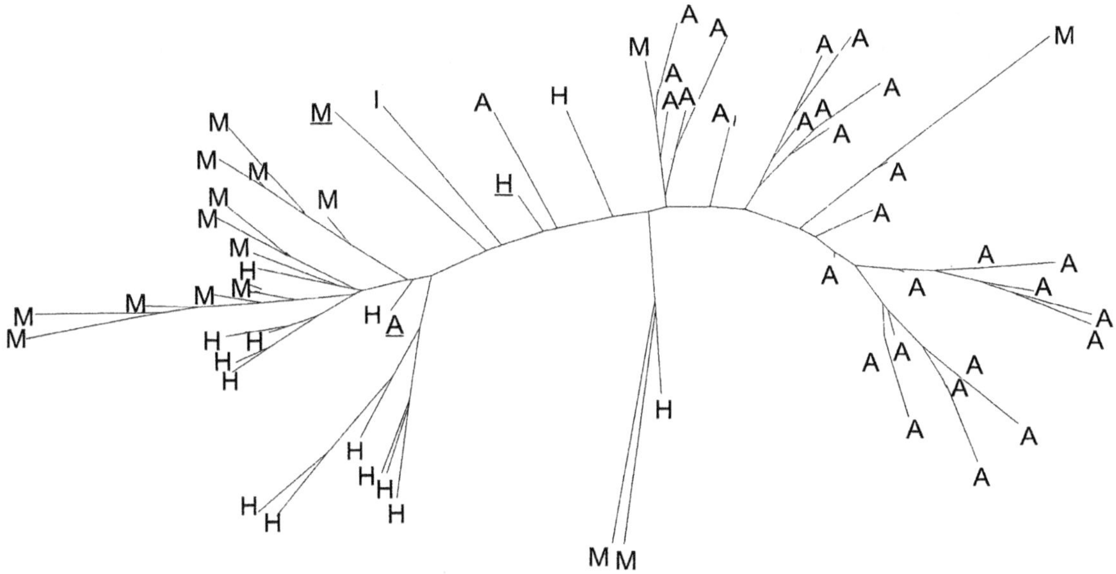

Figure 3. Phylogenetic tree generated using RAPDPLOT (Kambhampati et al., 1992) and PHYLIP (Felsenstein, 1993). RAPD data from Ravenscar animals was used to produce RAPDPLOT output (similarity matrix) which was then clustered using neighbour-joining with the NEIGHBOR program in PHYLIP. H and M refer to *L. saxatilis* H and M forms, I refers to *L. saxatilis* intermediate form. A is *L. arcana.* Underlined individuals were misclassified in the discriminant analysis.

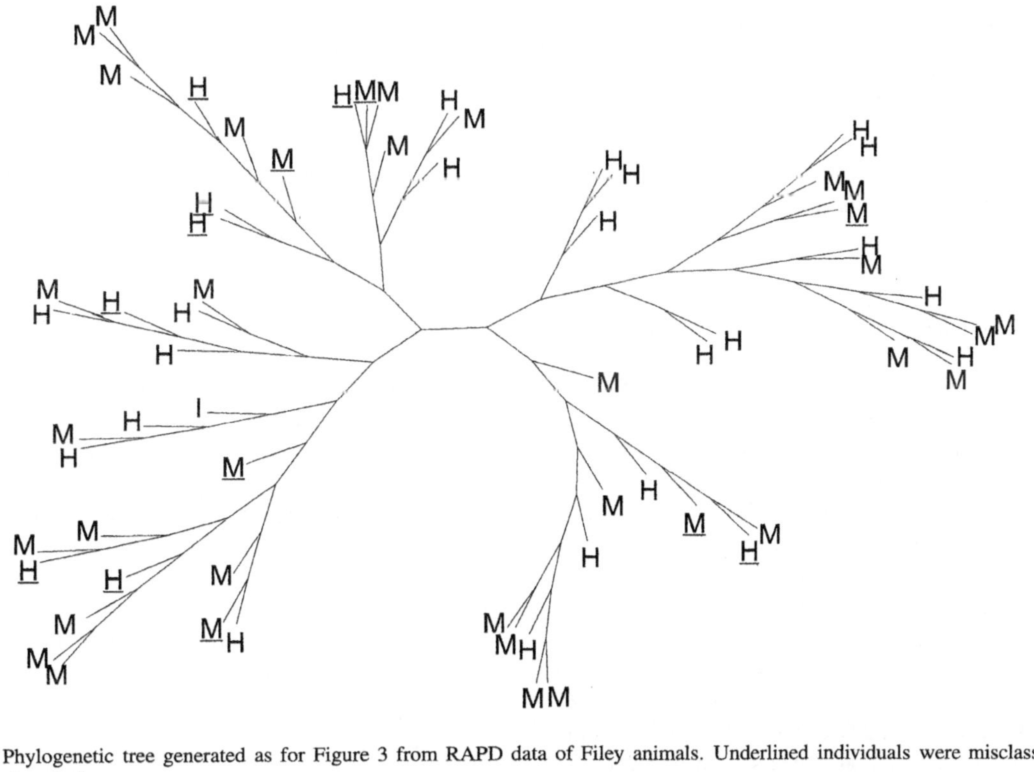

Figure 4. Phylogenetic tree generated as for Figure 3 from RAPD data of Filey animals. Underlined individuals were misclassified in the discriminant analysis.

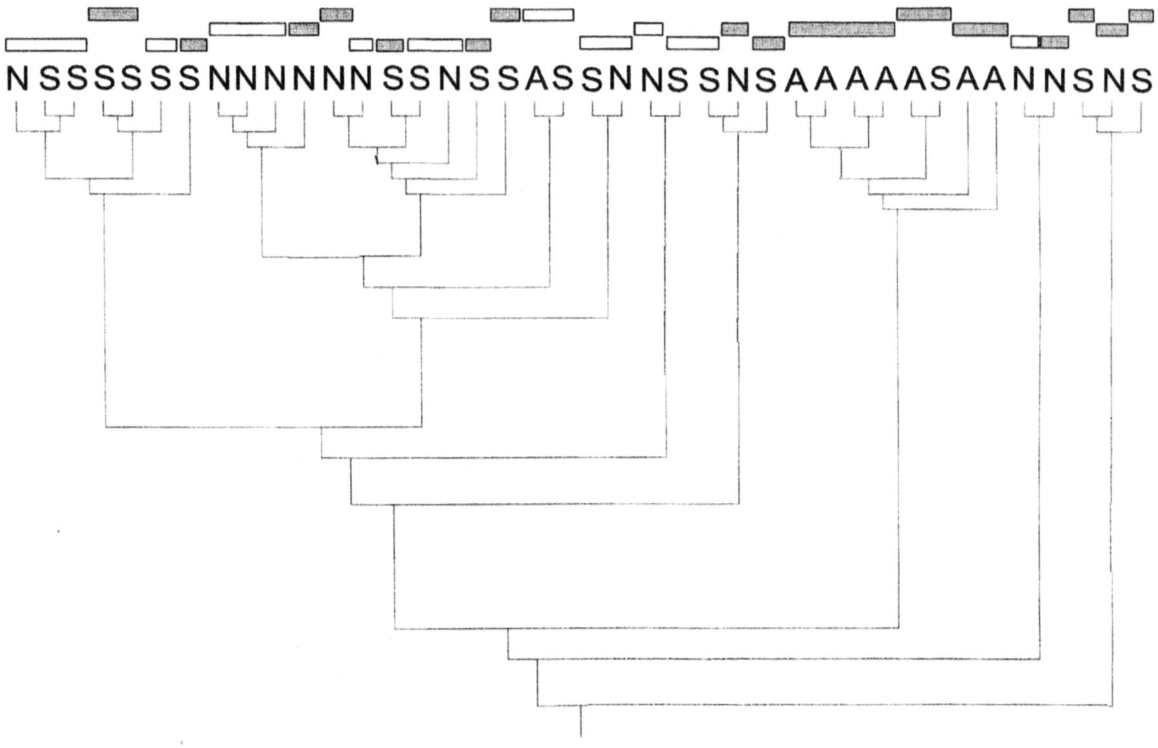

Figure 5. UPGMA dendrogram generated using RAPDPLOT (Kambhampati et al., 1992) and PHYLIP (Felsenstein, 1993) from RAPD data of Peak Steel animals. RAPDPLOT similarity matrix was clustered with the UPGMA option of NEIGHBOR in PHYLIP and drawn using DRAWGRAM in PHYLIP. S = *L. saxatilis* b, N = *L. neglecta*, A = *L. arcana*. Animals were collected from 3 locations on Peak Steel represented by three levels of horizontal bars, and from 2 adjacent patches at each location, represented by flecked or unflecked bars.

No band was diagnostic for species or locality although one band seen in the majority of *L. arcana* was virtually absent from other species. However distinct patterns were evident, at least from Ravenscar animals when similarity measures calculated from frequency differences of bands, were clustered. Animals from Ravenscar tended to cluster into *a priori* groupings (*L. arcana*, *L. saxatilis* 'H' and 'M') regardless of treatment of data. However when only bands that yielded an $N_e m$ estimate < 1 were utilised in production of the distance metrics clustering was more obvious (Figure 3). Filey animals did not display this pattern to such an extent (Figure 4) since after F_{ST} analysis and pruning of the dataset to bands with $N_e m < 1$ the data were reduced to only 12 bands and meaningless patterns were produced after reduction to a similarity matrix and clustering. These differences in pattern between collection sites were also suggested by the results of a discriminant analysis applied to the band data where 96.1% of animals at Ravenscar were correctly placed into their respective group ('H', 'M' or 'A', 1 of each group were misclassified) whilst only 78.1% of Filey samples were correctly placed (9 'H's and 5 'M' misplaced). Animals that were misclassified in the discriminant analysis are marked on Figures 3 and 4.

Figure 5 displays the dendrogram computed from similarity measures calculated between small, barnacle dwelling littorinids collected from Peak Steel. This shows that small barnacle dwelling individuals displayed some tendency to cluster by species but there was also a general tendency to cluster by collection patch. Discriminant analysis on these data with species as discriminator showed that 95% were correctly classified (one *L. saxatilis* classified as *L. neglecta* and one *L. arcana* classed as *L. saxatilis*). However, a similar level of classification was detected when collection patch (animals were taken from 6 patches in total) was used as discriminator from which 9.25% were incorrectly placed (3 individuals misclassified).

Discussion

Allozyme analyses have previously shown littorinids to be highly polymorphic with intraspecific variation often exceeding interspecific differentiation (e.g., Ward, 1990). The ability to differentiate *L. arcana* from *L. saxatilis* has been taken as a benchmark of the aptness of this technique and although there are no diagnostic loci, allele frequencies do separate these species (Sundberg et al., 1990). RAPD analysis has however been demonstrated capable of completely differentiating the closely related species *L. arcana* and *L. saxatilis* at the individual level (Crossland et al., 1993, 1996). However the ability to distinguish species is dependent on the primer used: analysis of band sharing indices with one primer may yield tightly clustered single species groups divergent from the other species, whilst the same samples analysed with an alternative primer fail to differentiate the species. Such between primer variation in utility has also been recognised in studies of *Ceratitis* fruitflies (Haymer & McInnis, 1994).

In the present study, the single primer RAPD-H, which has previously been used for examining intraspecific variation in *L. saxatilis* (Grahame et al., 1997) distinguished between almost all specimens of *L. saxatilis* and *L. arcana* analysed at Ravenscar. It is also notable from this study that *L. arcana* are considerably less polymorphic than brooding littorinids. Such reduced variability has been seen at the PNP locus (Knight & Ward, 1986), and for non-specific esterases (Mill & Grahame, 1992) in *L. arcana*. The same primer, used for samples from the same locality, also separates most *L. saxatilis* H from *L. saxatilis* M; notably, this differentiation is about as good between the H and M forms of *L. saxatilis* as it is between them and *L. arcana*. The tree showing relationships (Figure 3) together with the discriminant analysis result (96.1% of specimens correctly placed) show a high level of discrimination. This strongly suggests that the two morphs (H and M) might not share the same gene pool. Although it could be argued that such a result may occur if allopatric populations are studied, these populations cannot be considered truly allopatric, since although they are found predominantly in different zones on the shore, some animals are to be found in the 'wrong' location, and they are subject to much mixing after storms (Hull et al., 1996). Therefore if there is a barrier to gene flow, it must be intrinsic rather than extrinsic.

This primer however does not discriminate well between *L. saxatilis* b, *L. neglecta* or *L. arcana* on the barnacle flat at Old Peak (about 500 m away from the H and M collection sites) – see Figure 5 (although we note that discriminant analysis correctly placed 95.0% of the specimens correctly, a figure nearly as high as that for the Ravenscar samples). These observations are difficult to interpret together, since on the one hand there is evidence of lack of gene flow between forms regarded as conspecific (*L. saxatilis* H and M) while on the other hand, there is no evidence of a reproductive barrier between forms regarded as different species (*L. saxatilis* b and *L. arcana*). We can only speculate at this stage that ability or failure to differentiate is likely to be highly primer-specific (Crossland et al., 1993; Haymer & McInnis, 1994) and application of more primers is necessary.

At Filey, there is a further test of discrimination of *L. saxatilis* H and M. The tree (Figure 3) indicates less good differentiation between them than occurred at Ravenscar. Discriminant analysis placed 78.1% of individuals into their suggested group, this also confirms that differentiation at Filey was less good than at Ravenscar. This difference in pattern between Filey and Ravenscar suggests that the discriminant utility of this single primer (RAPD-H) is not maintained throughout the range of H and M. Further progress would require screening with multiple primers, and this is in progress at several sites in Britain.

An alternative approach to screening with multiple primers is to focus on a band or bands occurring at different frequencies (or even uniquely) in different populations, leading to the identification of sequence-characterised amplified regions (SCARS, see Xu et al., 1995; Bodénès et al., 1997). This is also being pursued.

These procedures are being followed in the light of the findings of Hull et al. (1996), that H and M forms of *L. saxatilis* have distinctly different reproductive characteristics, shell shapes and habitats, and that the observation of rare intermediate forms with a high incidence of abortion suggests the presence of a reproductive barrier. Allozyme investigations do not reveal consistent differences between the forms, therefore it is of interest that the morphological observations are supported by the RAPD data. At the same time, the evident sensitivity of the RAPDs approach to changes of sample location means that a primer useful in one context will not necessarily be useful in another. This is evident in the failure to discriminate between the small barnacle dwelling *L. saxatilis* b and either *L. arcana* or

L. neglecta (Figure 5) and the reduced differentiation of H and M at Filey in comparison with Ravenscar. Although the patterns we have uncovered are not clear cut they are evidence that the RAPD approach and related methodologies such as SCAR analysis will prove valuable in the study of *Littorina* population genetics and taxonomy.

Acknowledgements

This research was supported by the MAST-3 programme of the European Commission under contract number MAS3-CT95-0042 (AMBIOS).

References

Ashburner, M., 1989. *Drosophila* a laboratory manual. Cold Spring Harbor Laboratory Press, 434 pp.

Backeljau, T. & T. Warmoes, 1992. The phylogenetic relationships of 10 Atlantic Littorinids assessed by allozyme electrophoresis. In J. Grahame, P. J. Mill & D. G. Reid (eds), Proceedings of the Third International Symposium on Littorinid Biology. The Malacological Society of London: 9–24.

Bodénès, C., S. Joandet, F. Laigret & A. Kremer, 1997. Detection of genomic regions differentiating two closely related oak species *Quercus petraea* (Matt.) Liebl. and *Quercus robur* L. Heredity 78: 433–444.

Caley, K. J., J. W. Grahame & P. J. Mill, 1995. A geographically based study of shell shape in small rough periwinkles. Hydrobiologia 309: 181–193.

Clark, A. G. & C. M. S. Lanigan, 1993. Prospects for estimating nucleotide divergence with RAPDs. Mol. Biol. Evol. 10: 1096–1111.

Crossland, S., D. Coates, J. Grahame & P. J. Mill, 1993. Use of random amplified polymorphic DNAs (RAPDs) in separating two sibling species of *Littorina*. Mar. Ecol. Prog. Ser. 96: 301–305.

Crossland, S., D. Coates, J. Grahame & P. J. Mill, 1996. The *Littorina saxatilis* species complex- interpretation using random amplified polymorphic DNAs. In J. Taylor (ed.), Origin and Evolutionary Radiation of the Mollusca. Oxford University Press: 205–209.

Felsenstein, J., 1993. PHYLIP (Phylogeny Inference Package) version 3.5c. Distributed by the author. Department of Genetics, University of Washington, Seattle.

Grahame, J., P. J. Mill, S. L. Hull & K. J. Caley, 1995. *Littorina neglecta* Bean: ecotype or species? J. nat. Hist. Lond. 29: 887–899.

Grahame, J., S. L. Hull, P. J. Mill & R. Hemingway, 1997. Discovering unrecognised diversity among marine molluscs. In R. F. Ormond, J. D. Gage & M. V. Angel (eds), Marine Biodiversity: Patterns and Processes. Cambridge University Press: 293–318.

Grosberg, R. K., D. R. Levitan & B. B. Cameron, 1996. Characterization of genetic structure and genealogies using RAPD-PCR markers: a random primer for the novice and nervous. In J. D. Ferraris & S. R. Palumbi (eds), Molecular Zoology: Advances, Strategies and Protocols. Wiley-Liss: 67–100.

Hadrys, H., M. Balick & B. Schierwater, 1992. Applications of random amplified polymorphic DNA (RAPD) in molecular ecology. Mol. Ecol. 1: 55–63.

Haymer, D. S., 1994. Random amplified polymorphic DNAs and microsatellites: what are they, and can they tell us anything we don't already know? Ann. ent. Soc. Am. 87: 717–722.

Haymer, D. S. & D. O. McInnis, 1994. Resolution of populations of the Mediterranean fruit fly at the DNA level using random primers for the polymerase chain reaction. Genome 37: 244–248.

Hull, S. L., J. W. Grahame & P. J. Mill, 1996. Morphological divergence and evidence for reproductive isolation in *Littorina saxatilis* (Olivi) in northeast England. J. moll. Stud. 62: 89–99.

Johannesson, K. & B. Johannesson, 1990. Genetic variation with *Littorina saxatilis* (Olivi) and *Littorina neglecta* Bean: Is *L. neglecta* a good species? Hydrobiologia 193: 89–97.

Kambhampati, S., W. C. Black IV & K. S. Rai, 1992. Random amplified polymorphic DNA of mosquito populations (Diptera: Culicidae): techniques, statistical analysis, and applications. J. med. Ent. 29: 939–945.

Kimura, M. & T. Ohta, 1971. Theoretical Aspects of Population Genetics. Princeton University Press, Princeton, N.J., 219 pp.

Knight, A. J. & R. D. Ward, 1986. Purine nucleoside phosphorylase polymorphism in the genus *Littorina* (Prosobranchia: Mollusca). Biochem. Genet. 24: 405–413.

Mill, P. J. & J. W. Grahame, 1992. Clinal changes in esterase variability in *Littorina saxatilis* (Olivi) and *L. arcana* Hannaford Ellis in southern Britain. In J. Grahame, P. J. Mill & D. Reid (eds), Proceedings of the Third International Symposium on Littorinid Biology. The Malacological Society of London: 31–37.

Muralidharan, K. & E. K. Wakeland, 1993. Concentration of primer and template qualitatively affects products in random-amplified polymorphic DNA PCR. Biotechniques 14: 362–364.

Reid, D. G., 1993. Barnacle-dwelling ecotypes of three British *Littorina* species and the status of *Littorina neglecta* Bean. J. moll. Stud. 59: 51–62.

Reid, D. G., 1996. Systematics and evolution of *Littorina*. The Ray Society, London. 463 pp.

Reid, D. G., E. Rumbak & R. H. Thomas, 1996. DNA, morphology and fossils: phylogeny and evolutionary rates of the gastropod genus *Littorina*. Phil. Trans. Roy. Soc. Lond. B 351: 977–895.

Rumbak, E., D. G. Reid & R. H. Thomas, 1994. Reconstruction of phylogeny of 11 species of *Littorina* (Gastropoda: Littorinidae) using mitochondrial DNA sequence data. The Nautilus Suppl. 2: 91–97.

Scheirwater, B., 1995. Arbitrarily amplified DNA in systematics and phylogenetics. Electrophoresis 16: 1643–1647.

Smith, J. E., 1981. The natural history and taxonomy of shell variation in the periwinkles *Littorina saxatilis* and *Littorina rudis*. J. mar. biol. Ass. U.K. 61: 215–241.

Sunberg, P., A. J. Knight, R. D. Ward & K. Johannesson, 1990. Estimating the phylogeny in the mollusc *Littorina saxatilis* (Olivi) from enzyme data: methodological considerations. Hydrobiologia, 193: 29–40.

Ward, R. D., 1990. Biochemical genetic variation in the genus *Littorina*. Hydrobiologia 193: 53–69.

Xu, H., D. J. Wilson, S. Arulsekar & A. T. Bakalinsky, 1995. Sequence specific polymerase chain-reaction markers derived from randomly amplified polymorphic DNA markers for fingerprinting grape (*Vitis*) rootstocks. J. am. hort. Soc. 120: 714–720.

Hydrobiologia **378:** 79–88, 1998.
R. M. O'Riordan, G. M. Burnell, M. S. Davies & N. F. Ramsay (eds), Aspects of Littorinid Biology.

Assortative mating between two distinct micro-allopatric populations of *Littorina saxatilis* (Olivi) on the northeast coast of England

S. L. Hull
CERCI, University College Scarborough, Filey Road, Scarborough, YO11 3AZ, U.K.

Key words: Littorina saxatilis, polymorphism, assortative mating, reproductive isolation, reinforcement, sympatric speciation, mating behaviour

Abstract

Size assortative mating is a common invertebrate mating pattern and is usually accompanied by male and female sexual selection, and these three behaviours can contribute to reproductive isolation. Two distinct populations of the marine prosobranch *Littorina saxatilis*, H and M, occur within 15 m of each other on the same shore. Previous studies have demonstrated that these two forms have different reproductive strategies and that the rare hybrids between the two forms show evidence of reproductive dysfunction and hence are less fit than the assumed parental forms. In both populations, female shell height was shown to be a predictor of the number of embryos contained within the brood pouch. The mean shell height of the M population was significantly larger than that of the H population, and the M population matures at a larger shell size than the H population.

The two populations show complete assortative mating to type in the field, and occupy different microhabitats on the same shore. Therefore, laboratory-based experiments were performed to determine if assortative mating was maintained in sympatry and also to determine the effect of population density on mate choice. The males of both populations showed sexual selection for female size, choosing to mate with females approximately 10% larger than themselves from an assortment of female sizes. The M population showed complete assortative mating to type, irrespective of the density of H and M females, whereas at low densities the H males did occasionally mate with M females. The role of assortative mating and reinforcement (due to natural selection acting against the less fit hybrids), in maintaining the partial reproductive barrier between the two populations is discussed.

Introduction

One of the commonest mating patterns found within invertebrates is that of assortative mating by size, which is typically defined as a positive correlation between the size of mates within a population or sample (Arnqvist et al., 1996). This mating pattern is usually accompanied by sexual selection, and has been shown to have a profound effect on the demography and genetics of populations, and may help to preserve phenotypic and genetic variation (for review, see Crespi, 1989). According to Crespi (1989), assortative mating patterns in arthropods may be the result of more than one mechanism. These mechanisms can be broadly classified into three categories, mate choice, mate availability and mating constraints, and he suggests that no single mechanism could be the cause of all assortative mating patterns observed in arthropods. Arnqvist et al. (1996) described 'true' assortative mating as a linear relationship between male and female size with the observations being equally distributed around the regression line of female versus male size, and distinguish 'true' assortative mating from 'apparent' assortative mating in which the latter results from an increased or decreased variance in male size with respect to increased female size. Understanding the causes of assortative mating is essential to assess the relationship of assortative mating to sexual and natural selection, and also its role in promoting genetic differentiation and speciation (Crespi, 1989).

Along with assortative mating, behavioural and ecological studies of male sexual selection and female sexual selection are frequently used to investigate mate choice and success within a species, and these behaviours can also be applied to the study of incipient speciation (Johannesson et al., 1995). Differences in mating behaviour between diverging populations and the subsequent reinforcement of post-zygotic isolation by natural selection acting against less fit hybrids, may explain why some closely related species show greater reproductive isolation in sympatry than allopatry (Ridley, 1993). Reinforcement can only support allopatric speciation if the incipient species do not have complete post-zygotic reproductive isolation (Butlin, 1987), but is thought to be an essential component of sympatric speciation (Barton & Hewitt, 1985; Ridley, 1993) and speciation resulting from habitat divergence (Diehl & Bush, 1989).

Copulations between individuals of different species result in a reduction of fitness in the non-discriminating individuals, therefore the relative importance of species recognition systems depend upon the cost of interspecific matings to the non-discriminating individuals (Saur, 1990). Both interspecific and intersexual copulations have been reported to occur in the prosobranch genus *Littorina*. Raffaelli (1977) reported infrequent interspecific copulations between *Littorina nigrolineata* (=*Littorina compressa*) Gray and *L. saxatilis*, and Saur (1990) reported observing rare interspecific copulations between *L. saxatilis* and *Littorina littorea* (L.) in the field and suggested that size differences would contribute to mating incompatibility. Intrasexual copulations have also been recorded in the field. On a shore in Sweden, 33 out of 124 *L. saxatilis* and 9 out of 129 of *L. littorea* copulations were observed to be between males. In laboratory-based experiments, male *L. littorea* did not appear to be able to distinguish the sex of a snail prior to copulation, but copulation duration was much shorter in intrasexual copulations suggesting that mating was not complete (Saur, 1990). Erlandsson & Johannesson (1994) reported that *L. littorea* demonstrated size-assortative mating on one out of three shores, and sexual selection for female size on three out of the five shores investigated in Sweden. Not only did the males select larger females with which to mate, the males mated for longer periods with the larger females than with the small females. They suggested that during the longer matings the males could transfer more sperm. In trail following experiments, male *L. littorea* could distinguish the mucus trails of male and female conspecifics during the breeding season and the males would closely follow the trail of a female (Erlandsson & Kostylev, 1995). The authors suggested that the mucus trails contained information about the sex of the individual, and that male:male matings were the result of a male following a multiple-layered trail which would cause confusion during the identification of a prospective mating partner.

Johannesson et al. (1995) reported an example of incipient reproductive isolation in two sympatric populations of the intertidal gastropod *Littorina saxatilis* (Olivi) from Galicia, Spain. The two conspecific morphs occurred over a scale of 25 km, and consisted of a ridged and banded upper shore morph and a smooth, unbanded low shore morph, which had overlapping distributions in the midshore where hybrids between the two morphs accounted for approximately 20% of the total population (Johannesson et al., 1993). Allozyme variation at five polymorphic enzyme loci indicated that the two morphs shared the same gene pool. However, gene flow between the morphs was less than that within morphs suggesting that the two populations demonstrate incipient reproductive isolation (Rolan-Alvarez et al., 1996). They suggested that habitat selection, assortative mating and female sexual selection contributed to the maintenance of the partial reproductive barrier between the two morphs. The high shore morphs showed positive size-assortative mating which was suggested to be a consequence of non-random distributions of the different morphs between microhabitats. In a further study, Rolan-Alvarez et al. (1995) showed that sexual selection acted against the hybrids, as the hybrids mated less frequently than the pure morphs in regions where either of the pure morphs was present in high frequencies. Thus, they suggested that sexual selection acted to restrict gene flow between the two populations.

At Ravenscar, on the northeast coast of England, a similar situation has been reported where two distinct populations of *L. saxatilis* occur on the same shore, but as isolated populations only metres from each other. These populations are distinct in shape, occupy different microhabitats at different heights on the shore and have different reproductive strategies. The high shore *L. saxatilis* H population produced a small number of large embryos, whereas the *L. saxatilis* M population produces a large number of small young (Hull et al., 1996). Rare intermediates between the H and M forms occurred in the same habitat as the M population, but accounted for less than 2% of the M population. More interestingly, the brood pouches of the intermediates contain eggs of both sizes and there is evidence of reproductive dysfunction as high proportions of the embryos contained within the brood pouch were aborting. The two populations only appeared to mix after storms displaced animals from their normal habitat, but within a month the displaced animals had returned to their former habitats. An identical pattern occurs at another site, Filey Brigg, and analysis of allozymes at five polymorphic loci for both H and M *L. saxatilis* and also *Littorina arcana* Hannaford Ellis populations from the two sites, revealed little genetic differentiation between the populations (Grahame et al., 1997). There is little evidence of genetic divergence between the H and M populations at the allozyme level (Hull, 1994). Yet, the two populations are significantly different in shape, and show differences in habitat choice and reproductive strategy. DNA analysis may yield more promising results as small differences between H and M forms at Ravenscar have been revealed using RAPDs in which 2 out of 10 primers screened produced bands which were present in a small sample of *L. saxatilis* M and the intermediates, but not *L. saxatilis* H (Grahame et al., 1997).

The aims of the current study were to, firstly, determine if female shell height was an indicator of reproductive success, and if males of both the H and M populations demonstrated sexual selection for female size and size assortative mating. Secondly, experiments were performed in the laboratory to determine if the H and M populations showed assortative mating to type and data on mate choice were also collected from the field.

Method

Site description

The two populations of *L. saxatilis* occupy different microhabitats on the same shore, and rarely intermingle, thus samples of both populations were collected for laboratory-based studies of breeding behaviour. Samples of *L. saxatilis* H were collected from large (approximately 2–3 m in height) high-shore boulders at MHW level, and samples of *L. saxatilis* M were collected from smaller (approximately 0.5–1 m in height) mid-shore boulders 15 m horizontally down the same shore at Old Peak, Ravenscar (British National Grid reference NZ984021), northeast England. For full site description see Hull et al. (1996) for details.

Data from previous samples collected at these two sites were used to compile size frequency histograms of the two populations, and also to determine if female shell height was a predictor of the number of juveniles contained within the brood pouch. Field observations of copulating pairs were made during June/July 1996 to compare with the laboratory-based results. The samples of *L. saxatilis* H and *L. saxatilis* M were collected from the shore at Old Peak, during March and June 1996. The populations were sexed by allowing the animals to dry out, then re-submerging them in seawater. As the animals emerged from the shells the mature males could be distinguished from the females by the presence of a penis in the mantle cavity. The males and females were then kept separately in the aquarium in tanks with stones from their native habitat at 12 °C (±1 °C) until required for experiments (Warwick, 1982).

Mate choice experiments

The mate choice experiments were conducted in 15 cm diameter containers containing seawater at a depth of approximately 5–10 mm, into which a single male of known size was presented with females of known size either from his own population or from the other population. Notes were made on the mating behaviour of both males and females. The behaviour of the animals was then monitored until either mating took place (i.e. the male inserted his penis inside the mantle cavity of the female), or the experiment had run for 3 h. Initially, males of different sizes from each population were presented with 10 females of different sizes from their own populations to determine if female size was important in male mate choice. All shell measurements were of maximum shell height in mm. Each experiment was performed 10 times for both populations using males of different shell sizes. The same protocol was followed for the experimental determination of the effect of the density of the females from the two populations on male mate choice. Four separate experimental regimes were used according to the format below.

a. One male with one female of either his own type or the other population (H males $n = 17$, M males $n = 17$).
b. One male with one female of each population of approximately equal size (H males $n = 25$, M males $n = 33$).
c. One male with two females from each population (H males $n = 15$, M males $n = 15$).
d. One male with ten females from each population (H males $n = 12$, M males $n = 12$).

Finally, 10 males of each population were placed in a container to determine if the males attempted to copulate with other males of either population, and in a similar experiment 10 males and 10 females of the same population were kept in a container to determine if males could distinguish females from males. These experiments were performed in September 1996.

The data were an analysed using the Isolation indices developed by Ribi & Porter (1995), where,

Isolation index, (1)

$$I = \frac{((h, h) + (m, m)) - ((h, m) + (m, h))}{N}.$$

The Isolation index has a range of $I = -1$ to 1, where 0 = complete random mating, $I > 0$ positive assortative mating, $I < 0$ negative assortative mating where (h, h) = H males × H females, (m, m) = M males × M females, (h, m) = H males × M females, (m, h) = M males × H females.

$$\text{Asymmetry index, } A = \frac{(h, m)}{(h, m) + (m, h)}. \quad (2)$$

The range of A is 0 to 1, where $A = 0.5$ signifies that inter-population mating is symmetrical. If $A < 0.5$ then (m, h) copulations are more common.

Chi-square tests were used to determine if the males of both populations demonstrated significant deviations from complete random mating when presented with females from both populations. In the absence of positive assortative mating, the males would select females from either population at random with which to mate, thus the expected values for the Chi-square tests were estimated assuming that the males would mate with a female from either population with equal frequency.

Results

Shell height and number of juveniles

There was a significant difference in shell height between the two populations. The mean shell height of the *L. saxatilis* M population was significantly larger than that of the *L. saxatilis* H population (*L. saxatilis* H mean shell height = 7.99 mm, $n = 160$; *L. saxatilis* M mean shell height = 10.62 mm, $n = 89$; Z-test = 11.03, $p = 0.001$), as reflected by the size frequency histogram (Figure 1). Regression of log female shell height against log number of juveniles contained within the brood pouch demonstrated a significant relationship between the two variables in both populations (*L. saxatilis* H log number = −1.57 + 3.62 log height; ANOVA $F_{1,31} = 58.81$, $p > 0.0001$. *L. saxatilis* M log number = −0.71 + 2.75 log height; ANOVA $F_{1,24} = 7.49$, $p = 0.012$). A common slope could be fitted to the compete data set (ANCOVA, $F_{1,54} = 184.8$, $p < 0.001$), however, there was a significant difference in intercepts between the two data sets (ANCOVA, df = 1, $F = 215.01$, $p < 0.001$). Thus, female shell height could be used as a predictor of the number of juveniles contained in the brood pouch in both the H and M populations.

Mating behaviour experiments

Overall, 37 H/H copulations on the high-shore boulders and 27 M/M copulations on the mid-shore boulders were observed in the field, but no copulations were observed to occur between the two populations, nor were any H animals found in the M habitat and vice versa.

In the laboratory, the males crawled anti-clockwise over the shell of the females with their tentacles fully extended and apparently touching the surface of the female's shell. Before insertion of the penis into the mantle cavity of the female, the male positioned himself so that the penis could be inserted on the right-hand side of the females' shell between the substrate and the edge of the shell. Females appeared to be passive during male attempts to copulate, but on three occasions M females closed their opercula in response to an H male trying to insert his penis into the mantle cavity.

Initially, a range of different sized males from each population presented with ten females of assorted size from their own populations to determine if the size of the female had any effect on male mate choice (H population size range 4.75–10.15 mm; M population size range 7.95–14.05 mm). There was a significant correlation between the size of the male and the size of the female chosen in both populations (H population, $n = 10$, Spearman rank correlation coefficient = 0.867, $p < 0.01$. M population, $n = 10$, Spearman rank correlation coefficient = 0.945, $p < 0.01$). Regression of male size against size of female chosen by the male resulted in significant regressions for both the H and M populations (H population, female size = 1.37 + 0.864 male size, $r^2 = 71\%$, $F_{1,9} = 19.59$, $p = 0.002$, Figure 2 (a). M population; female size = 1.36 + 0.941 male size $r^2 = 89.1\%$,

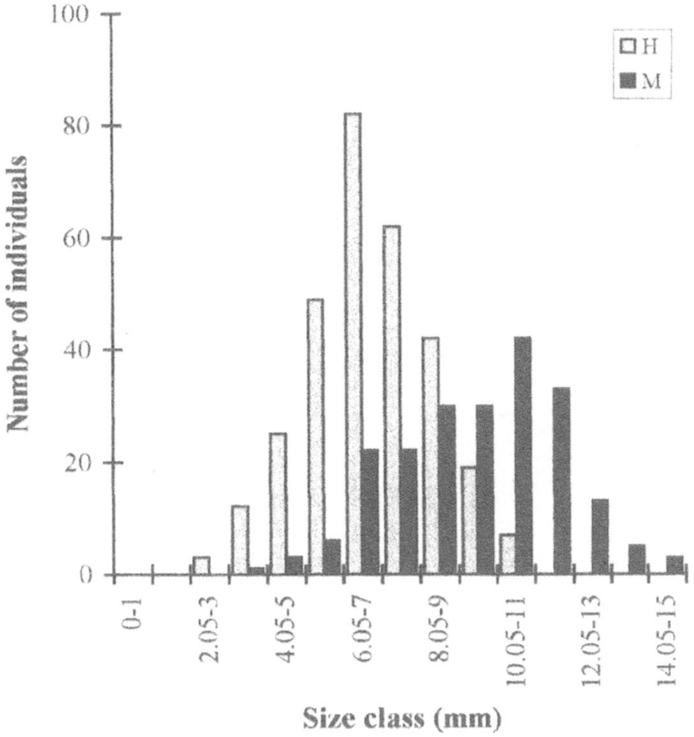

Figure 1. Size frequency histograms of the H and M populations of *L. saxatilis* at Old Peak.

Table 1. Results of the frequency of mate choice experiments to determine if males would mate with a single female from either population

	H population female with male		M population female with male	
Male population	No mating	Mated	No mating	Mated
H	0	17	16	1
M	15	0	0	16

$F_{1,9} = 65.12$, $p = 0.0001$, Figure 2 (b)). There was no significant difference between the slopes of the two regression lines of the two populations (t-test, $t = 0.15$, $p > 0.05$). The males of both populations chose to mate with females which were approximately 10% larger than themselves from a population of females of assorted sizes, although they did not necessarily choose the largest female within the population, therefore there appears to be sexual selection for female size exhibited by the males.

When presented with either an M or H female, the M males would not try to copulate with the H females (Table 1) (even though they were observed to crawl over them), but they did mate with the M females. The H males also mated with females of their own population, and on one occasion out of 17 replicate experiments, an H male attempted to mate with an M female. The M males demonstrated complete assortative mate choice, whereas in the H males, mate choice was almost completely assortative ($I = 0.944$). Preliminary investigations of inter-population mate choice indicated that it was very asymmetrical ($A = 1.0$) as, although one H male attempted to mate with an M female, on no occasion did one of the M males attempt to mate with an H female. However, as the sample size is so small no conclusions can be drawn from this event.

When a single male from the H population was presented with both H and M females, the males exhibited a significant deviation from random mate choice ($\chi^2 = 13.29$, df $= 1$, $p < 0.001$). In all instances but one, H males chose to mate with H females, indicating almost complete assortative mating between the two populations. The M males also demonstrated a signif-

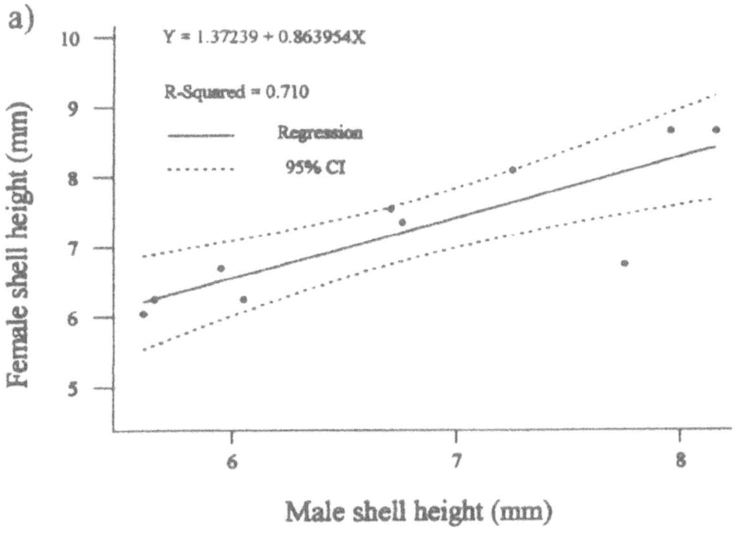

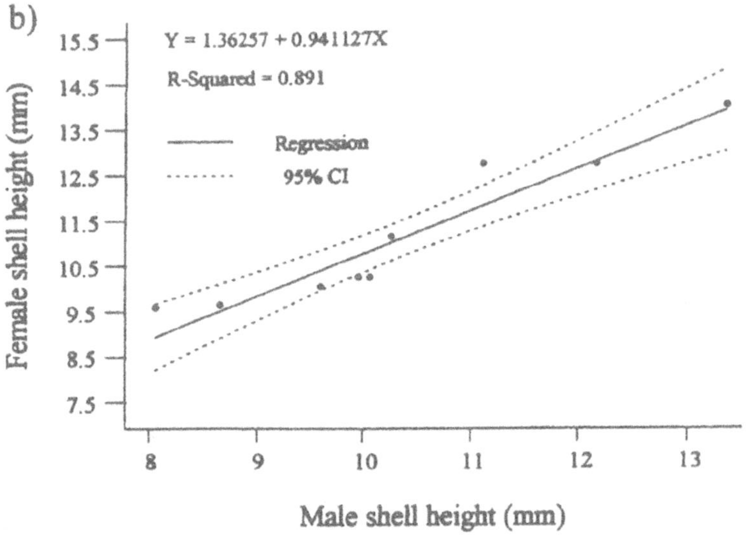

Figure 2. Regression of male shell height (mm) against shell height of female (mm) chosen from ten females of different sizes, (a) *L. saxatilis* H and (b) *L. saxatilis* M populations, (dotted lines denote 95% confidence limits).

icant deviation from random mate choice between the two females ($\chi^2 = 14.06$, df=1, $p < 0.001$), and in all cases M males mated with M females, demonstrating complete positive assortative mating. One H male mated with an M female, thus inter-population mating was asymmetrical ($A = 0$), and the males of both populations showed positive assortative mating ($I = 0.983$: Table 2).

When presented with two H and two M females, H males showed a significant deviation from random mate choice ($\chi^2 = 21.2$, df=1, $p < 0.001$), and on all occasions mated with a female from their own population (Table 3). The M males also chose mates from their own populations on all occasions, again showing a significant deviation from random mate choice ($\chi^2 = 31.02$, df=1, $p < 0.001$). A similar pattern was also seen at high female densities, i.e. ten H

Table 2. Results of the frequency of male mate choice experiments in which a male was placed in a container with a similarly-sized female from each of the two populations

Male population	Female population H	Female population M	Number of replicates
H	24	1	25
M	0	33	33

Table 3. Results of frequency of male mate choice experiments performed at higher densities of females

Female chosen from two H and two M females: Male	H	M	Female chosen from ten H and M females: Male	H	M
H	15	0	H	12	0
M	0	15	M	0	12

and ten M females. The H males showed a significant deviation from random mating ($\chi^2 = 26.12$, df=1, $p < 0.001$), selecting only H females with which to mate, and the M males only selected M females ($\chi^2 = 20.16$, df=1, $p < 0.001$) again demonstrating a significant deviation from random mating.

There were no attempted male:male copulations observed during the study, nor during the experiment in which ten males from each population were studied to find out if the males would try to copulate with other males either from their own or the other population. In the control experiments where ten males were placed in a container with ten females from their own population, copulations were observed. In the H population, two male:female attempted copulations were observed and in the M population one male:female attempted copulation was observed.

At high densities, the males from both the H and M populations attempted to mate with females from their own populations, indicating complete positive assortative mating with no mating asymmetry. The M males showed complete positive assortative mating irrespective of population density and did not attempt to mate with the H females even when they were of similar size to the M females. On one occasion an H male did attempt to mate with an M female in the absence of H females, but at higher population densities they demonstrated complete positive assortative mating.

Discussion

There was a significant relationship between shell height and number of juveniles contained within the brood pouch for the females of both the H and M *L. saxatilis* populations. In other studies, the number of juveniles contained within the brood pouch of mature *L. saxatilis* was observed to be linearly related to shell height (three Firth of Forth populations, Ross & Berry, 1991). Janson (1985) found a linear relationship between shell height and log embryo number and also size-specific differences in fecundity in populations of *L. saxatilis* from Sweden. Therefore, shell height can be regarded as an indicator of the fecundity of the mature female *L. saxatilis*. There was also a significant difference in mean shell height between the H and M populations. This pattern was also reflected in the size frequency histogram, with the M population showing a greater number of larger animals than the H population. Hull et al. (1996) reported that both sexes of the *L. saxatilis* H population matured at a smaller size than did those of the *L. saxatilis* M population. Despite their proximity (10–15 m apart on the same shore), the two populations are distinct in shape; the H populations being thin-shelled and large-apertured, whereas the M populations have thick shells, large shell lips and small apertures (Hull et al., 1996; Grahame et al., 1997). The more usual pattern of morphological change within *L. saxatilis* populations, is one of horizontal change along a shore. Extreme phenotypic morphs colonise the more exposed and sheltered habitats, and a range of phenotypically intermediate morphs occur within the transitional environments, (which only account for approximately 7% of the shoreline under investigation) (Janson, 1983; Janson & Sundberg, 1983). A similar situation, although lower on the shore, has been described in Galicia, Spain (Johannesson et al., 1993; Rolan-Alvarez et al., 1995, 1996), where two extreme morphological types occupy different microhabitats at different vertical heights on the same shore.

During the current study, observation of copulating pairs of snails in the field indicated complete positive assortative mating to type within the two populations. However, this could be a consequence of the non-random distribution of the two populations within different microhabitats, rather than mate choice (Rolan-Alvarez et al., 1995). Over the last five years, no copulations between H and M populations have been observed in the field (Hull, unpublished data). Occasionally, animals from the M population can be found on the high shore boulders after a storm (Hull et al., 1996), but the animals appear to return to their own microhabitats, as these patterns only persist for a matter of days. There is no zone of overlap between the two populations and H/M hybrids are rare (less than 2% of the M population), unlike the situation in Galicia, where an upper shore morph and lower shore morph appear to hybridise in a mid-shore zone where the hybrids are found alongside the parental populations (Johannesson et al., 1993; Rolan-Alvarez et al., 1995, 1996). Therefore, to ascertain if the H and M populations showed assortative mating to type in sympatry, laboratory-based studies had to be performed.

Both H and M males chose to mate with females larger than themselves when presented with females of a range of sizes from their own populations. Interestingly, the males from both populations did not select the largest female within the sample, but selected a female approximately 10% larger than themselves. This indicates that the males of both populations show true size assortative mating (Arnqvist et al., 1996). Size assortative mating has also been demonstrated in the Galician populations of *L. saxatilis*, especially in the upper shore zone (Johannesson et al., 1995). In a laboratory-based study, Erlandsson & Johannesson (1994) demonstrated that large *L. littorea* females mated more frequently, and for longer periods than small conspecific females, thus the males appeared to show sexual selection for female size. Whilst climbing onto the female shell, males appear to assess female size by using their tentacles (Saur, 1990; this study), therefore it is possible that male *L. saxatilis* assess the size of their prospective mates prior to copulation. In both H and M *L. saxatilis* populations, female shell size was positively related to the number of juveniles within the brood pouch. A male may be able to enhance his life-time reproductive success by choosing large, more fecund females with which to mate, thus sexual selection for large female size could have evolved. There could also be mating constraints acting to promote size assortative mating in the *L. saxatilis* populations (after Crespi, 1989). Large males may find it physically impossible to mate with small females due to the relative size of the penis to the opening in the mantle cavity (Saur, 1990), and during mating, small males may also be more readily displaced by large males (Erlandsson & Johannesson, 1994).

Male *L. saxatilis* M showed complete positive assortative mating to type, irrespective of the density of the H and M females, where as at low densities *L. saxatilis* H males did occasionally attempt to mate with the *L. saxatilis* M females. This discrepancy could possibly be explained by the sexual selection for female size exhibited by the males of both populations. As the M population was larger in shell height than the H population, mature *L. saxatilis* M females would be of equivalent size to, or larger than the *L. saxatilis* H males. However, *L. saxatilis* H females would generally be smaller than *L. saxatilis* M males. Therefore, the *L. saxatilis* H males would perceive the large M females as prospective mates, whereas the M males would reject the H females as they would be smaller than themselves. It is possible that the pattern of assortative mating to type observed during the current study in the laboratory could be the result of sexual selection for female size by the male. In Galicia, assortative mating to type was thought to occur due to the non-random micro-distribution and to type-assortative mating of the two morphs within the habitat (Johannesson et al., 1995). Further study indicated that, in the mid-shore region of overlap between the two morphs, sexual selection appeared to act against the hybrid forms, as they mated less frequently in the micro-patches dominated by either one of the true upper or lower shore morphs (Rolan-Alvarez et al., 1995). Palumbi (1994) noted that, in a review of speciation in the marine environment, assortative mating to type appeared to be more prominent in areas where populations were sympatric, than allopatric. In the current study, complete positive assortative mating to type has been observed in the field, which could be the result of sexual selection for female size by the males of two distinct populations which demonstrate a non-random distribution.

Assortative mating, along with sexual selection, is thought to be one of the mechanisms which contributes towards the evolution of reproductive isolation (Diehl & Bush, 1989; Gilbert & Starmer, 1985). The above could also explain why the hybrids between the two populations are found in such low frequencies (Hull et al., 1996). It is possible that gene flow between the H and M populations occurs predominantly in one direction due to the sexual selection for female size exhibited by the males. Hypothetically, a displaced mature M female would be of such a size that H males would attempt copulation. However, a displaced M male would perhaps not attempt to copulate with the smaller H females. Another point of note, is that animals of the M population mature and become reproductively active at a larger size than those of the H population, therefore small M males or females (< 6 mm) may not be mature enough to mate (Hull, 1994). Therefore, hybridisation would occur in predominantly one direction, and only rarely during the periods when animals had been displaced into the different micro-habitats. There is evidence of a partial reproductive barrier between the H and M populations, as the broods of the H/M hybrids contain many either aborting or deformed embryos, in fact up to 55% of the brood may be dysfunctional (Hull et al., 1996). The hybrids are therefore less fit than either of the two assumed parental types (the broods of which show little evidence of dysfunction). It is therefore possible that the two distinct populations are maintained through habitat selection, sexual selection for female size, assortative mating, and reinforcement due to natural selection acting against the hybrids which have been shown to be less fit than the parental populations (Butlin, 1987, 1989). Reinforcement can support speciation only if postzygotic isolation between two populations is not already complete (Butlin, 1987), and the pleiotropic effects of selection on other characters can produce similar patterns to those of reinforcement (Butlin, 1987, 1989). Nevertheless, the results that have been obtained do suggest that the H and M *L. saxatilis* populations have developed partial reproductive isolation, and that the divergence is mediated via differential habitat selection between the two morphs, size assortative mating involving sexual selection for female size and assortative mating within a population, and possible reinforcement of reproductive isolation due to natural selection acting against the less fit H/M hybrids.

Acknowledgements

The author would like to acknowledge the financial support provided by the Research and Postgraduate School at the University College Scarborough, which enabled the experimental work to be undertaken. The comments of two anonymous referees helped to improve the manuscript.

References

Arnqvist, G., L. Rowe, J. J. Krupa & A. Sih, 1996. Assortative mating by size: a meta-analysis of mating patterns in water-striders. Evol. Ecol. 10: 265–284.

Barton, N. H. & G. M. Hewitt, 1985. Analysis of hybrid zones. Ann. Rev. Ecol. Syst. 15: 133–164.

Butlin, R., 1987. Speciation by re-inforcement. TREE 2: 8–13.

Butlin, R., 1989. Reinforcement of Premating Isolation. In D. Otte & J. A. Endler (eds), Speciation and its Consequences. Sinauer, Sunderland, Mass., U.S.A.: 158–179.

Crespi, B. J., 1989. Causes of assortative mating in arthropods. Anim. Behav. 38: 980–1000.

Diehl, S. R. & G. L. Bush, 1989. The Role of Habitat Preference in Adaptation and Speciation. In D. Otte & J. A. Endler (eds), Speciation and its Consequences. Sinauer, Sunderland, Mass., U.S.A.: 345–365.

Erlandsson, J. & K. Johannesson, 1994. Sexual selection on female size in a marine snail, *Littorina littorea* (L.). J. exp. mar. Biol. Ecol. 181: 145–157.

Erlandsson, J. & V. Kostylev, 1995. Trail following, speed and fractal dimension of movement in a marine prosobranch, *Littorina littorea*, during a mating and a non-mating season. Mar. Biol. 122: 87–94.

Gilbert, D. G. & W. T. Starmer, 1985. Statistics of sexual isolation. Evolution 39: 1380–1383.

Grahame, J., S. L. Hull, P. J. Mill & R. Hemmingway, 1997. Discovering unrecognised diversity among marine molluscs. In Ormond, R. F. G., J. D. Gage & M. V. Angel (eds), Marine Biodiversity: Patterns and Processes. Cambridge University Press: 293–318.

Hull, S. L., 1994. Polymorphism, physiological ecology and the systematics of *Littorina*. Unpublished Ph.D. thesis, University of Leeds.

Hull, S. L., J. Grahame & P. J. Mill, 1996. Morphological divergence and evidence for reproductive isolation in *Littorina saxatilis* (Olivi) in northeast England. J. moll. Stud. 62: 89–99.

Janson, K., 1983. Selection and migration in two distinct phenotypes of *Littorina saxatilis* in Sweden. Oecologia 59: 58–61.

Janson, K., 1985. Variation in the occurrence of abnormal embryos in the females of the intertidal gastropod *Littorina saxatilis* Olivi. J. moll. Stud. 51: 64–68.

Janson, K. & P. Sundberg, 1983. Multivariate morphometric analysis of two varieties of *Littorina saxatilis* from the Swedish west coast. Mar. Biol. 74: 49–53.

Johannesson, K., B. Johannesson & E. Rolan-Alvarez, 1993. Morphological differentiation and genetic cohesiveness over a micro-environmental gradient in the marine snail *Littorina saxatilis*. Evolution 47: 1770–1787.

Johannesson, K., E. Rolan-Alvarez & A. Ekendahl, 1995. Incipient reproductive isolation between two sympatric morphs of the intertidal snail *Littorina saxatilis* (Olivi). Evolution 49: 1180–1190.

Palumbi, S. R., 1994. Genetic divergence, reproductive isolation and marine speciation. Ann. Rev. Ecol. Syst. 25: 547–572.

Raffaelli, D. G., 1977. Observations on the copulatory behaviour of *Littorina rudis* (Maton) and *Littorina nigrolineata* (Gray). Veliger 20: 75–77.

Ribi, G. & A. H. Porter, 1996. Mating between two hybridising species, *Viviparous ater* and *V. contectus* (Mollusca: Prosobranchia). Anim. Behav. 49: 1389–1398.

Ridley, M., 1993. Evolution. Blackwell Scientific Publications, New York, 670 pp.

Roberts, D. J. & R. N. Hughes, 1981. Growth and reproductive rates of *Littorina rudis* from three contrasted shores in North Wales, U.K. Mar. Biol. 58: 47–54.

Rolan-Alvarez, E., K. Johannesson & A. Ekendahl, 1995. Frequency and density-dependent selection in natural populations of *Littorina saxatilis* (Olivi). Hydrobiologia 309: 167–172.

Rolan-Alvarez, E., E. Rolan & K. Johannesson, 1996. Differentiation in radular and embryonic characters, and further comments on gene flow, between two sympatric morphs of *Littorina saxatilis* (Olivi). Ophelia 45: 1–15.

Ross, B. & A. J. Berry, 1991. Annual and lunar reproductive cycles in *Littorina saxatilis* (Olivi) and differences between breeding in the marine Firth of Forth and the Forth Estuary. J. moll. Stud. 57: 347–358.

Saur, M., 1990. Mate discrimination in *Littorina littorea* (L.) and *Littorina saxatilis* (Olivi) (Mollusca: Prosobranchia). Hydrobiologia 193: 261–270.

Warwick, T., 1982. A method of maintaining and breeding members of the *Littorina saxatilis* (Olivi) species complex. J. moll. Stud. 48: 368–370.

Hydrobiologia **378:** 89–94, 1998.
R. M. O'Riordan, G. M. Burnell, M. S. Davies & N. F. Ramsay (eds), *Aspects of Littorinid Biology.*

Natural selection on a vertical environmental gradient in *Littorina saxatilis*: analysis of fecundity

R. Cruz[1], E. Rolán-Alvarez[2] & C. Garcia[1]
[1] *Departamento de Bioloxía Fundamental, Facultade de Bioloxía, Universidade de Santiago de Compostela, E-15706 Spain*
[2] *Departamento de Bioloxía Fundamental, Facultade de Ciencias, Universidade de Vigo, E-36200 Spain*

Key words: Littorina, speciation, sympatry, natural selection, intertidal zone, environmental pressures, fecundity

Abstract

Two different morphs of the intertidal snail *Littorina saxatilis* are found in the wave-exposed coasts of Galicia, NW Spain. One lives on the upper shore and the other on the lower shore. They meet and produce phenotypically intermediate hybrids in a narrow transition band on the mid-shore.

To understand the maintenance in sympatry of such an extreme intraspecific polymorphism, we studied the change in fecundity in each morph along two vertical transects of the coast. We did not find a progressive reduction in fecundity in any morph as it approached the midshore, and conclude therefore that the observed spatial segregation of the morphs cannot be explained by divergent natural selection acting through fecundity. We also measured 16 environmental variables, but none explained an important proportion of fecundity variability.

Introduction

Littorina saxatilis (Olivi) is a dioecious intertidal gastropod, with internal fertilisation and direct development. It has a very low dispersal ability, and this often results in great genetic and phenotypic differences between populations (Janson, 1983, 1987; Ward, 1990; Reid, 1996).

A remarkable polymorphism exists in some *Littorina saxatilis* populations of the wave-exposed coasts of Galicia, NW Spain (Rolán et al., 1989; Johannesson et al., 1993). A ridged and banded morph (RB) of this snail is found on the upper shore, and a smooth and unbanded one (SU) lives on the lower shore. They meet and produce phenotypically intermediate hybrids in the midshore (Johannesson et al., 1995).

Genetic differences have been found between the morphs for several fitness-related traits, despite the fact that they seem to maintain an important amount of gene flow for some presumably neutral isozyme loci (Johannesson et al., 1993; Rolán-Alvarez et al., 1996; Johannesson et al., 1997).

The midshore band in which both morphs are found, along with the phenotypically intermediate individuals, is rather narrow in terms of the dispersal ability of the species, which is a few metres per month (Janson, 1983; Johannesson et al., 1993; Rolán-Alvarez et al., 1996). The forces that prevent each morph from invading the other's area of distribution, or both morphs from making a single genetic population through genetic flow, are still not completely clear. Johannesson et al. (1995) studied mating behaviour in these populations, and found that habitat selection, assortative mating and possibly sexual selection among females contributed to the partial reproductive isolation between the morphs.

The spatial and genetic separation between the morphs could also be due to a narrow adaptive specialisation of each morph to its own coastal band, its fitness being sharply reduced when approaching the other morph's habitat. To study this possibility, we registered the change in an important fitness component, female fecundity as measured by the number of viable embryos carried by females, in both morphs along two vertical coastal transects. Thus, if natural selection was operating through female fecundity to maintain the two morphs separated, we would find a clear decrease in the number of embryos carried by females of the two morphs when approaching the intermediate zone. Many studies have dealt with fecundity in *Littorina saxatilis* since the early nineteenth century, because the number of embryos in the brood pouch is easy to count (see a review in Reid, 1996). This number is correlated with maternal body size (Fish & Sharp, 1985).

Materials and methods

Snail sampling

We sampled snails in two coastal transects, transversal to the coastline, and separated by about 3 km of coast, in Galicia, NW Spain. The locations of these transects were chosen because there were especially wide bands occupied by phenotypically intermediate individuals. The samples were taken on May 25 and June 1 1995 for locality 1 and June 16 1995 for locality 2.

There were three coastal bands in each transect: the upper shore band, occupied by RB snails, the lower shore band, occupied by SU snails, and a mid-shore band, delimited by the highest area occupied by the SUs and the lowest area occupied by the RBs. We tried to get the same number of samples from every coastal band, despite the fact that they were of different widths. Thus, the spacing between samples was: 1.0, 2.4 and 7.0 m for the lower, mid and upper shore bands in the first transect, and 1.8, 1.6 and 3.2 m in the second one.

The sampling experimental unit was a 30 × 30 cm plastic sheet that was laid on the ground at the sampling points. A sample of 5 adult snails was taken from the area covered by the square, and a series of environmental variables was measured for this area. When there were not enough adult snails within the sampling area, we took the nearest ones. This happened only in the extreme upper shore, and the distances from the sampling area were rarely greater than 25 cm, and never greater than one half of the distance between two consecutive sampling areas.

Each sample of snails was put into a separate plastic bag, that was placed in boiling water for a few seconds and then stored at –22 °C. The brief scalding prevented the snails becoming too withdrawn within the shell when frozen, and thus facilitated the extraction of the body for dissection. The entire female (shell + body) was weighed before dissection. This dissection took place in artificial sea water (35 g of salt l^{-1}) and the fecundity of the females was measured as the number of embryos with normal appearance within the embryo brood pouch. Male individuals were not used in this study.

Measurement of environmental variables

The following environmental variables were measured at every sampling point. *Sample population density* was the number of animals present in the area defined by the extent of the plastic sheet. This sheet was equally divided into 16 'big' squares, that were in turn divided into four 'small' squares each, so that the sheet had a total of 64 small squares. *Mussel abundance 1* was the number of small squares that were completely occupied by mussels, and *mussel abundance 2* was the number of big squares covered by mussels. *Barnacle abundances 1* and *2* were measured in the same way. To get a measure of the continuity of the areas covered by mussels or barnacles, we defined the variable *mussel (or barnacle) patchiness* as the quotient: mussel (or barnacle) abundance 1/mussel (or barnacle) abundance 2. *Number of matings in the sample. Rock slope* was measured as the angle made by the rock surface and a horizontal plane. *Wave exposure* was measured with a compass, as the angle of view in which the sea could be seen from the sampling point (0° for a point completely covered, from which no extent of the sea can be seen, and 180° for a point completely exposed). *Vertical level* was a qualitative measure of habitat zonation, based on the presence of different species of animals and plants. Vertical level 1 corresponded to the lowest shore, in which there are stalked barnacles (*Pollicipes cornucopia* Leach), the alga *Corallina mediterranea* Areschough and sea anemones living outside pools. In vertical level 2, *Pollicipes* and *Corallina* disappear, and there is a dominance of mussels (*Mytilus galloprovincialis* Lamarck) over barnacles (*Chthamalus stellatus* Poli). In level 3 the barnacles predominate over mussels, and in level 4 the mussels disappear. Level 5 was characterised by the presence of the lichen *Verrucaria*. The *rock surface roughness* of the sites from which each animal was sampled, was qualitatively measured with a three-value index, from 1 (a flat rock) to 3 (a crevice). The average for the values corresponding to each animal was calculated for each sample. *Sun cover* was measured as the number of rock walls that were near the sampling point and that could shade it some time

during the day. *Percentage of bare rock, percentage of crevice and percentage of pool* were three-value indices (from 1 to 3) measuring the percentage of individuals in the sample that occupied these habitats.

In all the analyses that follow, individual fecundities were converted to relative fecundities by dividing them by the average number of embryos of the individual's morph and transect. This procedure corrected for transect differences in mean fecundity, which was useful to carry out joint analyses for both transects, and also allowed us to make clearer graphical comparisons between the morphs.

Data analysis

In order to see if there was a gradient of fecundity along the coastal transects, we calculated the polynomial regression of relative fecundity on coastal transect position. The model used was: Relative fecundity = position + position^2. Every transect position was expressed as a deviation from the average for its corresponding morph and transect before being squared. Thus, individuals near the centre of its morph's coastal band had low values for this independent variable, and those far from it had high values. The squared-position term in the model would enable us to detect any nonlinear relationship between fecundity and position.

To improve our understanding of the forces causing the spatial distributions of relative fecundity and body weight, we tried to ascertain if the position effects could be ascribed to any identified environmental factor. Thus, we carried out a multiple regression analysis of fecundity on the environmental variables described above. For every environmental variable measured, we introduced two independent variables in the model: the first was the environmental variable untransformed, and the second was the same variable expressed as a deviation from the corresponding morph and transect's average and squared, to detect nonlinear relationships with fecundity. Since there could be correlations between some of these environmental variables, we used the *Stepwise* method to select among them those variables that best explained the variability of the females' fecundity. Only variables significant at the 0.15 level were retained in the model. Separate analyses were carried out for each morph.

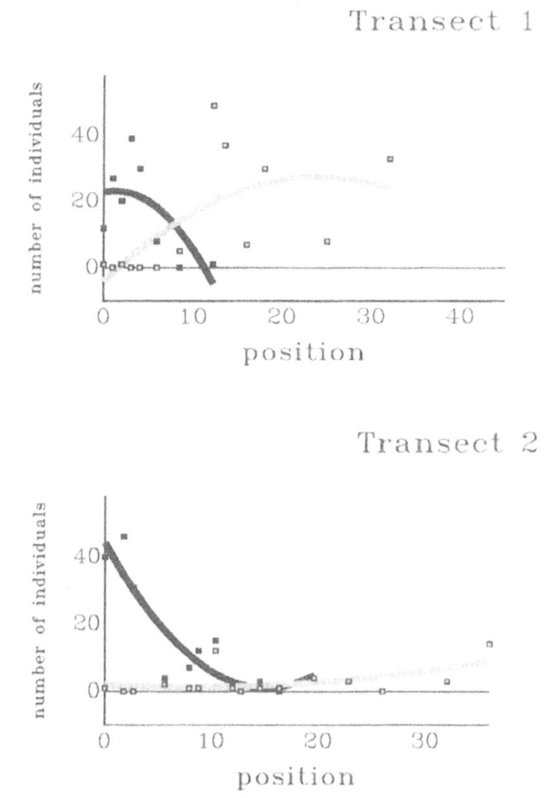

Figure 1. Regression of number of individuals in the sampling area on position in the transect. Grey observations and fitted curves correspond to the RB morph, and black ones, to the SU morph. The model R-square for the SUs and RBs was 0.47 and 0.42 in transect 1, and 0.83 and 0.29 in transect 2.

Results

Both morphs tended to decrease in number in the midshore (Figure 1). The average fecundity was higher for the RB morph than for the SU morph in both transects, and the hybrids had intermediate values (Figure 2). The fecundity differences between morphs were clearly significant (for transect 1, $t_{32} = 6.51$, $p < 0.0001$, and for transect 2 $t_{29} = 4.95$, $p < 0.0001$). Given the small number of observations available for hybrids in the experiment, these data were not further analysed.

No clear trends in fecundity were detected for the SU morph along the coast transects (Figure 3), as the whole model was not significant in a joint analysis of both transects ($F_{2,36} = 2.04$, $p > 0.14$). In a separate analysis of transect 1, however, we found a fecundity increase for this morph when approaching the midshore, the estimated regression coefficient

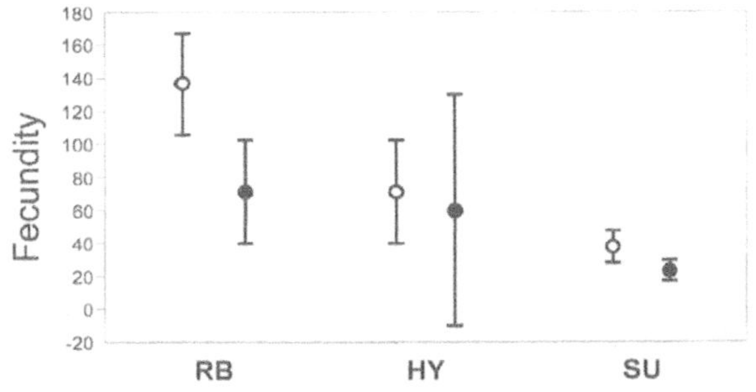

Figure 2. Average fecundity in pure morphs (SU, RB) and hybrids (HY) in transect 1 (black circles) and transect 2 (white circles), with 95% confidence intervals.

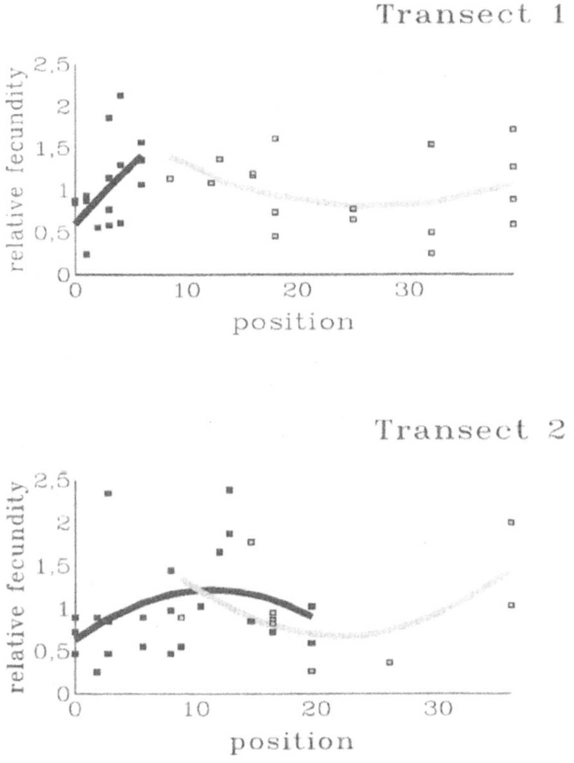

Figure 3. Regression of relative fecundity on position in the transect. Grey observations and fitted curves correspond to the RB morph, and black ones, to the SU morph. The model R-square for the SUs and RBs was 0.27 and 0.13 in transect 1, and 0.12 and 0.30 in transect 2.

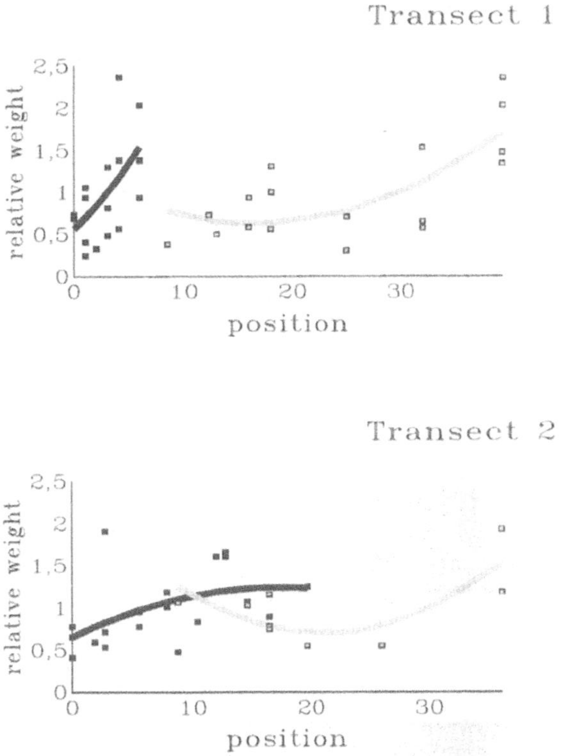

Figure 4. Regression of relative weigth of individuals in the sampling area on position in the transect. Grey observations and fitted curves correspond to the RB morph, and black ones, to the SU morph. The model R-square for the SUs and RBs was 0.32 and 0.55 in transect 1, and 0.26 and 0.60 in transect 2.

for the independent variable 'transect position' being significant and positive ($F_{1,14} = 5.17$, $p < 0.04$).

The joint analysis of the fecundities in both transects of the RB morph, on the other hand, found a significant and positive regression coefficient for the 'squared transect position' independent variable ($F_{1,23} = 4.73$, $p < 0.05$). Thus, the females being farther from the RB average transect position (i.e. in the uppershore borders) tended to have higher fecundity than those in the centre of the uppershore. No effect was found for the non-squared transect position ($F_{1,23} = 0.02$, $p > 0.8$).

This non-random distribution of fecundities along the transects could be related to the variation in body weight, as there was a clear correlation between these two variables ($r = 0.46$, $p < 0.02$ for the RBs and $r = 0.87$, $p < 0.0001$ for the SUs) in our data. Figure 4 shows the results of an analysis for the dependent variable relative weight that used the same model as shown in Figure 3. It can be seen that the spatial distribution of weights is very similar to the one found for the fecundities.

The environmental variables that explained most of the fecundity variation (i.e. those that were retained in the model) were different for the two morphs (Table 1). However, only one environmental variable (the

Table 1. Results of a stepwise multiple regression analysis of relative fecundity on all the environmental variables and squared environmental variables measured. The RB and SU data were analysed separately. The *F* value (with 1 and 24 degrees of freedom for the Rbs and 1 and 37 degrees of freedom for the SUs), probability and adjusted slope are given for these variables. WMPC is the Pearson correlation between each environmental variable and transect position, as calculated within the corresponding morph. JMPC is this same correlation as calculated in a joint analysis for both morphs. *, $p < 0.05$; **, $p < 0.01$; ***, $p < 0.001$

Environmental variables	*F*	Pr > *F*	Slope	WMPC	JMPC
RB upper shore					
Barnacles patchiness ^{2}	4.48	0.045	0.005	–0.05 ^{NS}	0.06 ^{NS}
Percentage of pond	3.85	0.062	–0.386	–0.01 ^{NS}	0.03 ^{NS}
SU lower shore					
Rock slope	3.16	0.084	0.012	0.10 ^{NS}	–0.23 ***
Wave exposure ^{2}	3.70	0.074	0.001	–0.07 ^{NS}	–0.01 ^{NS}
Rock slope ^{2}	2.89	0.098	–0.001	–0.21 ^{NS}	0.34 **
N° of matings in the sample ^{2}	3.38	0.063	–0.222	–0.46 **	–0.42 ***
Vertical level ^{2}	3.44	0.073	–0.326	0.45 **	0.35 **

square of barnacles patchiness in the RBs) in the final models was significant at the 0.05 level.

We tried also to detect any significant between-morph heterogeneity in adjusted slope for the environmental variables measured. Such heterogeneity could be due to divergent selection on fecundity in these morphs and could be important to explain their spatial distribution. A joint analysis of the effect on fecundity of all the environmental variables measured, along with their interaction with the morph factor, failed to obtain a significant *F* value for the model ($F_{29,35} = 1.17$, $p < 0.3261$). Then we tried an analogous model that included only the variables that had been selected for any morph in Table 1. We found a significant *F* value for the interaction of squared rock slope with morph ($F_{1,51} = 6.93$, $p < 0.02$) and the interaction of the square of vertical level with morph ($F_{1,51} = 5.04$, $p < 0.03$). These interactions, however, involved squared variables, and did not reveal therefore divergent selection, but different shapes in the fecundity-environmental variable relationship. The curve was concave for both squared rock slope and squared vertical level in the RBs, and convex in the SUs.

Discussion

A study of the variation in fecundity and its relationship with the spatial distribution of the two morphs of *Littorina saxatilis* was carried out. In contrast to what would be expected if the natural selection on fecundity was involved in the maintenance of the spatial segregation between morphs, we found that the fecundity in the RB morph increased when approaching the mid-shore, that is, the area of contact between the morphs. Therefore we can conclude that the spatial distribution of the morphs is not due to natural selection related to differences in fecundity. As seen in the Results, the changes in fecundity along the transect could be due to changes in body size. *Littorina saxatilis* is an indeterminate-growth animal, so that the observed spatial distributions of body sizes could in turn be related to non-random distributions of snail ages along the transects. In fact, age and size are closely related in age-structured *Littorina saxatilis* populations in the White Sea (Sergievsky et al., 1991). We could not estimate the age of the individuals but have three reasons to believe that this non-random distribution of ages, if it exists, cannot be too extreme. First, the linear distances between the extreme points of the upper shore band are rather large in terms of the likely life-long dispersal ability of the snails. Secondly, it is possible to find many juvenile snails in the extreme upper shore, where the largest snails are usually found, and thirdly, and as seen in the present work, there are mature females, and in similar proportions to the total number of individuals in the samples, at any point along the transect.

We found, anyway, some significant effects of the transect position on fecundity. It is interesting that we

were unable to find any environmental variable that could be termed the main factor responsible for these effects. Perhaps many of these variables contribute to the effect of position; their individual contribution being too small to be detected, using the amount of data that was available for this work. An alternative explanation could be that most of the observed spatial distribution of fecundity is due to a single environmental variable, and that we failed to identify and include it in our experimental design. For example, several workers (Emson & Faller-Fritsch, 1976; Raffaelli & Hughes, 1978; Reid, 1996) have found strong correlations between refuge size and snail size, so that refuge size could be important for the analysis of fecundity variability. The sometimes significant, but never very high, correlations found between transect position and the environmental variables are compatible with both alternatives.

The SU morph individuals are smaller than the RBs, and we found that they have also lower fecundities. However, the SUs carry larger embryos and produce therefore larger and perhaps more viable juveniles, which could be an adaptation to life in very wave-exposed habitats (Rolán-Alvarez et al., 1996, other interpretations are reviewed in Reid, 1996). There could be a trade-off between embryo number and embryo size, and the optimum combination of these traits could depend on the environmental variables. In that case, a more complete analysis of fecundity in the future should include measures of embryo size.

In conclusion, we think it is unlikely that natural selection, acting through differences in fecundity, may explain the spatial distribution of the morphs. Perhaps natural selection does act through another fitness component, such as viability. We intend to study this in the future. Rolán-Alvarez et al. (1997) have shown that selection on viability could explain most of the observed spatial distribution of the morphs. It is interesting to note that we found some evidence of gradual maladaptation in the between-morphs overlapping zone. In fact, both morphs' densities tended to decrease when approaching this zone. A better understanding of the evolutionary forces responsible for the situation of this polymorphism could be very useful for our knowledge of the process of parapatric speciation.

Acknowledgements

We thank Jorge Otero-Schmitt for help in sampling and Alberto Gayoso for help with the graphics. This work was supported by grant XUGA 20008B94 from Xunta de Galicia. R. C. thanks Xunta de Galicia and Santiago de Compostela University for a Ph.D. grant.

References

Emson, R. H. & R. J. Faller-Fritsh, 1976. An experimental investigation into the effect of crevice availability on abundance and size-structure in a population *of Littorina rudis* (Maton): Gastropoda: Prosobranchia. J. exp. mar. Biol. Ecol. 23: 285–297.

Janson, K., 1983. Selection and migration in two distinct phenotypes of *Littorina saxatilis* in Sweden. Oecologia 59: 58–61.

Janson, K., 1987. Allozyme and shell variation in two marine snails (*Littorina*, Prosobranchia) with different dispersal abilities. Biol. J. linn. Soc. 30: 245–256.

Johannesson, K., B. Johannesson & E. Rolán-Alvarez, 1993. Morphological differentiation and genetic cohesiveness over a microenvironmental gradient in the marine snail *Littorina saxatilis*. Evolution 47: 1770–1787.

Johannesson, K., E. Rolán-Alvarez & A. Ekendahl, 1995. Incipient reproductive isolation between two sympatric morphs of the intertidal snail *Littorina saxatilis*. Evolution 49: 1180–1190.

Johannesson, K., E. Rolán-Alvarez & J. Erlandsson, 1997. Growth rate differences between upper and lower shore ecotypes of the marine snail *Littorina saxatilis* (Olivi) (Gastropoda). Biol. J. linn. Soc. 61: 267–279.

Raffaelli, D. G. & R. N. Hughes, 1978. The effects of crevice size and availability on populations of *Littorina rudis* and *Littorina neritoides*. J. anim. Ecol. 47: 71–83.

Reid, D. G., 1996. Systematics and evolution of *Littorina*. The Ray Society, London. 463 pp.

Rolán, E., J. Otero-Schmitt & E. Rolán-Alvarez, 1989. Moluscos de la Ría de Vigo II. Poliplacóforos. Bivalvos. Escafópodos. Cefalópodos, *Thalassas*, Anexo 2. Universidad de Santiago de Compostela, Vigo. 276 pp.

Rolán-Alvarez, E., E. Rolán & K. Johannesson, 1996. Differentiation in radular and embryonic characters, and further comments on gene flow, between two sympatric morphs of *Littorina saxatilis* (Olivi). Ophelia 45: 1–15.

Rolán-Alvarez, E., K. Johannesson & J. Erlandsson, 1997. The maintenance of a cline in the marine snail *Littorina saxatilis*: the role of home site advantage and hybrid fitness. Evolution 51: 1838–1847.

Sergievsky, S. O., A. I. Granovitch & N. A. Mikhailova, 1991. The age structure of White Sea populations of *Littorina obtusata* and *L. saxatilis*. Trudy Zoologicheskogo Instituta 233: 79–126. (Russian)

Ward, R. D., 1990. Biochemical genetic variation in the genus *Littorina* (Prosobranchia: Mollusca). Hydrobiologia 193: 53–69.

Hydrobiologia **378:** 95–103, 1998.
R. M. O'Riordan, G. M. Burnell, M. S. Davies & N. F. Ramsay (eds), Aspects of Littorinid Biology.

Effects of habitat on growth and shape of contrasting phenotypes of *Bembicium vittatum* Philippi in the Houtman Abrolhos Islands, Western Australia

M. S. Johnson & R. Black
Department of Zoology, University of Western Australia, Nedlands, Western Australia 6907 Australia

Key words: translocation experiments, growth, shape, *Bembicium vittatum*, plasticity, genetic variation, size, shells

Abstract

Translocation experiments were used to test the effect of habitat on growth and shape of three contrasting phenotypes of *Bembicium vittatum*: dwarf, highly domed snails from an usually dry tidal pond; large, moderately domed snails from a sheltered, regularly inundated pond; and relatively flat snails from a vertical, exposed shore.

Snails from both ponds grew nearly twice as fast in the wet pond as in the dry pond, indicating a high degree of plasticity of growth. Associated with these changes in growth rates was convergence of shape. Under conditions of rapid growth, the dwarf snails became relatively flatter, and hence more similar to the native snails at that site. These results indicate that the dwarf phenotype is largely a plastic stunting in response to conditions of little submersion time.

The snails from the exposed shore also grew faster in the sheltered, wet pond than at their native site. However, they not only retained their flat shape, but actually became flatter (and hence divergent from the pond snails) when grown in the pond. Thus, variation in shell shape was due to interactions between source population and a common plastic association of flatter growth profile with more rapid growth. Previous experiments had demonstrated high heritability of the flat phenotype, while the present results show that the expression of the genetically different types is affected substantially by the conditions of growth, and that phenotypic differences among populations may either overestimate or underestimate the underlying genetic differences. This unpredictability of the relationship between variation in shell form and its underlying genetic basis complicates interpretations of geographical variation or palaeontological sequences based on shell form.

Introduction

Extensive phenotypic variation is characteristic of many intertidal gastropods, and much of that variation is associated with habitat. Fundamental to interpreting the evolutionary significance of this variation is the determination of how much it is a direct, plastic response to local environmental conditions and how much it results from underlying genetic variation. Experimental studies have revealed both environmental (e.g. Creese & Underwood, 1982; Fletcher, 1984) and genetic (e.g. Janson, 1982; Yamada, 1987, 1989) control of differences in growth rate. Similarly, variation in shape has been shown to have both environmental (e.g. Phillips et al., 1973; Kemp & Bertness, 1984; Boulding & Hay, 1993; Gibbs, 1993) and genetic (e.g. Newkirk & Doyle, 1975; Crothers, 1980; Palmer, 1985; Johannesson & Johannesson, 1996) components. Although the two traits have often been examined separately, shell shape may also be affected by growth rate (Vermeij, 1980; Kemp & Bertness, 1984; Boulding & Hay, 1993), complicating analyses of shape variation.

The littorine *Bembicium vittatum* Philippi is a direct-developing species endemic to southern Western Australia and South Australia (Reid, 1988). In Western Australia, it has a highly disjunct distribution, with extensive populations in the Houtman Abrolhos Islands (29° S, 114° E) and near Albany

(35° S, 118° E), separated by about 900 km, but with only a small population at Penguin Island (32° S, 116° E) in between. As expected for a direct developer, *B. vittatum* shows substantial genetic subdivision both among these three areas and within the Abrolhos archipelago (Johnson & Black, 1991, 1996; Parsons, 1996). Populations also differ substantially in shell size and shape (Johnson & Black 1996; Black & Johnson, 1997; Parsons, 1997a). In a common-garden breeding experiment involving tall snails from an Abrolhos site and relatively flat snails from Penguin Island, Parsons (1997b) showed that shape was determined strongly by genotype. In contrast, her analysis of growth rates showed genetic compensation over the large latitudinal distances between the Abrolhos and Albany: despite similar growth rates in their native environments, the parental types and their hybrids grew at very different rates in a common environment. These results fit the general expectation that direct-developing species are likely to show heritable phenotypic variation among populations.

While Parsons's (1997b) experiments revealed genetic differences affecting growth and shape among populations separated by 400 to 900 km, equally large phenotypic differences occur over distances of tens of metres to kilometres within the Abrolhos Islands. Some of these differences show clear associations with habitat. In this study, we have used translocation experiments to test for direct effects of habitat on growth and shape in populations representing three characteristic types: dwarf, highly domed shells from high-shore tidal ponds that are inundated for only short periods; large, domed shells from tidal ponds with persistent inundation; and relatively flat shells from vertical, relatively exposed shores.

Materials and methods

Study sites and material

All three study sites were in the southern, Pelsaert Group of the Abrolhos Islands. Site S75 is a vertical limestone shore on Middle Island. Although not exposed to strong waves, it does receive splash. Among habitats occupied by *B. vittatum* in the Abrolhos Islands, this site represents one extreme, and the snails have among the flattest shells found in this species. Another extreme habitat and set of phenotypes is provided by site P101, an isolated tidal pond 19 m long and 6 m wide, separated from the adjacent lagoonal shore by a berm of coral rubble 14 m wide and 1.5 m tall. There are more than 50 isolated tidal ponds in the Abrolhos Islands, with only subterranean connections to the sea (Black & Johnson, 1997). Pond 101 represents a set of high-intertidal ponds, which are shallow, and in which the snails are inundated only for a few hours at a time during spring tides, and not at all during neap tides, thereby restricting the opportunities for feeding. These ponds are characterised by small, highly domed *B. vittatum*. The third site, P81, is representative of the many more typical tidal ponds in the Abrolhos, in which frequent inundation and sheltered conditions are associated with large, relatively highly domed *B. vittatum*. Ponds 101 and 81 are about 200 m apart on Pelsaert Island, and are separated from S75, on Middle Island, by about 9 km.

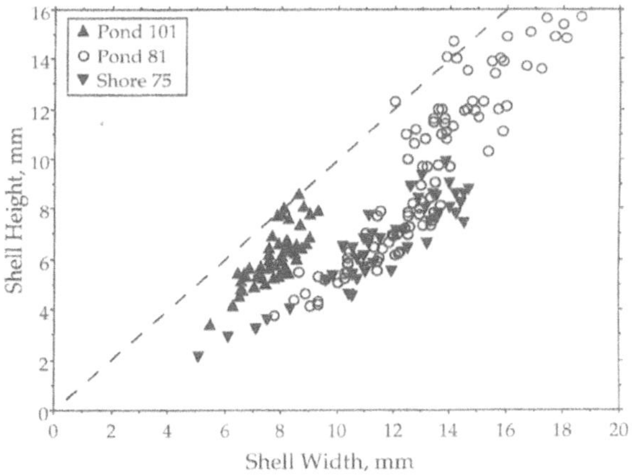

Figure 1. Comparison of shell height (maximum dimension, parallel to columella) and shell width (maximum diameter, perpendicular to columella) among populations of *B. vittatum* at the three study sites. The dashed line indicates height = width.

The contrasts amongst these three populations are summarised in Table 1, and illustrated by a plot of shell height against shell width in random samples from each site (Figure 1). Shell height increases allometrically with width at all sites, but the allometries differ among the sites (Table 1). Snails as small as 8 mm wide can be as tall as they are wide in P101, but the equivalent relative height is not reached until at least 12 mm width in P81, and is never reached at S75 (Figure 1). Figure 1 also illustrates the contrast in sizes of the snails: none wider than 9 mm were found at P101, whereas the large majority were > 12 mm at P81. Based on measurements of mating pairs, sexual maturity occurs at about 6 mm width in P101 and 9 mm in P81 and at S75.

Table 1. Summary of characteristics of the three populations of *Bembicium vittatum* used in this study. a and b are the coefficients in the allometric relationship between height and width: $H = bW^a$

Site	Description	Shell type	Width (mm) at maturity	a	b	H:W at maturity
S75	Vertical shore	Flat	9	1.349	0.238	0.51
P81	Wet pond	Large, domed	9	1.875	0.073	0.50
P101	Dry pond	Dwarf, domed	6	1.346	0.384	0.71

It seemed likely that growth and or shape could be constrained at both the dry pond P101 and the vertical shore S75, whereas conditions in the wet pond P81 are apparently more permissive. The likely constraint in P101 is the obvious lack of time for feeding and other activities, along with the decreased potential for growth of the algae that the snails eat. At S75, a selective constraint on shape is highly likely, as domed snails are easily washed off the vertical shore. In addition, splash from the rising tide appears to keep the snails inactive, thereby reducing the time for feeding at this site. In contrast, P81 is sheltered, while providing much time for feeding, and it is highly productive of algae.

Growth experiments

To test the effect of the contrasting habitats on growth, marked small snails were reciprocally translocated between Ponds 101 and 81. Snails from Shore 75 were also released *in situ* and in Pond 81; reciprocal transplants were not made, because earlier experiments had shown that relatively domed snails disappear within days when placed on the vertical shores. From each source, snails were measured in groups by sieves in 0.5 mm gradations. Each sieve class was divided equally among the two treatments: half to be returned to their native site, the other half to their translocation site. The snails were marked by spray-painting the dorsal surface of the shell with enamel paint (Taubman's Fiddly Bits), with a different colour to identify each source population. The snails from each source were captured, marked, and released within a two day period in April 1995, and the experiments were harvested in September, providing measurements of growth over a five-month period. Upon recapture, the final width, final height, and initial width (indicated by the spray paint) were measured to 0.1 mm with digital calipers. The radial growth was also estimated, with an accuracy of 5 to 10 degrees, and expressed as increment in whorls.

Analyses

Growth increments generally decrease as initial size increases. Consequently, growth rates were compared between treatments by analysis of covariance (ANCOVA) of increase in shell width or radial increment regressed against initial width. Unless otherwise indicated, slopes were homogeneous in any analysis, allowing comparisons of elevations over the entire range of initial widths.

Analysis of change in shape was complicated by three things. First, simple comparison of final shape (height against width) is confounded by the initial shape, so measures of the shape of new growth were required. Second, because we individually measured the snails only upon recapture, we could measure directly initial width, but not initial height; the snails grew too much to allow us to measure initial height on the recaptured snails, so we needed indirect measures of shape of new growth. Third, differences in size-specific survivorship amongst sites meant that the effective starting points differed among treatments, so some comparisons had to be made over restricted size ranges.

To obtain a measurable index of change in shape during the experiment, we determined the increase in width per whorl. Since both lateral and radial growth are related to initial width, comparisons between treatments were made by ANCOVA, with initial width as the covariate. Greater lateral growth should indicate growth along a flatter profile. To confirm that change in width per whorl was not simply an index of overall size increase (i.e. that both height and width might increase more per whorl in a particular treatment), we estimated initial height of each snail from the regression of log height against log width in its native source population. The r^2 values for these regressions were 0.87, 0.84, and 0.67, for S75, P81, and P101, respectively, so the the estimates of height were reasonably accurate. This approach allowed an indirect estimate of increase in height during the experiment. The difference between measured growth of width and estimated growth of height gave an estimate of the shape of new growth; a positive value indicated that growth was mainly lateral (i.e. flat), while a negative value indicated mainly vertical (i.e. tall) growth. This measure was then compared between treatments in the same manner as for increase in width.

Results

Recapture rates varied between 11% and 41%, being highest in the dry pond P101 and lowest in the wet pond P81 (Table 2). There were also clear differences in the size-specific recapture rates: average initial sizes of those recaptured in P81 were larger than their counterparts at the other sites for each source group.

Growth rates were highly responsive to habitat. In the reciprocal translocation between Ponds 81 and 101, both groups of snails grew much faster in P81 (Figure 2). Up to initial widths of 6 to 7 mm, both lateral and radial growth rates of P101 snails in P81 were as great as those of the natives. At larger sizes, however, the P101 snails grew substantially less. The resulting heterogeneity of slopes of the regressions of growth on initial size complicates the comparisons of growth rates over the full range of sizes.

Nevertheless, the 95% confidence intervals of the regression lines for the native and translocated snails from each site did not overlap over the included size ranges (Figure 2), so the contrasts in growth rates were clear. In P101, the lateral growth of snails from P81 was substantially lower than at their native site (about 3 mm, instead of 6 mm, for an initial width of 5 mm), but was still significantly greater than that of the native P101 snails (Figure 2a).

A similar pattern was seen in the comparisons of S75 snails grown *in situ* and in P81 (Figure 3). The S75 snails grew significantly faster both laterally and radially in P81 compared with their native site (ANCOVA), although there was a large variance in growth in both groups. The mean increase in width in P81 was indistinguishable by ANCOVA from that of the native P81 snails (Figure 3a), indicating a high degree of convergence. The responsiveness of radial growth was not as complete; the translocated S75 snails were intermediate to the growth rates of the two native groups (Figure 3b). Thus overall, there was a high degree of plasticity of growth rates in all experiments, but convergence in the same environment was not complete.

The slightly different contrasts observed for lateral and radial measures of growth indicate that shape was affected by both source population and growth environment. This is seen clearly in the comparisons of increase in width per increase in whorl (Figure 4). The change in shape was most striking for snails from the dry pond P101, which added on average about 1.5 mm more width per whorl, and hence became flatter, when grown in the wet pond P81 than at their native site (Figure 4a). Many of these translocated snails had a distinct 'sombrero' phenotype, with a high peak from their initial size and a relatively flat profile from post-translocation growth. For initial sizes less than 7 mm, the P101 snails in P81 grew on a shape trajectory that was indistinguishable from that of the native P81 snails (compare Figure 4a, b). For initial widths of 4.5 to 5.5 mm, for which the covariate could be ignored, the mean ($\pm$ SE) increase in width per whorl was 3.50 $\pm$ 0.06 mm for the P101 snails, compared with 3.74 $\pm$ 0.14 for the P81 snails ($t = 1.48$, $df = 33$, $p = 0.15$), indicating nearly complete convergence of the translocated snails. Only the larger individuals grew differently between these two groups, as the natives at P81 increased their lateral growth per whorl with increasing size, but those from P101 did not (Figure 4a,b). The reciprocal experimental snails from P81 showed a similar contrast between the two growth sites. Although they added signficantly less width per whorl (and therefore became more domed) when translocated to P101, the convergence with the P101 snails was incomplete; the translocated P81 snails were intermediate in their growth pattern between those of the two groups of native snails.

The indirect estimates of increase in height relative to width confirmed that the shape of new growth differed between both source pond populations and site of growth (Figure 5a,b). The transition from greater lateral than vertical growth at small sizes to greater vertical growth at larger sizes is a useful means of comparing the allometries of the different groups. At their native site, the snails from P101 made this transition at 6 mm width, but at P81, where they grew faster, the transition was delayed to about 8.5 mm width. Conversely, the snails from P81 made the transition at 9 mm at their native site, but at 7.5 mm when translocated to P101. Thus, snails from both sources

Table 2. Numbers and mean sizes (± SD in mm) of snails released and recaptured for each treatment

Source	Destination	Number released	Mean width	% Recapture	Mean width of recaptures	
					Initial	Final
P101	P101	379	5.06 ± 1.20	40.9	5.27 ± 1.10	7.28 ± 0.72
P101	P81	379	5.06 ± 1.20	16.4	5.87 ± 1.23	10.98 ± 1.10
P81	P81	488	4.09 ± 1.01	10.7	5.80 ± 1.15	11.59 ± 1.59
P81	P101	488	4.09 ± 1.01	30.1	4.66 ± 1.14	7.87 ± 0.89
S75	S75	198	6.09 ± 1.46	29.2	5.42 ± 1.20	8.20 ± 1.50
S75	P81	198	6.09 ± 1.46	24.2	7.47 ± 1.31	12.62 ± 1.94

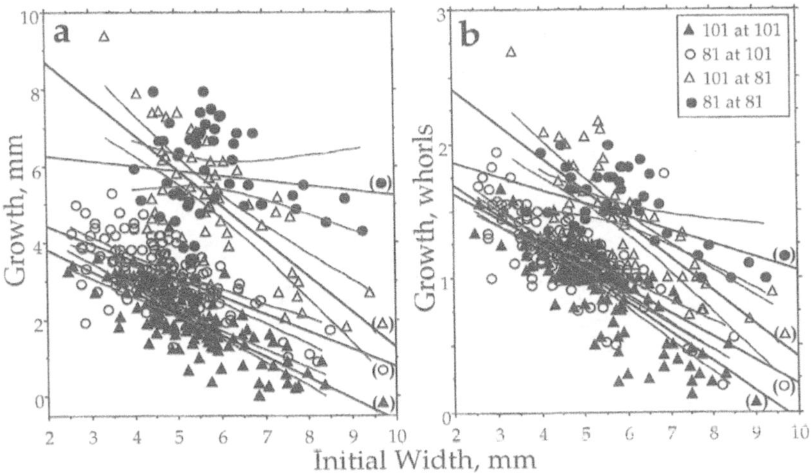

Figure 2. Comparison of lateral (a) and radial (b) growth, as functions of initial width, for recapture snails in reciprocal translocation experiments between Ponds 101 and 81. Regression lines and 95% confidence intervals of the mean are shown. The regression lines are labelled by symbols in brackets, which are not data points.

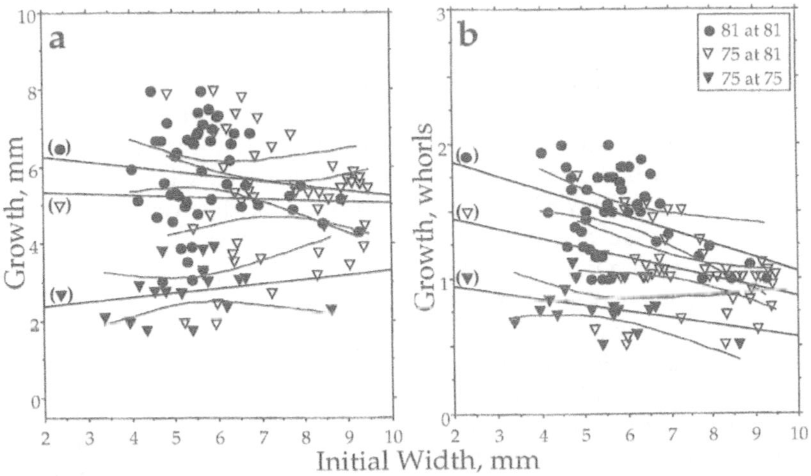

Figure 3. Comparison of lateral (a) and radial (b) growth, as functions of initial width, for recaptured snails in the translocation experiment between S75 and P81. Regression lines and 95% confidence intervals of the mean are shown. The regression lines are labelled by symbols in brackets, which are not data points.

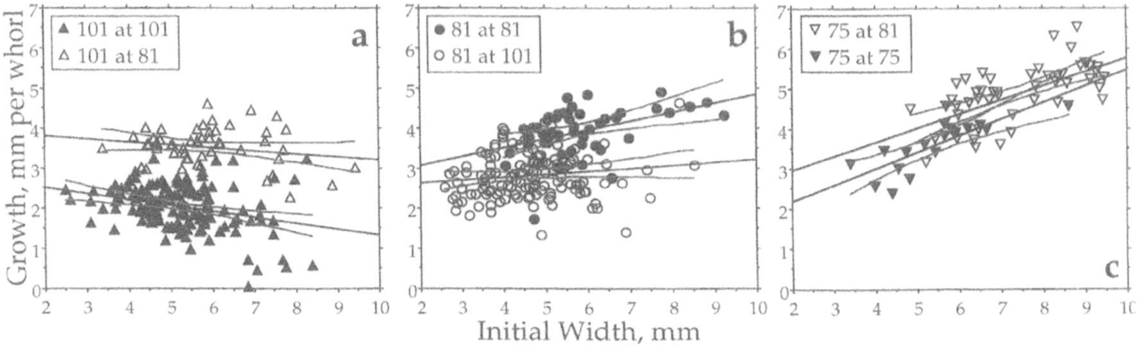

Figure 4. Comparisons of lateral growth per whorl, as a function of initial width, in translocated and native snails from the three study sites. Regression lines and 95% confidence intervals of the mean are shown.

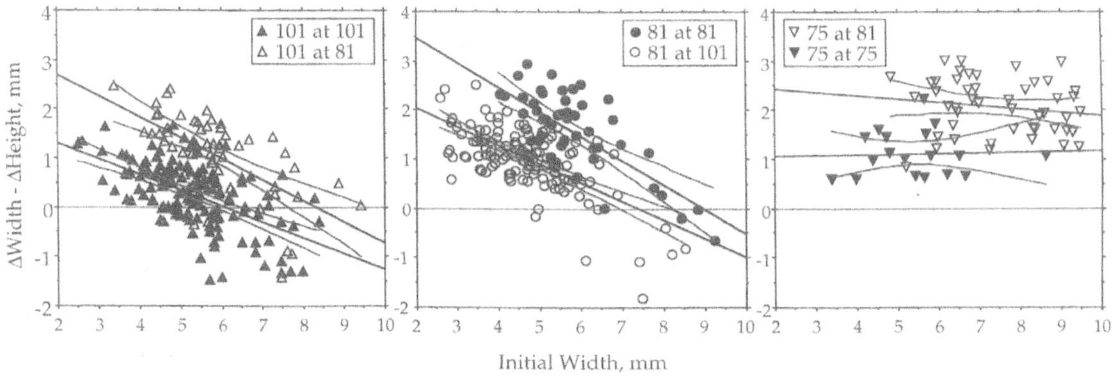

Figure 5. Comparisons of differences between lateral growth and vertical growth, as a function of initial width, in translocated and native snails from the three study sites. Regression lines and 95% confidence intervals of the mean are shown.

had taller growth trajectories at P101, where they grew more slowly. However, the convergence of translocated snails to the natives was incomplete. Thus, while there was substantial plasticity of growth form, there were residual differences due to source.

The patterns of growth of the flat snails from S75 revealed a different set of interactions. There was a significant positive regression of mm per whorl against initial width, regardless of growth site, as expected for this relatively flat phenotype (Figure 4c). As for the other groups of snails, greater increase in width per additional whorl was seen in P81, where growth was faster (Figure 4c). However, rather than resulting in convergence of shape with the native P81 snails, this caused the S75 snails grown in P81 to be even flatter than those at the S75 native site; hence, at larger sizes the S75 and P81 snails were even more distinct when grown together (compare Figure 5b, c). This effect was so great that it could be seen by a comparison of final height against final width (Figure 6). For that comparison, only snails with width in the range of 7 to 11 mm were used, as that was the range of overlap between the two groups of S75 snails; the significance of the differences between these two groups was confirmed by ANCOVA. Even though this direct comparison of final shape includes the influence of pre-experiment growth, the divergence of phenotypes in the common environment is clear.

Discussion

The consistent result from these experiments is that growth rate is highly plastic. As anticipated, the substantially higher growth rates under the permissive conditions of Pond 81 indicate that growth is constrained at both P101 and S75. Nevertheless, convergence between snails from different sources grown in

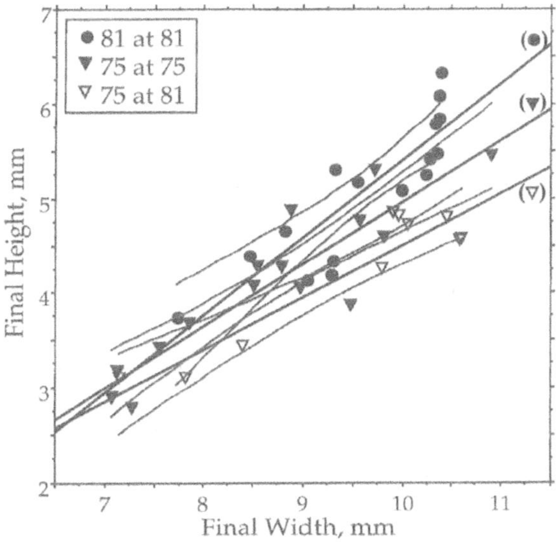

Figure 6. Comparisons of the relationship of height to width in recaptured snails in the translocation experiment between S75 and P81. Regression lines and 95% confidence intervals of the mean are shown. Only snails with final widths in the range 7–11 mm are included, as that is the range of overlap between the two groups of S75 snails. The regression lines are labelled by symbols in brackets, which are not data points.

the same environment was incomplete, and it is essential to consider whether the residual differences might be genetically determined.

In the case of the dwarf P101 snails grown at P81, the incomplete convergence seems to be due to differences in age and maturity between the two sources. Whereas almost all the snails originating from P81 were pre-reproductives at the time of marking, the individuals from P101 bigger than 6 or 7 mm were likely to be reproductively active. In addition, for a given initial width, the snails from P101 would have been older than those from P81, because growth is slower in P101. It was only these larger, older, reproductively mature snails from P101 that had slower growth than the natives at site P81. Although these individuals grew faster in the wet P81 than their stunted counterparts in their native P101, it is not surprising that their response to the favourable conditions was only moderate. In contrast, the lateral growth rate of the P81 snails was not as depressed in P101 as that of the native snails at that dry site. This could well be due to both a residual effect of physiological condition from their pre-translocation growth in the more favourable P81 and possibly an effect of their younger age. However, the incomplete convergence of lateral growth rates in P101 could also indicate retention of differences in shape, since the radial growth rates did converge completely. Taken together, these considerations suggest that the differences in growth between the dwarf snails of the dry P101 and the large snails of the wet P81 are largely plastic responses to growth conditions, but with possibly some small underlying genetic differences. In terms of lateral growth, the same applies to the slower growth at S75, which is determined by the growth environment.

The contrasts of detailed patterns of lateral and radial growth, however, emphasise the complexity of analyses of growth rates, and show that growth rate and shape are inextricably linked. Regardless of source, the snails were flatter when they grew in P81, where growth rates were higher. This association is similar to that found in *Littorina littorea*, in which faster growth leads to greater relative width (Kemp & Bertness, 1984). The opposite relationship, in which more rapid growth leads to narrower shells, was found in *Littorina subrotunda* (Boulding & Hay, 1993; see Reid, 1996), so the relationship of growth rate and shape is not consistent among species. The important point from these studies is that shape is determined in part by growth rate, and consequently may have a substantial associated plastic component. Our results also show that different contrasts of shape may have different causes, even within the same species. The highly domed dwarf shells at P101 are apparently largely a stunted phenotype determined directly by the harsh conditions of relatively limited access to food. The large height relative to width is mainly a consequence of the slow growth, which is determined by the habitat.

However, there may still be other components to this stunting. One of these is age or reproductive state. In both P101 and P81, several of the larger snails from P101 had distinctly smaller increases in width per whorl (Figure 4a). In these highly domed individuals, increases in width become very small as growth continues. These snails appear to be stuck on this trajectory, even when growth is greater, as in P81. More interesting, however, is the failure of the translocated snails from P81 to converge completely with the P101 snails in the relationship of shape to growth. This suggests that there may indeed be a genetic component to the ability to produce an extreme stunted phenotype. Unfortunately, with studies of growth of translocated animals, it is not possible to exclude completely the possible influence of early environment on subsequent development, so we cannot be certain of such a genetic influence.

In an elegant experiment that used allozyme markers to identify offspring of translocated adults, Parsons (1997b) was able to show that the relatively flat shells of *B. vittatum* from Penguin Island were the result of genetic differences from their tall counterparts from a site in the Abrolhos Islands. Our results for the flat snails from S75 are consistent, indicating that even on a more local scale this phenotype is genetically determined. Superimposed on this genetic difference, however, is the underlying common plastic response of even flatter shells associated with more rapid growth.

Taken together, these results are especially interesting in the context of Parsons's (1997b) comparisons over geographic distances of 400 to 900 km. Among the sites that she examined, growth rates were similar, despite the large latitudinal separations. However, when grown at a common, latitudinally intermediate site, the offspring of snails from different geographic regions had very different growth rates, and the inter-locality hybrids were intermediate. This indicated that, on a large geographic scale, there is genetic compensation; similar phenotypes (i.e. growth rates) are produced by different genotypes, in a countergradient response to different environments. In our study, we chose extreme sites within one region, in which the local variation in growth rates and habitats transcends the large-scale geographic variation studied by Parsons. In contrast to her results, these local differences in growth rates are not due to genetic differences, but almost entirely to the extreme local conditions; on this local scale, there is no evidence of countergradient selection for growth rate.

In the case of shape, however, our results on a local scale contrast with Parsons's (1997b) large-scale comparisons in the opposite way. Both experiments confirm that flat shells remain flat, regardless of habitat. However, in our study, it is also clear that there is genetic compensation, whereby different genotypes grown in different habitats may be more similar than when they are grown in the same habitat. Such genetic compensation, in which genetic differences counter the direct effects of environmental gradients, appears to be common for physiological traits, such as growth, respiration, and developmental rate (Conover & Schultz, 1995), but is thought to be rare for morphological traits (Levins, 1968; Bervan et al., 1979). Since shell form is affected by growth rate in *B. vittatum* and other littorines (e.g. Kemp & Bertness, 1984; Boulding & Hay, 1993; Boulding et al., 1993), it is clear that so-called physiological traits and morphological traits are not necessarily distinct.

The likely adaptive basis for the countergradient morphological variation in *B. vittatum* divergence is clear. The flat shells are associated with relatively exposed, vertical shores (Johnson & Black, 1996), where domed shells are easily dislodged. However, since the growth rates on those shores are not high, flat shells cannot be produced by the common plastic response of shape to growth rate. In the face of slow growth, there has apparently been selection favouring an inherently flatter growth form on the vertical shore. Only when these snails are transferred to an environment allowing rapid growth is the extent of the underlying genetic change evident.

Generally, morphological differences among populations are presumed to have covariant genetic and environmental components, so that the phenotypic differences overestimate the underlying genetic divergence. Recognition that substantial differences in shell form can be plastic, as exemplified by the dwarf, highly domed shells of *B. vittatum* in P101 and the large, less domed snails of P81, has called into question supposed palaeontological evidence for punctuated morphological change in gastropods associated with speciation (e.g. Kemp & Bertness, 1984; Palmer, 1985; McMahon, 1992; Trussell, 1996). The genetic compensation producing the flat shells of *B. vittatum*, however, shows that, contrary to the normal expectation, differences in shape among populations in some cases actually underestimate the underlying genetic differences between those populations. This unpredictability of the relationship between variation in shell form and its underlying genetic basis greatly complicates interpretations of geographic variation or fossil sequences based entirely on shell shape.

Acknowledgements

We thank John Clarke and Jane Prince for help in the field, and the Western Australian Fisheries Department for use of their field camp. The manuscript benefited from discussions with Karen Parsons and comments from members of the Marine Ecology Discussion Group. Financial support was provided by the Australian Research Council.

References

Bervan, K. A., D. E. Gill & S. J. Smith-Gill, 1979. Countergradient selection in the green frog, *Rana clamitans*. Evolution 33: 609–623.

Black, R. & M. S. Johnson, 1997. Tidal ponds: unusual habitats characteristic of the Houtman Abrolhos Islands. In F. E. Wells (ed.),The fauna and flora of the Houtman Abrolhos Islands, Western Australian Museum, Perth: 47–61.

Boulding, E. G., J. Buckland-Nicks & K. L. Van Alstyne, 1993. Morphological and allozyme variation in *Littorina sitkana* and related *Littorina* species from the northeastern Pacific. Veliger 36: 43–68.

Boulding, E. G. & T. K. Hay, 1993. Quantitative genetics of shell form of an intertidal snail: constraints on a short-term response to selection. Evolution 47: 576–592.

Conover, D. O. & E. T. Schultz, 1995. Phenotypic similarity and the evolutionary significance of countergradient variation. TREE 10: 248–252.

Creese, R. G. & A. J. Underwood, 1982. Analysis of inter- and intra-specific competition amongs intertidal limpets with different methods of feeding. Oecologia 53: 337–346.

Crothers, J. H., 1980. Further observations on the growth of the common dog-whelk, *Nucella lapillus* (L.), in the laboratory. J. moll. Stud. 46: 181–185.

Fletcher, W. J., 1984. Intraspecific variation in the population dynamics and growth of the limpet *Cellana tramoserica*. Oecologia 63: 110–121.

Gibbs, P. E., 1993. Phenotypic changes in the progeny of *Nucella lapillus* (Gastropoda) transplanted from an exposed shore to sheltered inlets. J. moll. Stud. 59: 187–194.

Janson, K., 1982. Genetic and environmental effects on the growth rate of *Littorina saxatilis*. Mar. Biol. 69: 73–78.

Johannesson, B. & K. Johannesson, 1996. Population differences in behaviour and morphology in *Littorina saxatilis*: Phenotypic plasticity or genetic differentiation? J. Zool. Lond. 240: 475–493.

Johnson, M. S. & R. Black, 1991. Genetic subdivision of the intertidal snail *Bembicium vittatum* (Gastropoda: Littorinidae) varies with habitat in the Houtman Abrolhos Islands, Western Australia. Heredity 67: 205–213.

Johnson, M. S. & R. Black, 1996. Geographic cohesiveness versus associations with habitat: genetic subdivision of *Bembicium vittatum* Philippi (Gastropoda: Littorinidae) in the Houtman Abrolhos Islands. Biol. J. linn. Soc. 58: 57–74.

Kemp, P. & M. D. Bertness, 1984. Snail shape and growth rates: evidence for plastic shell allometry in *Littorina littorea*. Proc. natn. Acad. Sci. U.S.A. 81: 811–813.

Levins, R., 1968. Evolution in Changing Environments. Princeton University Press, Princeton, 120 pp.

McMahon, R. F., 1992. Microgeographic variation in the shell morphometrics of *Nodilittorina unifasciata* from southwestern Australia in relation to wave exposure of shore. In J. Grahame, P. J. Mill & D. G. Reid (eds), Proceedings of the Third International Symposium on Littorinid Biology, The Malacological Society of London, London: 107–117.

Newkirk, G. F. & R. W. Doyle, 1975. Genetic analysis of shell-shape variation in *Littorina saxatilis* on an environmental cline. Mar. Biol. 30: 227–237.

Palmer, A. R., 1985. Quantum changes in gastropod shell morphology need not reflect speciation. Evolution 39: 699–705.

Parsons, K. E., 1996. The genetic effects of larval dispersal depend on spatial scale and habitat characteristics. Mar. Biol. 126: 403–414.

Parsons, K. E., 1997a. Role of dispersal ablility in the phenotypic differentiation and plasticity of two marine gastropods. 1. Shape. Oecologia 110: 461–471.

Parsons, K. E., 1997b. Contrasting patterns of heritable geographic variation in shell morphology and growth potential in the marine gastropod *Bembicium vittatum*: evidence from field experiments. Evolution 51: 784–796.

Phillips, B. F., N. A. Campbell & B. R. Wilson, 1973. A multivariate study of geographic variation in the whelk *Dicathais*. J. exp. mar. Biol. Ecol. 11: 27–69.

Reid, D. G., 1988. The genera *Bembicium* and *Risellopsis* (Gastropoda: Littorinidae) in Australia and New Zealand. Rec. Aust. Mus. 40: 91–150.

Reid, D. G., 1996. Systematics and evolution of *Littorina*. The Ray Society, London. 463 pp.

Trussell, G. C., 1996. Phenotypic plasticity in an intertidal snail: the role of a common crab predator. Evolution 50: 448–454.

Vermeij, G. J., 1980. Gastropod shell growth rate, allometry, and adult size: environmental implications. In D. C. Rhoads & R. A. Lutz (eds), Skeletal Growth in Aquatic Organisms. Plenum Press, NewYork: 379–391.

Yamada, S., 1987. Geographic variation in the growth rates of *Littorina littorea* and *L. saxatilis*. Mar. Biol. 96: 529–534.

Yamada, S., 1989. Are direct developers more locally adapted than planktonic developers? Mar. Biol. 103: 403–411.

Hydrobiologia **378:** 105–114, 1998.
R. M. O'Riordan, G. M. Burnell, M. S. Davies & N. F. Ramsay (eds), Aspects of Littorinid Biology.

Increasing precision in randomised field experiments: barnacle microtopography as a predictor of *Littorina* abundance

E. G. Boulding* & F. M. Harper[1]
Department of Zoology, University of Guelph, Guelph, Ontario, N1G 2W1, Canada
(author for correspondence)*
[1] *Present address: Department of Biology, Memorial University of Newfoundland, St. John's, Newfoundland, A1B 3V6, Canada*

Key words: barnacle, density-dependent, *Littorina*, microtopography, population density, refuges

Abstract

We investigated whether measurements of barnacle microtopography could be used as covariates to increase precision in randomised field experiments with intertidal *Littorina* populations. We used image analysis to quantify the microtopography of the barnacles within 100 cm^2 quadrats in July 1994 using as variables: number of barnacles, number of dead empty barnacles, mean distance to nearest neighbouring barnacle, total area covered by barnacles, and number of pockets of different size-classes [extra-small (< 0.15 cm^2), small (≥ 0.15 and < 0.30 cm^2), medium (≥ 0.30 and < 0.45 cm^2), large (≥ 0.45 and < 0.60 cm^2), and extra-large (≥ 0.60 cm^2)] at least 75% surrounded by barnacles. We then used these variables to predict the abundance of *Littorina* spp. in the same quadrat. Two variables: the number of extra-small pockets and the number of small pockets accounted for 50% of the variation in total littorinid snail abundance among quadrats for January 1995 and for 47% of the variation for August 1994 but for only 16% of the variation for June 1994. However the single variable, the number of barnacles, accounted for 29–36% of the variation in snail abundance on these dates and can be measured without special equipment. The best combination of covariates was the number of extra-small pockets alone. The number of extra-small pockets between barnacles may be influencing the snail abundance by providing a refuge from wave shock and heat stress/desiccation. We suggest that the availability of these refuges among the barnacles could act as a density-dependent regulator of the total population size of a particular species of *Littorina*.

Introduction

Rocky intertidal shores are well known as tractable systems for ecological field experiments (reviewed in Paine, 1994). However rocky shores have steep vertical environmental gradients that are often combined with complex topography. These factors cause high spatial heterogeneity in the distribution and abundance of intertidal organisms. Consequently a great number of replicates are usually required before even large differences among treatments of field experiments can be detected statistically. Unfortunately increasing the statistical power of experiments by increasing the number of replicates is often impossible because the area available at a particular tidal level on steep rocky shores is usually extremely limited.

Fisher (1932) suggests that an alternative method of increasing the statistical power of a field experiment is to correct for any pre-existing differences among quadrats with a covariate so that they all have similar means before the treatments are applied. He used the yields of tea bushes before an experiment began as a covariate; this corrected for the fact that some treatments would have by chance a more productive set of tea bushes than other treatments and reduced the experimental error (= mean square error). This makes it more likely that a small difference among treatments will be statistically significant for a given number of replicates.

The density of littorinid gastropods varies considerably among adjacent areas of the rocky intertidal zone of the Northeastern Pacific. On shores in Britain where extensive wave action occurs field observations have indicated a close relationship between the abundance and size structure of littorinid populations and the availability of crevices (Emson & Faller-Fritsch, 1976; Raffaelli & Hughes, 1978). Crevice-dwelling may decrease exposure to heat, desiccation and reduce the possibility of dislodgment by wave action – the three main causes of littorinid mortality (Faller-Fritsch & Emson, 1986). Barnacles may provide refuges similar to crevices, by altering the complexity of the substratum (Jernakoff, 1985). Indeed Emson & Faller-Fritsch (1976) found a correlation between the number of empty barnacle shells and the number of juvenile *Littorina 'rudis'*. But Fish & Sharp (1986) sampled a vertical transect and found no relationship between the number of empty barnacle shells and the number of *Littorina neglecta*.

Physical factors may be responsible for the regulation of intertidal populations if they result in density dependent mortality (Haldane, 1953). For example dislodgement by accelerating waves of any gastropods not in refuges could regulate the population density of littorinid gastropods if the availability of suitable refuges is limited. Therefore physical features which provide refuge from wave shock or desiccation may influence gastropod abundance.

Microtopographical features, such as crevices and barnacles have been previously measured using a variety of non-destructive sampling techniques. These include visually estimating the percent coverage of barnacles (e.g. Dethier, 1984), point quadrat sampling using grids with points to obtain a percent of those points which cover or contact a feature (e.g. Jernakoff, 1983; Dungan, 1986), and photo-quadrat sampling using slides of quadrats projected on to a grid of dots (e.g. Connell, 1970; Bohnsack, 1979), see review by Foster et al., 1991). Image analysis improves on these methods because each topographical feature can be individually quantified. This might be useful if *Littorina* respond to the microtopography of barnacles differently to that of crevices because of their habitat selection behaviour (Jones & Boulding, unpubl. data).

The systematics of Northeastern Pacific *Littorina* have recently been revised and we now know that four species can be found on wave-exposed shores: *Littorina scutulata* Gould, 1849, *Littorina plena* Gould, 1849, *Littorina sitkana* Philippi, 1846, and *Littorina subrotundata* (Carpenter, 1864) (Reid & Golikov, 1991; Boulding et al., 1993; Reid, 1996). Identifying these four species using shell morphology alone can be challenging (see figures in Reid, 1996) and previously there was some debate as to whether the wave-exposed form of *L. subrotundata* was an undescribed species (Reid & Golikov, 1991; Boulding et al., 1993) and several papers have referred to it as *Littorina* sp. (Boulding et al., 1993; Boulding & Van Alstyne, 1993; Boulding & Hay, 1993; Kim & DeWreede, 1996). However, recent mitochondrial DNA sequences obtained from different Northeastern Pacific populations of this species (Kyle & Boulding, 1998) support the hypothesis of Reid (1990) that the wave-exposed form should be assigned to *L. subrotundata* because of the similarity of the pallial oviducts of the females (Reid & Golikov, 1991; Reid, 1996). These four Northeastern Pacific species show two different types of reproductive modes. *L. scutulata* and *L. plena* have planktonic egg capsules and a planktonic larval period of at least three weeks whereas *L. sitkana* and *L. subrotundata* lay benthic egg masses which hatch directly into juvenile snails (reviewed by Reid, 1996). As the adult snails move little (Boulding & Van Alstyne, 1993) the recruitment to particular populations of snails would be expected to more dependent on the local adult density for the two direct-developing species than for the two planktonic-developing species.

We hypothesised that because Northeastern Pacific gastropods require refuges from heat stress/desiccation and wave action, there could be a large effect of microtopography on littorinid density. We predicted that quadrats with a large number of those types of microhabitats which provide suitable refuges would show a higher density of these gastropods than quadrats with a low number. Our goal was to identify variables that predicted which quadrats would have high initial densities of littorinid snails, for use as covariates in the linear models used to analyse manipulative field experiments using littorinids (E. G. Boulding, unpubl. data).

Methods

Snail abundance

Field work was carried out near Bamfield Marine Station (48°50′ N, 126°6′ W) on the west coast of Vancouver Island, B.C., Canada. A total of fifty 10 cm × 10 cm quadrats were established at three study sites on wave-exposed rocky shores in October, 1993. The location of the quadrats was chosen at random within the barnacle zone (*Balanus glandula* Darwin, 1854) subject to the constraint that the location was relatively flat and relatively bare of macroalgae. The locations of the quadrats were permanently marked by placing a stainless steel screw in the upper right-hand corner and marking the other three corners with epoxy.

Seven times between October 1993 and January 1995 all four species of littorinid gastropods, comprising *Littorina scutulata*, *L. plena*, *L. sitkana*, and *L. subrotundata*, larger than 2.0 mm were counted. This was done by using forceps to remove all the snails larger than 1.5 mm in shell length from each quadrat and measuring their shell lengths to the nearest 0.5 mm. Snails inside the wall plates of dead barnacles were not counted as they could not be removed without breaking the wall plates, but these were not common at these sites. After sampling, all snails were returned to the quadrats. The numbers of each size-class of each species of *Littorina* were recorded but the data for *L. scutulata* and *L. plena* were not entered separately as we did not then realise that both were present at this site. Recent molecular systematics work at our Nudibranch Point site has shown that of a sample of 81 high-spired snails in the *L. scutulata* species complex, 47% were *L. scutulata* s.s. and 53% were *L. plena* (Kyle & Boulding, 1998).

To quantify microtopography the quadrats were videotaped between June 23, 1994 and July 14, 1994 with a Hi8 camcorder. We videotaped the entire 10 cm × 10 cm quadrat as well as each of the four 5 cm × 5 cm sub-quadrats and ensured the quadrat frame was visible in the image.

Image analysis

Recorded images of the quadrats were analysed using video analysis software (JAVA, Jandel Scientific, Corte Madera, California, U.S.A.) on an IBM PC computer with an installed video digitising board (TARGA, Truevision, Inc., Indianapolis, Indiana, U.S.A.). This system allows video images to be frozen on the screen while distances and areas are measured.

Video images of the 100 cm^2 quadrats were used to assess the size and number of crevices present in the rock substrate. If the crevices were deeper than 2 mm the quadrats were excluded from analysis because crevices are three-dimensional features which cannot be quantified using a two-dimensional image analysis system. The thirty-seven quadrats chosen for microhabitat analysis had a total of three crevices and these were too small for snails > 2.0 mm shell length to take refuge in and therefore would not influence our analysis.

Actual measurements of the microtopographic variables were performed on four 25 cm^2 sub-quadrats which together comprised the 100 cm^2 quadrat. As the video camera was not held a uniform distance away from the rock surface, each image was calibrated for each individual quadrat using the outline of the 5 cm × 5 cm quadrat marker. An attempt was made to hold the long axis of the camera lens perpendicular to a line tangent to the shore surface. But even if the line parallel to the shore surface deviates from perpendicular by the angle θ, the ratio of the true distance to the measured distance on the video image is the cosine of θ which is close to one when theta is close to zero. Video images were not used to count snails as they were not visible in the black and white images.

The microtopographic variables measured included the number of live and dead barnacles present; the total area of the quadrat covered by live barnacles and the total sheltered area found between barnacles. The number of live and dead barnacles were counted because it had been previously observed that where there were more barnacles, there appeared to be more littorinid snails. In each quadrat the percent coverage of barnacles was determined. Sheltered areas of bare substrate in pockets between barnacles were also measured as we suspected they might provide protection from wave action and heat/desiccation stress. Pockets between barnacles were defined as areas 75% enclosed by surrounding barnacles and which were of an approximate minimum dimension of 2.0 mm (Figure 1). The total area of these pockets were summed for each quadrat. The number of pockets within certain size ranges were tallied into five size-classes: extra-small (< 0.15 cm^2); small (≥ 0.15 cm^2 and < 0.30 cm^2); medium (≥ 0.30 cm^2 and < 0.45 cm^2); large (≥ 0.45 cm^2 and < 0.60 cm^2); extra-large (≥ 0.60 cm^2). We divided the pockets into these size-classes to see if we could determine the sizes preferred by different size-classes of littorinid snails.

The index of aggregation (R), and mean distance to nearest neighbour r, were also calculated to determine whether the distribution of the barnacles in the quadrats affected snail abundance. We used the method described by Krebs (1989: 127) and defined a boundary strip around the inside edge of the image of our sub-quadrats. We calculated, $R = r_a/r_e$, where r_a is the measured mean distance to the nearest barnacle and $r_e = 1/(2\nu\rho)$ is the expected mean distance where

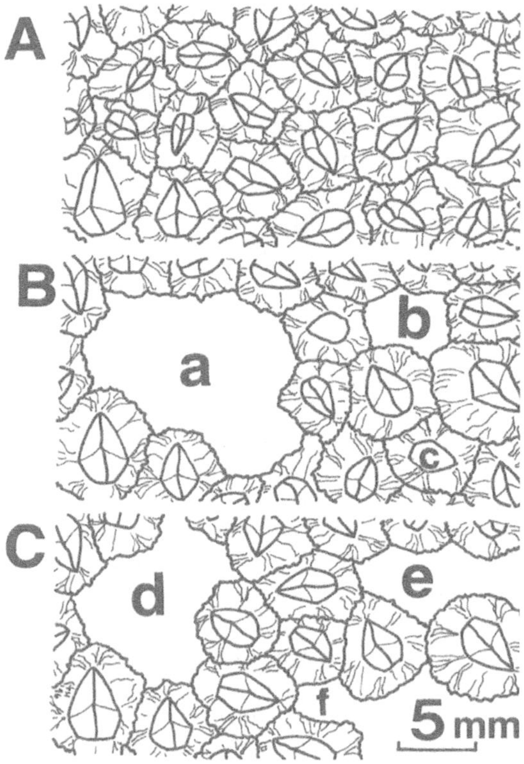

Figure 1. Classification of barnacle microtopography. A) Closely-packed, even-aged, barnacles with few refuges for snails. B) Gaps created by missing barnacles shown in b and dead barnacles with intact wall plates (= empty) shown in c provide refuges for snails. B/C) Pockets created by uneven-aged or missing barnacles are shown in a (1.4 cm^2, so not counted), d (extra-large pocket, 0.8 cm^2), e (large pocket, 0.32 cm^2), and f (extra-small pocket, 0.08 cm^2).

ρ is the number of barnacles inside the boundary strip divided by the area of the portion of the sub-quadrat inside the boundary strip (Krebs, 1989).

Statistical analysis

All statistical analyses were performed using a computerised statistical package (SYSTAT, Version 5.0, Inc.). Pearson correlation coefficients for single pairwise comparisons of the variables quantified through image analysis were calculated.

The independent variables which were significantly correlated with the dependent variables were used in forward stepwise multiple regression. Either total abundance of all snail species on a particular date or the sum of the 2 mm and 3 mm size-classes on a particular date was used as the dependent variable. From this, the subset of independent variables that best explained the variability in snail abundance was determined.

We then used the independent variables which occurred most frequently in the subsets chosen by the stepwise regression package as covariates in a repeated measures ANCOVA. The first linear model fitted was:

$$\mathrm{T1+T2+T3+T4+T5+T6=SITE+a \times P0015+b \times P0160+c \times BARNN+k}, \quad (1)$$

where T1 to T6 are six different counts of the same quadrat on six different dates, SITE is one of three sites, P0015, P0160, and BARNN are 3 barnacle microtopography measurements and a, b, c, and k are constants. We also fit other models with all possible combinations of pairs of these three covariates. We also fit a version of this model with only one covariate, P0015.

$$\mathrm{T1+T2+T3+T4+T5+T6=SITE + a \times P0015+k}. \quad (2)$$

Finally we tested the homogeneity of slopes between SITES assumption by fitting the model:

$$\mathrm{T2+T3+T4+T5+T6=SITE + a \times P0015+SITE \times P0015+k}. \quad (3)$$

Results

Barnacle distribution and abundance

In total we quantified the microtopography of 37 quadrats at three study sites. The mean number of barnacles per 100 cm^2 was 117 (Table 1). The mean area covered by each barnacle was 0.21 cm^2 and the mean distance to its nearest neighbouring barnacle was 0.602 cm. Empty barnacles made up only about 6% of the total number of barnacles. The barnacles had a mean index of aggregation, $R = 2.17$ which means they were uniformly distributed.

Snail abundance

The majority of snails in the quadrats were *L. subrotundata* and *L. plena – L. scutulata* (Figure 2) but a few *L. sitkana* were found especially at site 3 where they

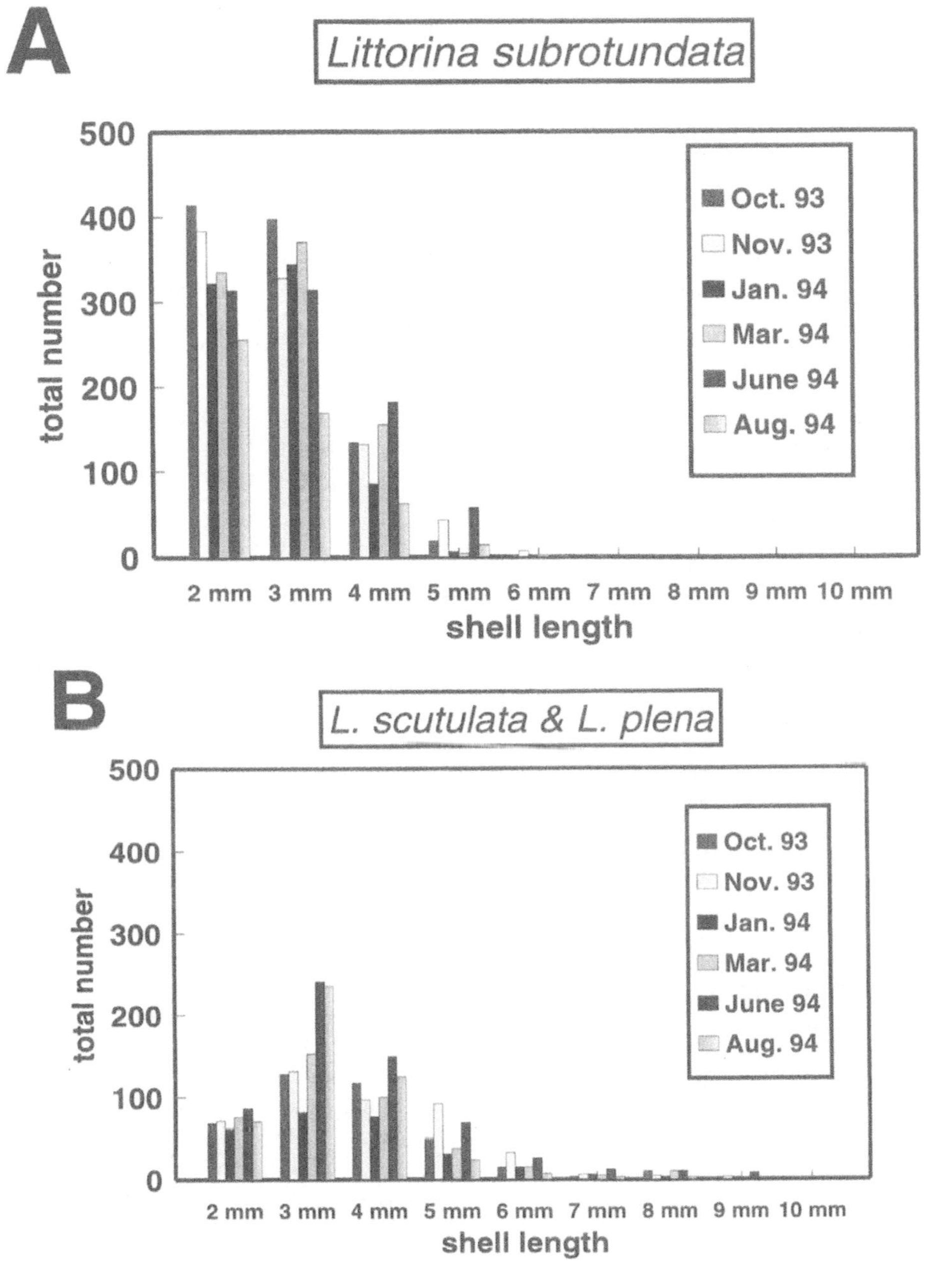

Figure 2. Histograms of different size-classes after the counts from all 37 quadrats were combined. (A) *Littorina subrotundata*; (B) *Littorina plena* and *Littorina scutulata*.

Table 1. Names, abbreviations, means and standard deviations of variables describing barnacle microtopography ($N = 37$)

Variable name	Abbreviation	Mean	Standard deviation
Number of barnacles	BARNN	117.0	41.2
Number of empty barnacles	EBNB	5.81	5.26
Total barnacle area (cm^2)	TBA	25.7	11.6
Mean barnacle size (cm^2)	MEANBA	0.210	0.040
Number of pockets < 0.15 cm^2	P0015	15.0	17.1
Number of pockets > 0.15 and < 0.30 cm^2	P0160	11.9	9.0
Number of pockets > 0.30 and < 0.45 cm^2	P0304	3.57	3.25
Number of pockets > 0.45 and < 0.60 cm^2	P0450	1.14	1.58
Number of pockets > 0.60 and < 1.00 cm^2	P0610	0.892	1.54
Mean distance to nearest neighbour	NNMEAN	0.602	0.060
Index of aggregation	*R*	2.17	0.320

Table 2. Names, abbreviations, means and standard deviations of littorinid gastropod density per 100 cm^2 quadrat for all four species combined ($N = 37$ except for January 1995 when $N = 32$)

Variable name	Abbreviation	Mean	Standard deviation
January 1994 all size classes	JAN94T	28.3	20.2
January 1994 2 & 3 mm size classes	JAN9423	22.2	15.2
March 1994 size classes	MAR94T	68.9	44.6
March 1994 2 & 3 mm size classes	MAR9423	25.6	17.4
June 1994 all size classes	JUN94T	33.6	19.8
June 1994 2 & 3 mm size classes	JUN94T23	25.9	17.0
August 1994 all size-classes	AUG94T	28.0	19.1
August 1994 2 & 3 mm size classes	AUG9423	21.1	14.9
January 1995 all size-classes	JAN95T	21.5	17.9
January 1995 2 & 3 mm size classes	JAN9523	13.2	14.9

made up less than 5% of the total. The most abundant size-classes were the 2 mm and the 3 mm for *L. subrotundata* and the 3 mm for *L. plena – L. scutulata*. The mean number of snails of all species and size-classes ranged from 13.2 to 68.9 during the period from Jan. 1994 to January 1995 (Table 2). There was considerable variation in the total count of snails from quadrat to quadrat with coefficients of variation up to 120%.

Pearson correlation matrix

The independent variables which were significantly correlated ($a = 0.05$, $df = 35$) with the total abundance of all *Littorina* spp. for most dates were: BARNN, P0015, P0160, and R (Table 3). Although there were significant pairwise correlations among the independent variables the r^2 never exceeded 0.50 (Table 3).

Multiple Regression

Multiple regression equations with independent variables incorporating the variables describing barnacle microtopography could account for up to 60% of the variance in counts of littorinid snails among quadrats (Table 4). Unfortunately different independent variables worked better on different dates. Nevertheless, P0015 and P0160 were most often chosen followed by BARNN (Table 4).

Selection of covariates for linear models

When no covariates were used in the linear model there was a significant effect of SITE ($p < 0.002$) which comes about because SITE 3 has a density of snails that is about twice that of the SITES 1 and 2. Adding all three covariates, P0015, P0160, and

Table 3. Pearson correlation coefficients for variables quantifying the microtopography of the barnacles and the abundance of the snails ($N = 37$, values > 0.33 or < -0.33 are significant at $\alpha = 0.05$)

	BARNN	EBN	P0015	P0160	P0304	P0450	P0610	*R*	JAN-94T	JUN-94T	AUG-94T	JAN-9423	JU94-T23	AUG 9423
BARNN	1.00													
EBN	0.57	1.00												
P0015	0.43	0.20	1.00											
P0160	0.75	0.31	0.50	1.00										
P0304	0.46	0.32	−0.42	0.16	1.00									
P0450	0.19	0.17	−0.43	−0.11	0.64	1.00								
P0610	0.42	0.32	−0.30	0.05	0.57	0.40	1.00							
R	0.86	0.41	0.37	0.73	0.48	0.18	0.35	1.00						
JAN94T	0.36	0.20	0.66	0.56	−0.17	−0.16	−0.13	0.34	1.00					
JUN94T	0.54	0.14	0.33	0.36	−0.03	−0.08	0.17	0.37	0.04	1.00				
AUG94T	0.60	0.16	0.57	0.62	−0.13	−0.11	0.08	0.48	0.41	0.72	1.00			
JAN9423	0.31	0.20	0.59	0.50	−0.15	−0.18	−0.09	0.29	0.97	0.01	0.33	1.00		
JU94T23	0.56	0.14	0.64	0.51	−0.08	−0.17	0.08	0.45	0.38	0.74	0.76	0.34	1.00	
AUG9423	0.55	0.17	0.51	0.55	−0.10	−0.06	0.14	0.42	0.41	0.67	0.98	0.32	0.75	1.00

Table 4. Results of Multiple Regression analysis using barnacle microtopography to predict snail abundance (N = 37 except Jan. 95 when N = 32). The constant and the coefficients estimated by forward stepwise multiple regression are shown under their independent variable names with '–' meaning that variable was not included in the final subset, r^2 is the multiple squared correlation coefficient. c = constant

Variable	BARNN	EBNB	P0015	P0160	P0304	P0450	P0610	c	r^2
JAN94T	–	–	0.6	0.7	–	–	–	11.0	0.51
MAR94T	–	–	1.9	–	–	11.0	–	26.0	0.45
JUN94T	0.5	–	−0.4	–	−3.8	–	–	1.9	0.43
AUG94T	–	–	0.4	0.9	–	–	–	11.0	0.47
JAN95T	–	–	0.4	0.8	–	–	–	5.0	0.50
JAN9423	–	–	0.4	0.5	–	–	–	11.0	0.40
MAR9423	–	–	0.6	4.0	–	–	–	11.0	0.32
JUN9423	0.2	–	0.7	0.5	–	–	–	−0.4	0.54
AUG9423	–	–	0.37	0.54	–	–	2.4	6.9	0.43
JAN9523	–	–	0.36	0.75	−1.8	2.5	–	2.6	0.61

BARNN, into the model successfully removed this site effect but resulted in a significant interaction between TIME and P0160. Interactions between our covariates and TIME would be predicted if a covariate had a significant effect on some sampling dates but not others as we might have expected from our stepwise regression subsets. We solved this by using only P0015 as the covariate. This successfully removed the differences among sites but did not result in an interaction between time and P0015 although there was still a time $*$ SITE interaction (Table 5). When we tested for homogeneity of slopes we found that the SITE $*$ P0015 interaction was not significant ($p = 0.31$) suggesting the assumption of equal slopes among sites was not violated.

Discussion

Our analysis showed that variables describing barnacle microtopography can be used as covariates to increase the precision of linear models for a given number of replicates. Interestingly, one covariate, the number of extra small pockets between barnacles, P0015, was

Table 5. Example of repeated measures ANOVA to test for differences in the abundance of littorinid gastropods over time with P0015 as the covariate

MEANS over time					
OCT93T	NOV93T	JAN94T	MAR94T	JUN94T	AUG94T
74.6	62.4	28.3	68.9	33.7	28.0

BETWEEN SUBJECTS SOURCE	SS	df	MS	F	*P*
SITE	1833.0	2	916.0	0.507	0.607
P0015	20290.0	1	20290.0	11.2	0.002
ERROR	59625.0	33	1807.0		

WITHIN SUBJECTS SOURCE	SS	df	MS	F	P	G-G	H-F
time	19166.0	5	3833.0	5.26	0.000	0.002	0.001
time * SITE	16408.0	10	1640.0	2.251	0.017	0.040	0.028
time * P0015	4896.0	5	979.0	1.34	0.248	0.263	0.258
ERROR	120262.0	165	729.0				

GREENHOUSE-GEISSER EPSILON: 0.6456
HUYNH-FELDT EPSILON: 0.7888

more successful at removing the differences among sites without adding interactions than were three covariates. This finding is important as it increases the statistical power of manipulative field experiments making them more feasible in the often limited space of the wave-exposed rocky intertidal shore.

Our work suggests that gastropod abundance is limited by the barnacle microtopography of wave-exposed rocky shore habitats. As the number of barnacles, extra-small pockets and small pockets increased, snail abundance increased irrespective of season. Our data showed that winter densities of *L. subrotundata* on wave-exposed shores were not necessarily lower than those in summer. This contrasts with the findings of Boulding & Van Alstyne (1993) who found that *L. subrotundata* living in a turf of a red alga *Mazzaella (= Iridaea) cornucopiae* had substantially lower winter densities in the most wave-exposed areas and also with the findings of Kim & DeWreede (1996) who found significantly lower winter densities of this snail in *M. cornucopiae* turf in a site close to ours. These lower winter densities of snails in the *M. cornucopiae* turf likely occur because this alga becomes compressed against the rock in severe winter storms reducing its value as a refuge for snails. In contrast, refuges among the barnacles are rigid and the barnacles usually persist throughout the most severe storms.

Wave-exposed littorinid populations can be crevice-limited (Raffaelli & Hughes, 1978; Hughes & Roberts, 1980), with competition for crevices possibly regulating snail abundance in a density-dependent manner (Faller-Fritsch & Emson, 1986). Although our study excluded quadrats with crevices from analysis, our results suggest a similar limiting role by small sheltered areas between barnacles. Indeed the density of the barnacle *Tesseropora rosea* has been shown to influence the distribution and abundance of the snail *Littorina acutispira* (Underwood & McFadyen, 1983).

The size distribution of gastropods on a wave-exposed shore may reflect the availability of refuges of suitable sizes. Although the shell lengths of the three littorinid species ranged from 2.0 mm to 10.0 mm, most of the snails were between 2.0 mm and 4.0 mm. Pockets smaller than 0.15 cm^2 accounted for more than 43% of the variation in snail abundance alone, suggesting the microtopography of the barnacle zone influenced the size structure of the three littorinid species. There were few pockets large enough to provide protection for snails of shell length greater than approximately 4.0 mm. Results found by Emson & Faller-Fritsch (1976) suggest where barnacles and small crevices are abundant and large crevices are scarce, small size-classes of snails are favoured. Further investigation is required to assess the importance of pocket size in determining the size-distribution of the four littorinid species in our study (*L. scutulata, L. plena, L. sitkana, L. subrotundata*) and to determine whether there are differences among species in preferences for pockets of particular sizes.

Our study used a correlation method to test the hypothesis that microtopography influences littorinid density. Manipulative experiments are required to prove whether an increase in the number of barnacles or extra-small pockets causes an increase in snail abundance. The removal of barnacles from an experimental area has been shown to result in a decline in *Littorina unifasciata* densities within 24 hours (Underwood & McFadyen, 1983). Indeed *L. unifasciata* moves further on smooth substrates without crevices and barnacles than on more topographically complex substrates (Underwood & Chapman, 1989). It seems likely that the correlations between the littorinid density and microtopography we have found are the results of active microhabitat selection by the snails, something that has also been demonstrated for *Littorina sitkana* (Jones & Boulding, unpubl. data).

Littorinid gastropods risk mortality from dislodgement by wave action, desiccation and heat stress (Faller-Fritsch & Emson, 1986). Barnacles may provide protection from all of these. When barnacles or other rough elements cover more than 1/12 of the substrate the waves flow over the tops of the barnacles reducing the water flow at the rock surface (Nowell & Church, 1979). Water may become trapped between the barnacles as the tide recedes reducing desiccation stress. Barnacles may also provide shade, reducing heat stress.

Video image analysis allows a novel examination of the microtopography of a wave-exposed shore area. This technique is non-destructive, fast and accurate. Images recorded on videotape retain a permanent record of data which can later be further analysed. We believe that this is the first time image analysis has been used to quantify microtopography although Roberts et al. (1994) have recently described a jump camera technique for quantifying subtidal macrobenthos at 100 m. The drawback to standard video image analysis is that only two dimensional surfaces can be digitised. Other workers have successfully quantified microtopography in three dimensions using ‘contour gauges’ (Underwood & Chapman, 1989; Kostylev, 1996).

Acknowledgements

We thank K. Bartlett, A. Barton, M. Berniez, S. Buchanan, C. Deming, C. Doyle, J. Ferris, E. N. Hay, T. K. Hay, G. Hodgins, M. Holst, K. M. Jones, B. Kerin, V. Pilon, and G. Taylor for help with censusing the permanent quadrats, and the Director and staff of Bamfield Marine Station for field support and the Huu-ay-aht First Nation for access to our study sites. Financial assistance was provided by N.S.E.R.C. (Canada) research and equipment grants to E. G. Boulding.

References

Bohnsack, J. A., 1979. Photographic quantitative sampling of hard-bottom benthic communities. Bull. mar. Sci. 29: 242–252.

Boulding, E. G., J. Buckland-Nicks & K. L. Van Alstyne, 1993. Morphological and allozyme variation in *Littorina sitkana* and related *Littorina* species from the Northeastern Pacific. Veliger 36: 43–68.

Boulding, E. G. & T. K. Hay, 1993. Quantitative genetics of shell form in an intertidal snail: constraints on short term response to selection. Evolution 47: 576–592.

Boulding, E. G. & K. L. Van Alstyne, 1993. Mechanisms of differential survival and growth of two species of *Littorina* on wave-exposed and on protected shores. J. exp. mar. Biol. Ecol. 169: 139–166.

Connell, J. H., 1970. A predator-prey system in the marine intertidal region. I. *Balanus glandula* and several predatory species of *Thais*. Ecol. Mon. 40: 49–78.

Dethier, M. N., 1984. Disturbance and recovery in intertidal pools: maintenance of mosaic patterns. Ecol. Mon. 54: 99–118.

Dungan, M. L., 1986. Three-way interactions: barnacles, limpets, and algae in a Sonoran Desert rocky intertidal zone. Am. Nat. 127: 292–316.

Emson, R. H. & R. J. Faller-Fritsch, 1976. An experimental investigation into the effect of crevice availability on abundance and size-structure in a population of *Littorina rudis* (Maton): Gastropoda: Prosobranchia. J. exp. mar. Biol. Ecol. 23: 285–297.

Faller-Fritsch, R. J. & R. H. Emson, 1986. Causes and patterns of mortality in *Littorina rudis* (Maton) in relation to intraspecific variation: a review. In P. G. Moore & R. Seed (eds), The Ecology of Rocky Coasts: Essays Presented to J. R. Lewis. Hodder & Stoughton, London: 157–177.

Fish, J. D. & L. Sharp, 1986. The ecology of the periwinkle, *Littorina neglecta* Bean. In P. G. Moore & R. Seed (eds), The ecology of rocky coasts: Essays Presented to J. R. Lewis. Hodder & Stoughton, London: 143–156.

Fisher, R. A., 1932. Statistical methods for research workers, 4th edn. Oliver & Boyd, Edinburgh.

Foster, M. S., C. Harrold & D. D. Hardin, 1991. Point vs. photo quadrat estimates of the cover of sessile marine organisms. J. exp. mar. Biol. Ecol. 146: 193–203.

Haldane, J. B. S., 1953. Animal populations and their regulation. New Biology 15: 9–24.

Hughes, R. N. & D. J. Roberts, 1980. Reproductive effort of winkles (*Littorina* spp.) with contrasted methods of reproduction. Oecologia 47: 130–136.

Jernakoff, P., 1983. Factors affecting the recruitment of algae in a midshore region dominated by barnacles. J. exp. mar. Biol. Ecol. 67: 17–31.

Jernakoff, P., 1985. An experimental evaluation of the influences of barnacles, crevices and seasonal patterns of grazing on algal diversity and cover in an intertidal barnacle zone. J. exp. mar. Biol. Ecol. 88: 287–302.

Kim, J. H. & R. E. DeWreede, 1996. Distribution and feeding preferences of a high intertidal littorinid. Bot. mar. 39: 561–569.

Kostylev, V., 1996. Spatial Heterogeneity and Habitat Complexity Affecting Marine Littoral Fauna. Unpublished Dissertation, Göteborg University, Göteborg, Sweden, 120 pp.

Krebs, C. J., 1989. Ecological Methodology. Harper Collins Inc., New York. 654 pp.

Kyle, C. J. & E. G. Boulding, 1998. Molecular genetic evidence for parallel evolution in a marine gastropod, *Littorina subrotundata*. Proc. R. Soc. Lond. B. 265: 303–308.

Nowell, A. R. M. & M. Church, 1979. Turbulent flow in a depth-limited boundary layer. J. Geophys. Res. 84: 4816–4824.

Paine, R. T., 1994. Marine Rocky Shores and Community Ecology: an Experimentalist's Perspective. Excellence in Ecology Series 4. Ecology Institute, Oldendorf/Luhe, Germany. 152 pp.

Raffaelli, D. G. & R. N. Hughes, 1978. The effects of crevice size and availability on populations of *Littorina rudis* and *Littorina neritoides*. J. anim. Ecol. 47: 71–84.

Reid, D. G., 1996. Systematics and evolution of *Littorina*. The Ray Society, London. 463 pp.

Reid, D. G. & A. N. Golikov, 1991. *Littorina naticoides*, new species, with notes on the other smooth shelled species from the northwestern Pacific. Nautilus 105: 7–15.

Roberts, D. E., S. R. Fitzhenry & S. J. Kennelly, 1994. Quantifying subtidal macrobenthic assemblages on hard substrata using a jump camera method. J. exp. mar. Biol. Ecol. 177: 157–170.

Underwood, A. J. & M. G. Chapman, 1989. Experimental analyses of the influences of topography of the substratum on movements and density of an intertidal snail, *Littorina unifasciata*. J. exp. mar. Biol. Ecol. 134: 175–196.

Underwood, A. J. & K. E. McFadyen, 1983. Ecology of the intertidal snail *Littorina acutispira* Smith. J. exp. mar. Biol. Ecol. 66: 169–197.

Hydrobiologia **378:** 115–127, 1998.
R. M. O'Riordan, G. M. Burnell, M. S. Davies & N. F. Ramsay (eds), Aspects of Littorinid Biology.

Temporal variation in parasite loading in relation to life history patterns of *Littorina obtusata* and *L. fabalis*

Gray A. Williams[1,3,*] & T. J. Brailsford[2]
[1] *Port Erin Marine Laboratory, The University of Liverpool, Port Erin, Isle of Man, U.K.*
[2] *Department of Life Sciences, University of Nottingham, University Park, Nottingham, NG7 2RD, U.K.*
[3] *Present address: The Department of Ecology & Biodiversity and The Swire Institute of Marine Science, The University of Hong Kong, Hong Kong*
(author for correspondence)*

Key words: life history, *Littorina*, parasite, temporal variation, trematode

Abstract

Parasitic infection of *Littorina obtusata* (L.) and *Littorina fabalis* Turton by larval digeneans was studied in summer (August) and winter (February) on a sheltered shore at St Michael's Island, Isle of Man. Sampling was timed to investigate parasite loading at different stages of the littorinids' life history. Nine species of digeneans were identified and infection rates were similar between the species, with a slightly higher prevalence in females for *L. obtusata* and males in *L. fabalis*. Temporal differences in infection were small, although metacercariae and degenerate infections were greater in female *L. obtusata* in February. Analysis of shell parameters and size of the penis in males and the pallial oviduct in females revealed morphological differences between parasitised and uninfected individuals. Parasitic stunting of the penis was evident in both species and the pallial oviduct was also smaller in parasitised females. There was also a tendency for parasitised individuals to have larger shells; although size also varied with season for *L. fabalis*.

Introduction

Gastropods, especially littorinids, are the primary intermediate hosts of many larval digenean trematodes (see review by Lauckner, 1987). The exact effects of parasitic infection are a matter of some debate; variation in size of infected host species has been shown to increase with infection (so called 'gigantism', Minchella, 1985) to be reduced (e.g. Huxham et al., 1993) or no noticeable size difference between infected and uninfected individuals (e.g. *L. mariae* = *fabalis*, Williams & Brailsford, 1990); reduced penis size has been recorded in infected individuals (so called 'parasitic castration', Cheung, 1967; Elner & Raffaelli, 1980; Huxham et al., 1993; Mouritsen & Jensen, 1994); egg production has been found to decrease in infected snails (e.g. Huxham et al., 1993; Mouritsen & Jensen, 1994) and changes in behaviour patterns have also been recorded (Robson & Williams, 1970; Williams & Ellis, 1975; Curtis, 1990). Recent debate has centred on the effects of the timing of infection relative to the host life-histories, especially reproduction. If the host is rendered unable to reproduce by the infection then energy should be directed to growth (i.e. gigantism, Minchella, 1985) whereas if castration is not absolute, relatively more energy should be channelled into reproduction, resulting in decreased growth (Sousa, 1983). The timing of infection relative to the snails' life history stage is therefore of great importance.

Littorina obtusata L. and *Littorina fabalis* Turton are epiphytic on fucoid algae (usually *Ascophyllum nodosum* Le Jolis and *Fucus serratus* L. respectively; Williams, 1990). *L. obtusata* is a perennial species, living for 2–3+ years, whereas *L. fabalis* is annual (Williams, 1992). Both species are known to be intermediate hosts to larval digenean parasites (James, 1968 as *L. littoralis*; Guiterman, 1970; Goodwin,

1975; Sergievsky, 1985; Williams, 1990; Williams & Brailsford, 1990). Previously, Williams and Brailsford (1990) suggested that the relative parasite loading of the flat winkles would be related to their life histories; *L. obtusata,* living for 2–3+ years, should have a greater loading than the annual *L. fabalis*. This hypothesis was investigated in summer at Sawdern Point in Wales, UK, where parasite loading was relatively greater in *L. obtusata*, therefore supporting this theory. *L. obtusata* also had stunted penes whereas *L. fabalis* never showed any signs of deformation. Williams and Brailsford estimated that *L. fabalis* used in this survey were 6–8 months old whereas *L. obtusata* were 1–2+ yrs old. If this difference in parasite loading was related to age of the animals (and hence exposure time to infection) then, as *L. fabalis* can live for 1+ yrs, older individuals of this species would be expected to have a higher incidence than younger individuals. The generality of this result, and of the annual life cycle of *L. fabalis*, has only been investigated at one site and at one time. Subsequent studies on the Isle of Man have revealed that whilst *L. fabalis* may live for a slightly longer period, the life-cycle was very similar to the Welsh population, and is basically annual in nature (Williams, 1996).

Infection incidence has been shown to vary seasonally in many species (e.g. *Littorina littorea*, Hughes & Answer 1982; *L. rudis* and *L. nigrolineata*, Elner & Raffaelli, 1980) being maximal in the summer. This temporal variation has been related to the incidence of the definitive hosts (principally seabirds, Hughes & Answer, 1980; Elner & Raffaelli, 1980) but also the life history stage of the littorinids. Infection rates in *L. littorea*, for example, are thought to increase after the snail has spawned (Robson & Williams, 1970, 1971) so that snails < 10 mm are largely uninfected (Hughes & Answer, 1982). Infection in *Littorina obtusata* has been shown to vary spatially, which is related to the degree of wave exposure and seabird populations (Goodwin, 1975) and also seasonally (Guiterman, 1970) being greater during November-January (Goodwin, 1975). Infection rates have been studied in some detail in the White Sea region, where variation between populations is very high (Sergievsky, 1985).

The aims of this study are to investigate parasite loading in *Littorina obtusata* and *L. fabalis* at a different geographic location (therefore allowing comparison with the results of Williams & Brailsford, 1990) and also at two different times at one site, to investigate temporal changes associated with the varying life histories of the two species.

Materials and methods

Sample site and timing

Collections of *Littorina obtusata* and *L. fabalis* were made from St Michael's Island, a sheltered rocky shore on the Isle of Man (O.S. Grid reference SC 296 675). This shore supports dense stands of fucoid algae, especially *Ascophyllum nodosum* and *Fucus serratus* (see Williams, 1996 for more details). Random collections of > 200 adult littorinids (as denoted by the thickening of the aperture lip, Goodwin, 1975) of both species were made in August 1988 (when populations of both species were at maximum numbers; *L. fabalis* individuals are estimated from Williams, 1992; 1996, to be 6–8 months old and *L. obtusata* 1+ yrs old) and February 1989 (when mature *L. fabalis* individuals would be ~ 1 year old, Williams, 1996).

Shell and soft body measurements

Animals were returned to the laboratory where the shell and soft body morphology was measured and the presence/absence of parasites scored. To describe shell morphology three parameters, a (overall shell length), b (aperture length) and c (shell height, see Williams & Brailsford, 1990) were measured (±0.05 mm using vernier calipers). Animals were then cracked open, sexed and the digestive gland squashed to investigate parasite infection. This method may slightly underestimate infection prevalence by metacercariae which may also be found in the pallial cavity, ctenidia and tissues of the head and foot. Parasites were scored to species level using James (1968). As well as identifiable parasite species, degenerate infections and metacercarial cysts were also recorded. The former were recorded when damaged remains of sporocysts were observed, but no intact and identifiable larvae were found. Metacercariae were recorded as a single group as accurate identification is extremely difficult because there are very few morphological differences between species and there is no recognised key available. The penis of male littorinids was removed and the length of the tip (t) and gland (g) area drawn using a *camera lucida* and the number of glands counted (see Williams & Brailsford, 1990). The pallial oviduct of the females was divided into three sections; the albumen gland (ag), the capsule (cg) and jelly gland (jg;

see Reid, 1990) and was drawn. All measurements were subsequently converted to mm.

Shell measurements were made of all snails, which were then cracked and investigated for parasite loading. Measurements of the soft body features of uninfected individuals were made until $n \sim 30$, after which, to increase processing time of infected animals, no further measurements of uninfected animals were made. Sometimes soft features of animals were damaged and unable to be measured when the shells were cracked, and this has resulted in unequal sample sizes.

Statistical analysis

Infection incidence for individual species, using sex and time as variables, was analysed using $2 \times 2 \times 2$, χ^2 contingency tables (Zar, 1996). Where significant non-independence between variables was detected partial independence was tested for. Where independence between variables was detected, sex and time were tested as single variables against infection rate by 2×2, χ^2 contingency tables using Yates correction for continuity (Zar, 1996).

Multivariate Discriminant Analysis was used to investigate morphological variation between infected and uninfected snails at the two sample dates. Shell and soft body measurements were log transformed (see Janson & Sundberg, 1983; Sundberg, 1988) and analysed using the Discriminant Analysis routine in Statistica (CSS, StatSoft Inc). Discriminant analysis generates a set of linear functions which combine measured variables so as to maximise differences between groups (see Janson & Sundberg, 1983; Sundberg, 1988; Lewis & Thorpe, 1994). In this study separate analyses were performed for the sexes and the species, as it is already known that differences exist in specific shell shape and that there is sexual dimorphism within species (see Williams, 1990, 1994). Fours analyses were therefore run; one for each sex/species combination to attempt to discriminate between infected and uninfected individuals and the different time periods.

Results

Parasitic infection and general littorinid morphology

Nine species of parasite were identified from both *Littorina obtusata* and *L. fabalis* and only two cases of a mixed infection (Table 1). The incidence of infection was similar between the two species. Infection incidence in *L. obtusata* was not independent of sex or time ($\chi^2 = 14.62$, 4 df, $p < 0.05$) although in the case of *L. fabalis* sex and time were mutually independent ($\chi^2 = 6.24$, 4 df, $p > 0.05$). Infection rate, therefore, in *L. obtusata* varied with time and sex, probably as a result of the high incidence of metacercariae recorded in female *L. obtusata* in February, and the increase in degenerate infections recorded in both species at this time. In the case of *L. fabalis*, there was no difference in infection rate with time ($\chi^2 = 0.95$, 1 df, $p > 0.05$) although there was a higher prevalence in males ($\chi^2 = 3.943$, 1 df, $p < 0.05$).

The dominant parasite species was *Microphallus similis* which was present in both littorinid species at both sample dates. The next most abundant infection was *Microphallus pygmaeus*, although this species was more prevalent in August (Table 1). *Cercaria lebourii* and *Cercaria littorinae obtusatae* were only recorded in *L. obtusata* in February and *Cercaria buccini* was only recorded from *L. fabalis*. Infection rates for *L. fabalis* with *Cercaria littorinae saxatilis* (IV) were very high for males in August. There was no other general pattern for the remaining, low, prevalence of the parasites between either species, sex or time of sampling.

In general, as would be expected, *Littorina obtusata* were larger than *L. fabalis* (Table 2) and females were larger than males for both species. Both species were larger in February than in August. There was little variation in size with infection for *L. obtusata* although *L. fabalis* size did appear to differ between infected and uninfected individuals (Table 2). The penes of male littorinids of both species were much smaller when the snails were infected (especially in the gland area for *L. obtusata*, Table 3) although there was little variation in the number of glands. The pallial oviduct size was also reduced in infected snails of both species (Table 4).

Discriminant analysis

The discriminant model for all four analyses was highly significant with the first function accounting for most of the between-group variance in all cases (Tables 5–8). The relatively high Wilks' Lambda values illustrate a certain degree of within group variation; however 52–68% of the total discriminant score variance was accounted for by the between-groups variance.

The first two discriminant functions accounted for 72% and 26% of the total between-groups variance for male *Littorina obtusata* (Table 5). The variables shell

Table 1. Prevalence of infection by digenean parasites in *Littorina obtusata* and *L. fabalis* in two different seasons, from St. Michael's Island, Isle of Man. Early, old and unidentified infections are grouped together under 'Unidentified infections'. Degenerate infections have not been used to calculate % infection

	Littorina obtusata				*Littorina fabalis*			
	August		February		August		February	
Digenean species	Male	Female	Male	Female	Male	Female	Male	Female
Microphallus similis	4.1	2.6	1.2	1.7	2.7	2.9	3.0	1.0
Microphallus pygmaeus	2.7	1.3	0	0	5.4	1.5	1.0	0
Cercaria lebouri	0	0	3.5	1.7	0	0	0	0
Cercaria buccini	0	0	0	0	1.4	0	2.0	1.0
Cercaria littorinae obtusatae	0	0	0	0.9	0	0	0	0
Cercaria littorinae saxatilis (IV)	0	0	1.2	0.9	2.8	0	0	0
Cryptocotyle lingua	1.4	1.3	1.2	3.5	0	1.5	0	0
Podocotyle atomon	1.4	0	0	0	1.4	0	0	0
Mixed infection	0	0	0	0.9	0	0	0	1.0
Unidentified species	0	0	2.4	1.7	1.4	0	0	0
Metacercariae	1.4	2.6	0	13.0	0	0	4.0	2.0
Degenerate	0	0	0	7.82	0	0	2	6.1
TOTAL infection	11.0	7.8	9.4	23.5	14.9	5.9	10	5.1
TOTAL individuals	73	77	85	115	74	68	100	99

Table 2. Mean shell measurements ± SD (a, shell length: b, aperture length and c, shell height) for infected and uninfected *Littorina obtusata* and *Littorina fabalis* in August 1988 and February 1989

Species	Sex	Infected	*n*	a (mm)	b (mm)	c (mm)
August 1988						
Littorina obtusata	M	No	65	13.98 ± 0.91	10.35 ± 0.57	12.86 ± 0.91
Littorina obtusata	F	No	71	14.36 ± 0.86	10.68 ± 0.86	13.08 ± 0.87
Littorina obtusata	M	Yes	8	13.66 ± 1.36	10.22 ± 0.78	12.70 ± 1.80
Littorina obtusata	F	Yes	6	14.35 ± 1.49	10.62 ± 0.92	13.00 ± 1.39
February 1989						
Littorina obtusata	M	No	77	14.64 ± 0.65	10.56 ± 0.48	13.27 ± 0.61
Littorina obtusata	F	No	88	14.82 ± 0.62	10.76 ± 0.47	13.44 ± 0.66
Littorina obtusata	M	Yes	8	14.89 ± 0.82	10.75 ± 0.53	13.61 ± 0.63
Littorina obtusata	F	Yes	27	14.73 ± 0.55	10.62 ± 0.56	13.27 ± 0.51
August 1988						
Littorina fabalis	M	No	63	10.24 ± 1.11	7.78 ± 0.81	9.00 ± 1.03
Littorina fabalis	F	No	64	10.49 ± 1.09	8.03 ± 0.86	9.15 ± 1.01
Littorina fabalis	M	Yes	11	9.80 ± 1.32	7.43 ± 0.99	8.70 ± 1.20
Littorina fabalis	F	Yes	4	12.25 ± 1.16	8.99 ± 0.78	10.99 ± 1.21
February 1989						
Littorina fabalis	M	No	90	10.51 ± 1.03	7.98 ± 0.76	9.37 ± 0.94
Littorina fabalis	F	No	94	11.15 ± 1.01	8.44 ± 0.77	8.44 ± 0.77
Littorina fabalis	M	Yes	10	11.09 ± 0.81	8.29 ± 0.55	9.90 ± 0.89
Littorina fabalis	F	Yes	5	10.93 ± 1.12	7.89 ± 0.65	9.63 ± 0.84

Table 3. Mean penis measurements ± SD (NG = number of glands; g = gland length; t = tip length) for infected and uninfected *Littorina obtusata* and *Littorina fabalis* in August 1988 and February 1989

Species	Infected	*n*	NG	g (mm)	t (mm)
August 1988					
Littorina obtusata	No	40	28.48 ± 4.44	3.46 ± 0.72	0.93 ± 0.27
Littorina obtusata	Yes	7	24.14 ± 10.67	1.81 ± 1.50	0.72 ± 0.17
February 1989					
Littorina obtusata	No	40	30.15 ± 5.96	4.15 ± 0.75	1.02 ± 0.19
Littorina obtusata	Yes	6	32.67 ± 4.27	2.93 ± 1.32	1.02 ± 0.22
August 1988					
Littorina fabalis	No	39	6.29 ± 2.02	2.67 ± 0.44	1.82 ± 0.49
Littorina fabalis	Yes	10	7.60 ± 7.40	1.56 ± 0.77	0.92 ± 0.49
February 1989					
Littorina fabalis	No	50	9.86 ± 1.54	2.52 ± 0.53	1.85 ± 0.56
Littorina fabalis	Yes	6	10.50 ± 1.38	1.89 ± 0.72	1.68 ± 0.92

Table 4. Mean pallial oviduct measurements ± SD (ag = albumen gland; cg = capsule gland and jg = jelly gland) for infected and uninfected *Littorina obtusata* and *Littorina fabalis* in August 1988 and February 1989

Species	Infected	*n*	ag (mm)	cg (mm)	jg (mm)
August 1988					
Littorina obtusata	No	28	1.16 ± 1.16	1.30 ± 1.30	3.16 ± 0.92
Littorina obtusata	Yes	6	0.61 ± 0.23	0.85 ± 0.34	2.02 ± 0.57
February 1989					
Littorina obtusata	No	41	1.72 ± 0.52	1.62 ± 0.31	3.32 ± 0.66
Littorina obtusata	Yes	12	1.35 ± 0.46	1.14 ± 0.44	2.25 ± 0.80
August 1988					
Littorina fabalis	No	33	1.45 ± 0.42	1.26 ± 0.29	2.73 ± 0.82
Littorina fabalis	Yes	4	0.94 ± 0.55	0.81 ± 0.17	1.96 ± 0.28
February 1989					
Littorina fabalis	No	39	1.48 ± 0.38	1.09 ± 0.31	2.90 ± 1.16
Littorina fabalis	Yes	5	1.26 ± 0.45	0.91 ± 0.40	2.20 ± 0.76

length (a); shell height (c) and the number of glands and the length of the gland area of the penis were all significantly different between the groups (Table 5, ANOVA). The standardised coefficients revealed that the greatest discriminatory power was attributed to the shell length (a) followed by the shell height (c) and the gland length of the penis (g). Use of the standardised coefficients to detect discriminatory power, however, should be made with caution as the measured variables are likely to be correlated. The correlations between the variables and the function scores (pooled within-group correlations) showed a strong negative relation-

Table 5. Summary statistics for discriminant analysis for male *Littorina obtusata.* F = Function (see Table 3 for description of letters used)

F	Eigen value	% Variance	Cumulative % variance	Canonical correlation	After function excluded	Wilk's Lambda	χ^2	df	*p*
1	0.992	72.05	72.05	0.706	0	0.3594	89.02	18	<0.001
2	0.352	25.57	97.62	0.510	1	0.7160	29.06	10	0.0012
3	0.032	2.38	100	0.178	2	0.9680	2.805	4	0.591

Analysis of variance for variables tested in the model (df 3, 96)

Variable	Wilks Lambda	Partial Lambda	*F* ratio	*p*
a	0.4606	0.7802	7.885	0.0001
b	0.3856	0.9321	2.039	0.1147
c	0.4050	0.8874	3.552	0.0178
NG	0.4216	0.8526	4.843	0.0037
g	0.5691	0.6315	16.338	<0.0001
t	0.3911	0.9191	2.466	0.0678

Standardised coefficients for canonical variables

Variable	Function 1	Function 2	Function 3
a	–1.3231	1.5747	–1.2045
b	0.43205	–0.6289	0.4193
c	0.9099	–0.5969	0.4035
NG	0.2297	0.8411	0.7743
g	–0.9007	–0.9198	–0.0352
t	–0.1939	0.5229	0.4024

Pooled-within group correlations (correlation of Discriminant function score and original variables)

Variable	Function 1	Function 2	Function 3
a	–0.3702	0.4419	–0.4352
b	–0.1721	0.1827	–0.2078
c	–0.1832	0.2951	–0.2529
NG	–0.2394	0.3014	0.7056
g	–0.8185	–0.2373	0.3934
t	–0.3559	0.2359	0.3289

ship with gland size for the first discriminant function, and a weaker correlation with shell length for the second function. The first function clearly separated the uninfected snails from the infected individuals (Figure 1). The second function was dominated by the same parameters and appeared to separate the different sample periods (Figure 1).

The first discriminant function for female *Littorina obtusata* accounted for 59% of the between-groups variance and was best expressed by shell length (a) and the length of the capsule gland (Table 6). ANOVA revealed that all parameters for the pallial oviduct measurements varied significantly between groups as did the overall shell length and these measurements were strongly negatively correlated with function 1 (pooled within-group correlations, Table 6). The second discriminant function, which accounted for 40% of the between-group variance, was explained by shell

Table 6. Summary statistics for discriminant analysis for female *Littorina obtusata* (see Tables 2 & 4 for explanation of letters used)

F	Eigen value	% Variance	Cumulative % variance	Canonical correlation	After function excluded	Wilk's Lambda	χ^2	df	*p*
1	0.897	72.05	59.69	0.688	0	0.322	91.73	18	<0.0001
2	0.549	39.60	96.29	0.596	1	0.611	39.88	10	<0.0001
3	0.056	3.71	100	0.229	2	0.947	4.39	4	0.3560

Analysis of variance for variables tested in the model (df 3, 96)

Variable	Wilks Lambda	Partial Lambda	*F* ratio	*p*
a	0.3978	0.8100	6.096	0.0009
b	0.3698	0.8713	3.840	0.0128
c	0.3364	0.9579	1.142	0.3376
ag	0.4089	0.7879	6.998	0.0003
cg	0.3558	0.9057	2.707	0.0509
jg	0.3683	0.8749	3.714	0.0149

Standardised coefficients for canonical variables

Variable	Function 1	Function 2	Function 3
a	1.1333	1.2626	–1.2329
b	–0.5333	–0.6332	–0.5610
c	–0.4147	–0.3329	1.1477
ag	–0.3404	0.8398	0.5138
cg	–0.6177	–0.3008	–0.3656
jg	–0.4295	–0.5055	0.1054

Pooled-within group correlations (correlation of Discriminant function score and original variables)

Variable	Function 1	Function 2	Function 3
a	–0.2019	0.4458	–0.6425
b	–0.2698	0.0464	–0.6938
c	–0.2164	0.2922	–0.3246
ag	–0.6792	0.6972	0.1275
cg	–0.7435	0.2859	–0.3695
jg	–0.7095	–0.1273	–0.0903

length and the length of the albumen gland (high coefficients and correlation, Table 6). As for male *L. obtusata*, the first discriminant function clearly separated infected and uninfected individuals whereas the second function separated the different sampling times (Figure 1).

Differences between the groups of male *L. fabalis* were explained by the shell aperture (b) and the length of the gland area of the penis (Table 7) which dominated the first function and accounted for 64% of the variation. Overall the length of the gland and tip of the penis and number of glands varied significantly between groups (ANOVA, Table 7). The second function accounted for 33% of the variance and was contributed to by the shell length (a), aperture (b) and, again, the length of the gland area of the penis, although the number of glands showed a strong positive correlation with the second discriminant function (Table 7). The

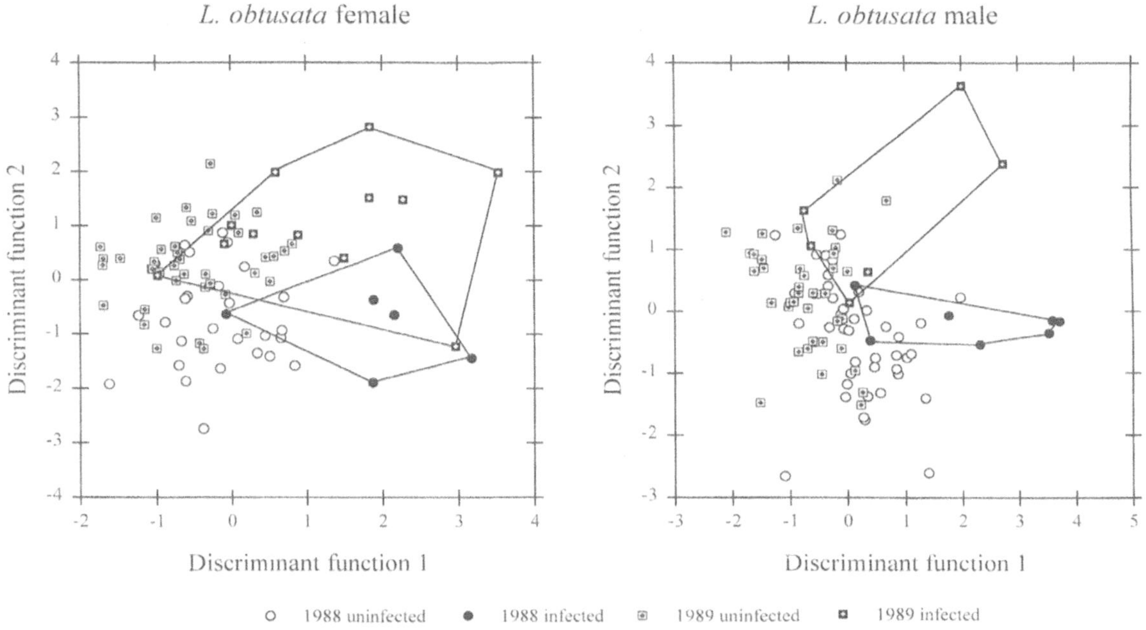

Figure 1. Projection of the first two discriminant functions for *Littorina obtusata* males and females in August 1988 and February 1989. Infected animals are contained within the polygons.

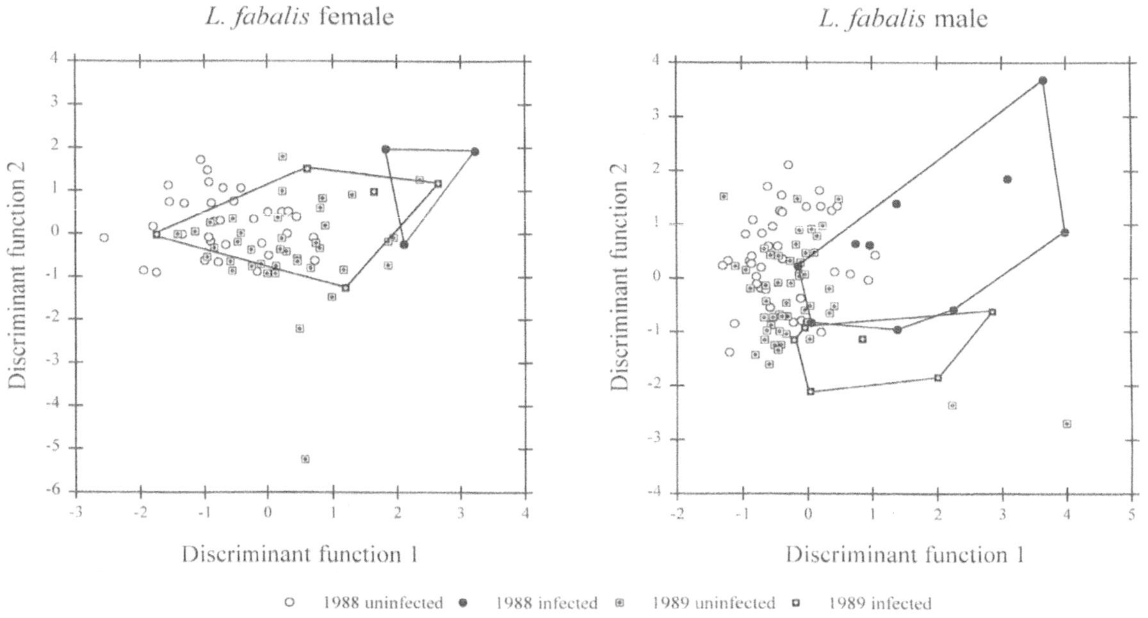

Figure 2. Projection of the first two discriminant functions for *Littorina fabalis* males and females in August 1988 and February 1989. Infected animals are contained within the polygons.

Table 7. Summary statistics for discriminant analysis for male *Littorina fabalis* (see Table 5 for explanation of letters used)

F	Eigen value	% Variance	Cumulative % variance	Canonical correlation	After function excluded	Wilk's Lambda	χ^2	df	p
1	0.587	64.38	64.38	0.6085	0	0.4725	74.230	18	<0.001
2	0.300	32.86	97.24	0.4800	1	0.7500	28.045	10	0.0015
3	0.025	2.76	100	0.1560	2	0.9750	2.469	4	0.6502

Analysis of variance for variables tested in the model (df 3, 96)

Variable	Wilks Lambda	Partial Lambda	*F* ratio	p
a	0.4988	0.9472	1.783	0.1555
b	0.4904	0.9635	1.213	0.3091
c	0.4981	0.9486	1.734	0.1652
NG	0.5269	0.8967	3.688	0.01458
g	0.5711	0.8273	6.681	0.00038
t	0.5222	0.9048	3.366	0.0218

Standardised coefficients for canonical variables

Variable	Function 1	Function 2	Function 3
a	0.3032	1.544	–3.718
b	–0.8998	–0.3353	1.1613
c	0.6619	–1.344	2.1599
NG	0.1159	–0.7859	–0.4314
g	–0.8052	0.9993	0.5467
t	–0.3113	–0.8933	–0.2676

Pooled-within group correlations (correlation of Discriminant function score and original variables)

Variable	Function 1	Function 2	Function 3
a	–0.1600	–0.2457	–0.5906
b	–0.1824	–0.3762	–0.4073
c	–0.1057	–0.3800	–0.3823
NG	–0.2920	–0.5511	–0.1540
g	–0.9129	0.0489	–0.0298
t	–0.8134	–0.2917	–0.1910

first function clearly separated infected and uninfected animals whilst the second function separated out the different sampling dates although the uninfected individuals showed little difference with sample times (Figure 2).

The first discriminant function for female *L. fabalis* accounted for a large amount of the variation between groups (73%, Table 8) and was largely contributed to by shell height (c) and the length of the capsule gland; which differed significantly between the groups (ANOVA, Table 7). The second function, accounting for 19% of the variation, was mainly composed of the shell length (a) and the length of the albumen gland (ag). The first discriminant function separated the different sample dates for infected and uninfected snails (Figure 2). There was no clear separation along the second axis.

Table 8. Summary statistics for discriminant analysis for female *Littorina fabalis* (see Tables 2 & 4 for explanation of letters used)

F	Eigen value	% Variance	Cumulative % variance	Canonical correlation	After function excluded	Wilk's Lambda	χ^2	df	*p*
1	0.652	72.66	72.66	0.628	0	0.481	54.10	18	<0.0001
2	0.171	19.08	91.74	0.382	1	0.795	16.97	10	0.0749
3	0.074	8.26	100	0.263	2	0.931	5.29	4	0.2593

Analysis of variance for variables tested in the model (df 3, 96)

Variable	Wilks Lambda	Partial Lambda	*F* ratio	*p*
a	0.5095	0.9448	1.382	0.2553
b	0.5087	0.9464	1.341	0.2679
c	0.5363	0.8977	2.697	0.0523
ag	0.5251	0.9168	2.149	0.1016
cg	0.5599	0.8598	3.859	0.0129
jg	0.4977	0.9672	0.803	0.4965

Standardised coefficients for canonical variables

Variable	Function 1	Function 2	Function 3
a	0.1128	1.2622	1.0735
b	–0.5201	–0.6050	–0.7337
c	1.1714	–0.6298	–0.7891
ag	0.0147	–0.8272	0.7779
cg	–0.6599	0.4466	–0.9027
jg	–0.2051	–0.3644	–0.1697

Pooled-within group correlations (correlation of Discriminant function score and original variables)

Variable	Function 1	Function 2	Function 3
a	0.4698	0.0826	–0.3920
b	0.3895	–0.2224	–0.6324
c	0.6546	–0.1734	–0.5261
ag	–0.2709	–0.5890	0.1548
cg	–0.5290	0.0065	–0.4176
jg	–0.1836	–0.4439	–0.2618

Discussion

This study attempted to compare parasite loading in *Littorina obtusata* and *L. fabalis* at different stages of their life histories. Populations sampled in August will be contributed to by newly recruited individuals of both species. In the case of *L. fabalis* it is likely that most of the population will be new recruits approaching reproduction in the winter months with a few adults persisting from the previous years recruits (see Williams, 1992; 1996). In contrast, *L. obtusata* populations will be made up of this year's adult recruits joining the established adult population and, therefore, be a mixed population of different aged individuals. In the winter (February) the populations of both species will be composed of older individuals and are likely to have reproduced, or be close to spawning. After this period the majority of *L. fabalis* will gradually die off to be replaced by new recruits, and so this sample represents the oldest individuals of this species which can be collected in any numbers, although a few individuals may persist. An increase in overall size was evident for both species, being larger in the February sample as compared to the August sample, which supports the proposed demographic changes.

Nine species of digenean parasites were recorded in the present study as compared to the 5 species recorded by Williams and Brailsford (1990). Goodwin (1975) recorded 8 species of parasite in *L. obtusata*. *Cercaria lebourii* and *Cercaria littorinae obtusatae* were only recorded from *L. obtusata* in February, and *Cercaria buccini* was only recorded from *L. fabalis*; all the other species were recorded in both snails. Only two mixed infections were recorded in the survey. Infection incidence was within the recorded range for *L. obtusata* (10–25%, Guiterman, 1970; Goodwin, 1975; Williams & Brailsford, 1990) although higher values have been recorded for *L. obtusata* in the White Sea, USSR (Sergievsky, 1985). Infection rates in *L. fabalis* were higher than previously recorded (5.5% Williams & Brailsford, 1990) and there was a clear male bias in infection incidence, a condition noted by Goodwin (1975) and Williams and Brailsford (1990). The exact reasons for this are not understood, although differential susceptibility or longevity have been suggested (Williams & Brailsford, 1990).

There was very little seasonal difference in infection incidence for *L. fabalis*, however the infection rate of female *L. obtusata* was much higher in February, due mainly to metacercariae, as compared to August. The lack of temporal change in *L. fabalis* may be a result of some of the individuals collected in August being second year animals and not newly recruited adults, although population frequencies from this site suggest these adults would be scarce (Williams, 1996). In both species metacercariae and degenerate infections were recorded in February but not in August. Degenerate infections are likely to be old infections or rejected infections. Metacercariae recorded also represented mature infections (i.e. actively producing the next stage of the life cycle) and are the final stage, when the snail is host to primary and secondary intermediate stages, before transfer to the definitive host. As such it would be expected for these older or more advanced infections to be prevalent in relatively older host individuals.

Both species were larger in February than in August, as would be expected from their life history patterns (Williams, 1990; 1992; 1996). In February both *L. obtusata* and *L. fabalis* individuals will be > 1 yr old (up to 3+ for *L. obtusata*) and will be mature and have spawned, or be close to spawning, during the winter. Infection rates are known to increase in post-spawning individuals of *Littorina littorea* (Robson & Williams, 1970) and high rates have been recorded for *L. obtusata* in November-January (Goodwin, 1975). In contrast, however, Elner and Raffaelli (1980) recorded relatively higher incidences of infection in *L. rudis* and *L. nigrolineata* in June (summer) as compared to November (early winter), although this only represents an interval of 4 months. The presumably older population of *L. fabalis* did not show an increased infection rate as compared to the recently recruited adults in August although degenerate and metacercariae infections were higher in February, suggesting increased exposure times. Many authors have linked temporal variation in infection incidence with variation in the abundance of the definitive hosts. In this case shorebirds and fish are thought to be the definitive hosts and correlations between seagull abundance, for example, and snail infection rates (Robson & Williams, 1970, 1971; Hughes & Answer, 1982) have been recorded.

Parasitic infection altered the morphology of both *L. obtusata* and *L. fabalis*. Changes in shell morphology were variable, infected animals sometimes having overall larger shells than uninfected animals (male snails and female *L. fabalis* in August), although infected male *L. fabalis* tended to be smaller than uninfected individuals in August. There were also temporal differences in size of uninfected individuals between winter and summer, illustrating natural growth patterns of these species, with individuals growing and thickening their shells with time (Goodwin, 1975, 1978). Evidence for gigantism was not clear in this study although increases in overall size, shell width and height were recorded.

Evidence for reduction in the reproductive organs for parasitised animals was, in contrast, very clear. Parasitised individuals of both species, in summer and winter, had reduced gonads. Both male *L. obtusata* and *L. fabalis* had stunted penes, especially the glandular areas for *L. fabalis*, in February. Stunting of the penis has been recorded in *L. obtusata* (Goodwin, 1975; Williams & Brailsford, 1990) but not previously for *L. fabalis*. The size of the pallial oviduct was also reduced in parasitised females of the two species (especially the capsule gland length for *L. fabalis*). This has not previously been recorded for these two species, but indicates that reproductive output of the females may be reduced due to infection as has been described for *L. rudis* and *L. nigrolineata* (Elner & Raffaelli, 1980) and *L. littorea* (Huxham et al., 1993).

In summary both *L. obtusata* and *L. fabalis* were parasitised by larvae of digenean trematodes. Infection rates were similar between the species, although *L. obtusata* females were more heavily infected in winter and infection rate was higher in male *L. fabalis* as compared to females. Infections in winter were dominated by metacercariae and degenerate infections. Infected animals had reduced or stunted gonads (both penes and pallial oviducts in the two sexes) which presumably will reduce reproductive output. Variation in shell morphology was observed, although no clear indication of gigantism was observed. This study presents a coarse scale of temporal resolution of the interaction between parasite infections and the life history patterns of littorinids. More thorough temporal investigations, at various life history stages of these species, are clearly needed to understand the dynamics of this relationship and the subsequent influence of parasite infection on the ecology of *L. obtusata* and *L. fabalis*.

Acknowledgements

This work was conducted whilst G. A. W. was in receipt of a NERC Post-Doctoral Fellowship. Mak Yiu Ming and Rupert Lewis (The University of Hong Kong) advised and helped with the Discriminant Analysis – thanks! Ms Kelly, Lau Lai Yip prepared the figures. Attendance at the conference was supported by a CRCG grant from The University of Hong Kong to G. A. W.

References

Cheung, T. C., 1967. Marine molluscs as hosts for symbioses, with a review of known parasites of commercially important species. Adv. Mar. Biol. 5: 424.

Curtis, L. A., 1990. Parasitism and the movements of intertidal gastropod individuals. Biol. Bull. 179: 105–112.

Elner, R. W. & D. G. Raffaelli, 1980. Interactions between two marine snails, *Littorina rudis* Maton and *Littorina nigrolineata* Gray, a predator, *Carcinus maenas* (L.), and a parasite, *Microphallus similis* Jägerskiold. J. exp. mar. Biol. Ecol. 43: 151–160.

Goodwin, B. J., 1975. Studies on the biology of *Littorina obtusata* and *L. mariae* (Mollusca: Gastropoda). Unpublished Ph.D. thesis, University College of Wales, Aberystwyth.

Goodwin, B. J., 1978. The growth and breeding cycle of *Littorina obtusata* (Gastropoda: Prosobranchiata) from Cardigan Bay. J. moll. Stud. 44: 231–242.

Guiterman, J. D., 1970. The population biology of *Littorina obtusata* (Gastropoda: Prosobranchiata). Unpublished Ph.D. thesis, University College of North Wales, Bangor.

Hughes, R. N. & P. Answer, 1982. Growth, spawning and trematode infection of *Littorina littorea* (L.) from an exposed shore in North Wales. J. moll. Stud. 48: 321–330.

Huxham, M., D. G. Raffaelli & A. Pike, 1993. The influence of *Cryptocotyle lingua* (Digenea: Platyhelminthes) infections on the survival and fecundity of *Littorina littorea* (Gastropoda: Prosobranchia); an ecological approach. J. exp. mar. Biol. Ecol. 168: 223–238.

James, B. L., 1968. The distribution and keys of the species in the family Littorinidae and of their digenean parasites, in the region of Dale, Pembrokeshire. Field Stud. 2: 615–650.

Janson, K. & P. Sundberg, 1983. Multivariate morphometric analysis of two varieties of *Littorina saxatilis* (L.) from the Swedish west coast. Mar. Biol. 74: 49–53.

Lauckner, G., 1987. Ecological effects of larval trematode infestation on littoral marine invertebrate populations. Int. J. Parasit. 17: 391–444.

Lewis, R. I. & J. P. Thorpe, 1994. Temporal stability of gene frequencies within genetically heterogenous populations of the queen scallop *Aequipecten (Chlamys) opercularis*. Mar. Biol. 121: 117–126.

Minchella, D. J., 1985. Host life-history variation in response to parasitism. Parasitology 90: 205–216.

Mouritsen, K. N. & K. T. Jensen, 1994. The enigma of gigantism: effect of larval trematodes on growth, fecundity, egestion and locomotion of *Hydrobia ulvae* (Pennant) (Gastropoda: Prosobranchia). J. exp. mar. Biol. Ecol. 181: 53–66.

Reid, D. G., 1990. Note on the discrimination of females of *Littorina mariae* Sacchi et Rastelli and *L. obtusata* (Linnaeus). J. moll. Stud. 56: 113–114.

Robson, E. M. & I. C. Williams, 1970. Relationships of some species of Digenea with the marine prosobranch *Littorina littorea* (L.). I. The occurrence of larval Digenea in *L. littorea* on the North Yorkshire coast. J. Helminth. 44: 153–168.

Robson, E. M. & I. C. Williams, 1971. Relationships of some species of Digenea with the marine prosobranch *Littorina littorea* (L.). II. The effect of Digenea on the reproduction of *L. littorea*. J. Helminth. 45: 145–159.

Sergievsky, S. O., 1985. Populational approach to the analysis of the periwinkle *Littorina obtusata* (L.) invasions with the trematode parthenitae. Helminthologia 22: 5–14.

Sousa, W. P., 1983. Host life history and the effect of parasitic castration on growth: a field study of *Cerithidea californica* Haldemann (Gastropoda: Prosobranchia) and its trematode parasites. J. exp. mar. Biol. Ecol. 73: 273–296.

Sundberg, P., 1988. Microgeographic variation in shell characters of *Littorina saxatilis* Olivi – a question mainly of size? Biol. J. linn. Soc. 35: 169–184.

Williams, G. A., 1990. The ecologies of the flat periwinkles, *Littorina obtusata* (L.) and *L. mariae* Sacchi et Rastelli. Field Stud. 7: 469–482.

Williams, G. A., 1992. The effect of predation on the life histories and ecologies of two species of flat periwinkles – *Littorina obtusata* (L.) and *L. mariae* Sacchi et Rastelli. J. mar. biol. Ass. UK. 72: 403–416.

Williams, G. A., 1994. Variation in populations of *Littorina obtusata* and *L. mariae* (Gastropoda) in the Severn Estuary. Biol. J. linn. Soc. 51: 189–198.

Williams, G. A., 1996. Seasonal variation in a low shore *Fucus serratus* (Fucales, Phaeophyta) population and its epiphytic fauna. Hydrobiologia 326/327: 191–197.

Williams, G. A. & T. J. Brailsford, 1990. The relationship between flat periwinkle life histories and digenean infections. Hydrobiologia 193: 183–190.
Williams, I. C. & C. Ellis, 1975. Movements of the common periwinkle *Littorina littorea* (L.) on the Yorkshire coast in winter and the influence of infection with larval digenea. J. exp. mar. Biol. Ecol. 17: 47–58.
Zar, J. H., 1996. Biostatistical Analysis. 3rd edition, Prentice Hall, New Jersey, 662 pp.

Hydrobiologia **378:** 129–132, 1998.
R. M. O'Riordan, G. M. Burnell, M. S. Davies & N. F. Ramsay (eds), Aspects of Littorinid Biology.

Gills and size in some small *Littorina*

D. H. Black, J. Grahame* & P. J. Mill
Department of Biology, The University of Leeds, Leeds LS2 9JT, U.K.
(author for correspondence)*

Key words: gill, *Littorina saxatilis*, *Littorina arcana*, *Littorina neglecta*, allometry, rough periwinkles

Abstract

Gills were examined in *Littorina saxatilis*, *L. arcana* and *L. neglecta*, using animals of similar size. The form of the gills in the three groups is very similar, although there is an indication of qualitative differences between them. The relationship of number of gill lamellae to shell length shows that *L. saxatilis* and *L. arcana* share the same allometric relationship, while in *L. neglecta* the slope is the same but the intercept is less: at equivalent shell sizes, these animals have about three fewer lamellae.

Introduction

The status of *Littorina neglecta* Bean is controversial. It is a small rough periwinkle, typically living low on rocky shores in the crevices between barnacles and in the dead shells of barnacles. It matures at a small size and (like *Littorina saxatilis* (Olivi)) reproduces ovoviviparously. *L. saxatilis* typically attains larger size and breeds when larger than *L. neglecta*, but it must pass through sizes characteristic of *L. neglecta* as it grows. The discovery that the egg-laying species *Littorina compressa* Jeffreys and *Littorina arcana* Hannaford Ellis possess dwarf barnacle-dwelling forms which seem to be ecotypes has reinforced the suggestions of various authors that *L. neglecta* is an ecotype of *L. saxatilis* (Reid, 1993; see also Reid, 1996). However the ecotypes of *L. compressa* and *L. arcana* are rare (or perhaps only rarely reported), while *L. neglecta* is geographically widespread and very consistent in shell shape over its wide range (Caley et al., 1995). In contrast to this, there is a periwinkle which is rare, ovoviviparous, barnacle-dwelling and dwarfed: this was called *L. saxatilis* 'B' by Grahame et al. (1995). It is to date known from only two sites on the Yorkshire coast of England, and has been studied at one of these. It occurs strictly sympatrically with *L. neglecta* on the shore at Peak Steel, Yorkshire (British National Grid Reference NZ 984021), and the two are distinct in terms of shell shape (Grahame et al., 1995) and also allozymes (Hull, pers. comm.). Also occurring in the same habitat are juveniles of *L. arcana*. The presence of all these small forms offers the opportunity to study allometric relationships in a set of animals over the same size range. A fundamental error in assigning animals to taxa is to mistake variation in form which is a reflection of allometric growth for that having phylogenetic significance. Here we can address this problem.

We chose to examine gill morphology and gill lamella number in *Littorina neglecta*, *L. arcana* and *L. saxatilis* 'B' from Peak Steel, using animals of about the same size ranges. A small sample of *L. neglecta* only was available from Golam Head, County Galway, Ireland (INGR L 826 214), and these were used for comparison.

Methods

Only females were used. Identification was by shell characters, *Littorina neglecta* and *L. saxatilis* 'B' being quite distinct at Peak Steel (Grahame et al., 1995). *L. arcana* from this site were distinguished again by shape, and confirmed using the form of the pallial oviduct. Of a total of 103 specimens examined from Peak Steel, 4 (3.9%) were considered indeterminate

Table 1. Sample size, coefficient of determination and regression statistics for the relationship of gill lamella number and columella length. Although 103 specimens were examined, not all preparations were good enough for counting, therefore the total sample size here is 71

	n	r^2	Intercept	Slope
L. saxatilis	20	0.735	1.028	0.646
L. arcana	22	0.772	0.927	0.820
L. neglecta (Peak Steel)	15	0.561	0.927	0.722
L. neglecta (Golam Head)	12	0.069	1.077	0.389

Table 2. ANCOVA statistics for the relationship between gill lamella number and columella length, data transformed to base 10 logarithms

	Probability LS $\text{mean}_i = \text{LS mean}_j$			Intercept
	L. saxatilis	*L. arcana*	*L. neglecta*	
L. saxatilis	–	0.3793	0.0016	0.9829
L. arcana		–	0.0162	0.9652
L. neglecta			–	0.9177

and not used. The Irish sample contained nothing other than *L. neglecta*.

Measurements of columella length (Grahame & Mill, 1989) were made using an eyepiece micrometer in a dissecting microscope. Animals were carefully dissected under sea water, viewed with a binocular microscope. First the anterior and dorsal shell was carefully removed, then the mantle cut along the right side and reflected back so as to expose the gill with minimal distortion. Entomological pins were used to secure the preparation in a Sylgard™-filled dish. Once revealed the gill filaments were counted. Some preparations were fixed for a minimum of 60 h in osmium tetroxide vapour, dehydrated in a graded alcohol series and critical point dried. They were coated with gold and observed using a scanning electron microscope.

Results

All three forms shared a common gill structure, such variation as was observed was overlapping. The gill consists of an axis on which is borne a series of roughly triangular lamellae (Figure 1). Each lamella was continued posteriorly as a vascularised fold, these folds were particularly variable between individuals. Smaller animals had fewer lamellae, and these often appeared to be relatively thick, especially in *L. neglecta* and *L. arcana*. The lamellae in *L. saxatilis* were generally rather thinner and in both this species and *L. arcana* the vascularised folds were longer than in *L. neglecta*. No systematic survey of these apparent differences was attempted.

Data on shell length and lamella number were transformed to base 10 logarithms, then analysed; basic statistics are shown in Table 1. All the individual correlation and regression coefficients were significant ($p = 0.05$), except for Golam Head *L. neglecta*. We will assume that an allometric relationship might be expected here as well, and that unusual variation in the sample is masking such a putative relationship. Therefore the data were next analysed by analysis of covariance (ANCOVA), pooling all *L. neglecta*. This showed that a common slope could be fitted ($p = 0.0001$), and that there were no differences between the separate slopes ($p = 0.5019$). There were differences among the intercepts ($p = 0.0043$). Table 2 shows the ANCOVA statistics for probability that least squares $\text{mean}_i = \text{least squares mean}_j$, and also the adjusted intercepts. The adjusted common slope is 0.7301.

The results of the analysis show that there is no difference in intercept between *L. saxatilis* and *L. arcana*, but both are different from *L. neglecta* (after multiple comparison correction, $\alpha = 0.05$). This conclusion is true whether the Golam Head *L. neglecta* are included in the analysis (as for Table 2) or not.

We are therefore justified in summarising the data with two least squares regression lines: one calculated for *L. saxatilis* pooled with *L. arcana*, and a second calculated for *L. neglecta*. These lines are shown in Figure 2, using the pooled regression coefficient and adjusted intercepts from an ANCOVA, analysing just two groups – *L. saxatilis* and *L. arcana* pooled, with *L. neglecta* as a second group.

Discussion

The observations made on the form of the gills are very similar to those reported by Caley (1994). While there may be differences in the form of the gill lamellae and vascularised folds between the three forms examined in this study, these have not been quantified and remain to be rigorously explored. The vascularised folds are an adaptation to an amphibious way of life, the animals respiring both in air and water (Hyman, 1967; see

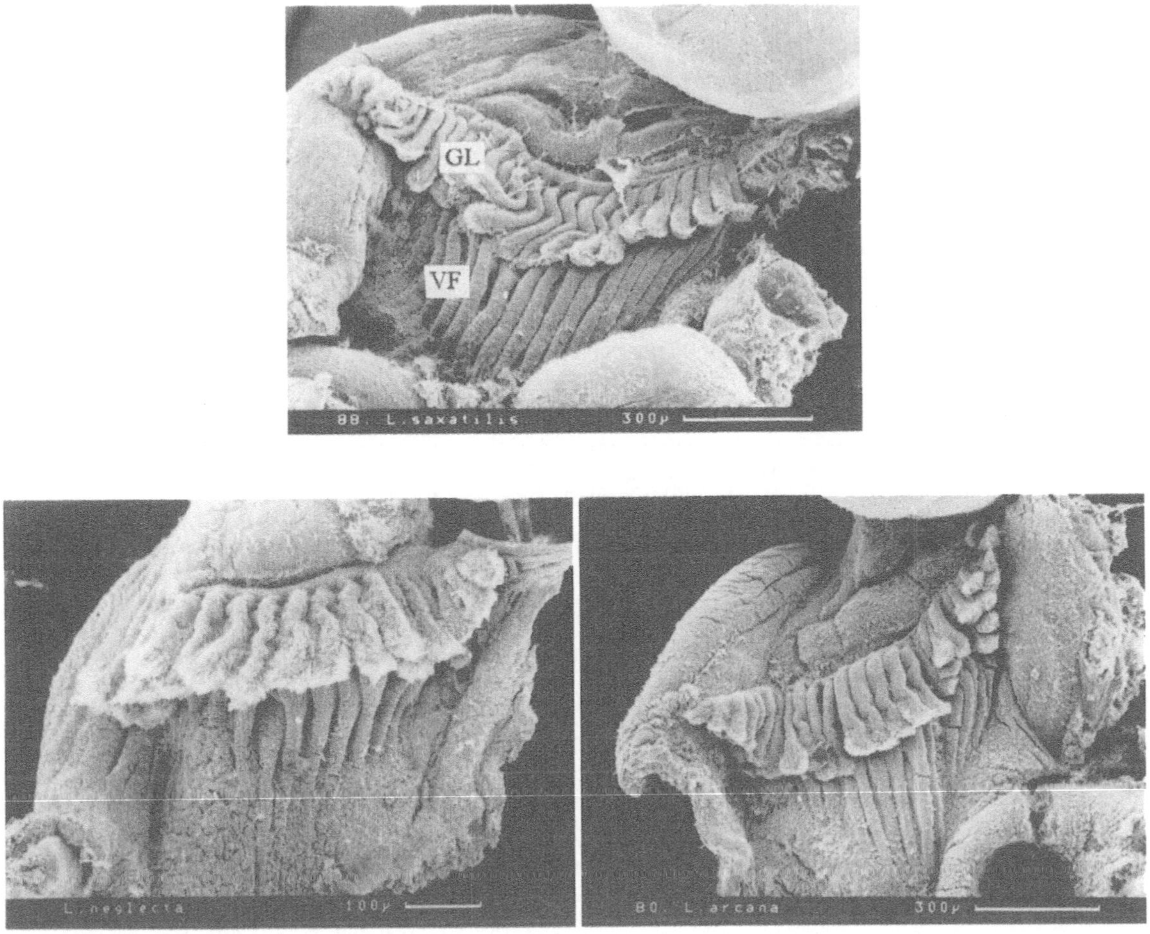

Figure 1. Scanning electron micrographs of the gills of *L. saxatilis*, *L. arcana* and *L. neglecta*. Gill lamellae (GL) and vascularised folds (VF) are indicated for the first of these specimens. The lamellae and folds in the case of *L. neglecta* appear thicker than in the other two specimens, but there was sufficient variation between individuals as to suggest that this should be quantitatively investigated before any conclusions are drawn on the point of possible differences.

also Sandison, 1966). That there is no clear difference in the vascularised folds between *L. neglecta* (a low shore animal) and *L. saxatilis* or *L. arcana* (both of which typically live higher up the shore) may reflect phenotypic variation: all the material used here came from the low shore, the two 'larger' species being represented by juveniles (*L. arcana*) or the dwarf 'B' form of *L. saxatilis*.

Quantitatively, there is a clear difference between the forms. ANCOVA shows that juvenile *Littorina arcana* should be treated as essentially the same as *L. saxatilis* 'B' in terms of the number of gill lamellae related to size. However, *L. neglecta* differs from both of these – there are slightly fewer lamellae (amounting to a difference of between 3.2 to 3.5 across the common size range of 2 to 4.5 mm). This differentiation is another piece of evidence which may be used to separate *L. neglecta* from other rough periwinkles, albeit on morphological grounds. It is also notable that while the *L. arcana* in this study may be regarded as juveniles in a nursery zone, the *L. saxatilis* 'B' are evidently not so, as they mature and breed in the barnacle habitat just as do *L. neglecta*. We are therefore confronted by two different forms of ovoviviparous rough periwinkle living in the barnacle habitat of Peak Steel,

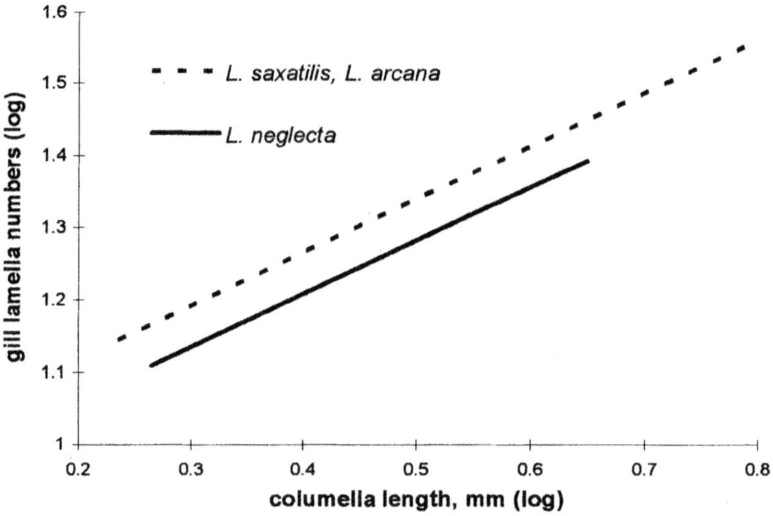

Figure 2. Regression lines expressing the summarised relationship of number of gill lamellae with columella length (mm), both expressed as logarithms to base 10.

only one of which seems truly, to resemble *L. saxatilis* (see also Grahame et al., 1995).

Acknowledgements

This research was supported in part by the MAST 3 programme of the European Commission under contract number MAS3-CT95-0042 (AMBIOS).

References

Caley, K. J., 1994. The taxonomic status and ecology of *Littorina neglecta*. Unpublished Ph.D. thesis, The University of Leeds.

Caley, K. J., J. Grahame & P. J. Mill, 1995. A geographically-based study of shell shape in small rough periwinkles. In P. J. Mill & C. D. McQuaid (eds), Advances in Littorinid Biology. Kluwer Academic Publishers, Dordrecht. Hydrobiologia 309: 181–193.

Grahame, J. & P. J. Mill, 1989. Shell shape variation in *Littorina saxatilis* and *L. arcana*: a case of character displacement? J. mar. biol. Ass. U.K. 69: 837–855.

Grahame J., P. J. Mill, S. L. Hull & K. J. Caley, 1995. *Littorina neglecta* Bean: ecotype or species? J. nat. Hist., Lond. 29: 887–899.

Hyman, L. J., 1967. The invertebrates, Volume VI. Mollusca I. McGraw-Hill Book Company, New York, vii + 792 pp.

Reid, D. G., 1993. Barnacle-dwelling ecotypes of three British *Littorina* species and the status of *Littorina neglecta* Bean. J. moll. Stud. 59: 51–62.

Reid, D. G., 1996. Systematics and evolution of *Littorina*. The Ray Society, London. 463 pp.

Sandison, E. E., 1966. The oxygen consumption of some intertidal gastropods in relation to zonation. J. Zool. Lond. 149: 163–173.

Hydrobiologia **378:** 133–141, 1998.

A hot summer, cold winters, and the geographical limit of *Trochocochlea lineata* in Somerset

J. H. Crothers
Field Studies Council at Nettlecombe Court, Williton, Taunton, Somerset TA4 4HT, U.K.

Key words: common topshell, northern limit, Somerset coast

Abstract

The common or toothed topshell, *Trochocochlea lineata* (da Costa, 1778), is at its northeastern geographical limit in the south and west of the British Isles. It is likely that these limits are related to some feature of air or water temperature.

Along the Somerset coast, as elsewhere in southern Britain, the summer of 1995 was warmer for longer than most of its predecessors and the 1995–1996 winter was colder for longer than any since 1978. This combination of circumstances prompted an investigation into the relative effects of these two seasons. Recruitment of the 1995 cohort was unusually successful at some sites but the subsequent cold winters have removed much of that advantage on Gore Point, probably the easternmost permanent population in the Bristol Channel. The two most easterly populations, either side of Helwell Bay (Watchet), did not survive the winter of 1996–1997.

The conclusion from these observations is that a hot summer can encourage recruitment but that the geographical limit in Somerset may be maintained by cold weather in winter.

Introduction

The common or toothed topshell, *Trochocochlea lineata* (da Costa, 1778),[1] is at its northeastern geographical limit in the south and west of the British Isles (Lewis, 1964).

Several authors have speculated about the factors affecting northern and eastern limits – notably, Garwood & Kendall (1985), Hawthorne (1965), Kendall (1987), Lewis (1986), Lewis et al. (1982), McMillan (1944), Preece (1993) and Williams (1965) – and most feel (if only intuitively) that they are related to temperature. It is usually concluded that the length of time temperatures remain below some critical value (e.g. winters too cold for too long or summers not warm enough for long enough) may be more significant than maxima or minima *per se*. However, the legacy of 1963 demonstrated that winter cold can be limiting (Crisp, 1964) and can affect distribution patterns for a surprisingly long time. No comparable effect relating to summer temperatures has been published to date.

Along the Somerset coast, as elsewhere in southern Britain, the summer of 1995 was warmer for longer than most of its predecessors and by contrast the 1995–1996 winter was colder for longer than is usual.[2] The present paper arose out of an interest to see whether:

- the 1995 settlement of *T. lineata* was more successful than usual,

[1] In the version presented at the Symposium, this species was referred to as *Monodonta lineata*, as was the custom in Britain at the time. Cesari (1986) suggested that the genus *Monodonta* (sensu lato) was better considered as the sub-family Monodontinae Cossmann, 1916 which he regarded as composed of four genera: *Monodonta* (sensu stricto) Lamarck, 1799, from Papua New Guinea and Australia; *Diloma* Philippi, 1845, from Japan and New Zealand; *Austrocochlea* Fischer, 1855, from Australia; and *Osilinus* Philippi 1847 from the Eastern Atlantic and the Mediterranean. He would have called the species mentioned in this paper *Osilinus*.

Gofas & Jabaud (1997) supported the dismemberment of *Monodonta* but showed that *Osilinus* is the senior synonym of *Austrocochlea*. The next available name for the Atlantic/ Mediterranean genus is *Trochocochlea* Mörch, 1852, and the generic name has been changed accordingly in the published version of the paper.

[2] Weather data are not collected at any of the sites mentioned in this paper. The nearest climatological station to Helwell Bay is at Nettlecombe Court (4 miles inland and at an altitude of some 96 m) which has maintained a log of daily observations at 0900 GMT since 1 January 1968.

- mortality following the 1995/1996 winter was more severe than usual.

Or, if both these statements were true,

- recruitment outweighed mortality, or vice-versa.

Fretter & Graham (1977, 1994) summarised the general biology of *T. lineata*. Garwood & Kendal (1985) investigated breeding cycles and Crothers (1994), Kendall (1987), Little et al. (1986), Williams (1965) and Williamson & Kendall (1981) have all investigated growth patterns. In Britain, the breeding season is confined to late July/early August and the new cohort appears on Somerset shores in late August or early September. Growth is rapid, initially (apart from the winter cessations), but slows appreciably as the animal reaches sexual maturity. Individuals have been known to reach 34 mm in shell height (measured in Skomer, North Haven, 30th June 1996) and 15 y of age (Stanbury, 1974) but these are exceptional figures. Unusually in gastropods, it is possible to relate size to age in this species and investigate population structure, growth rates, mortality rates and longevity.

Materials and methods

This paper concentrates on the population at Gore Point, west of Porlock Weir, Somerset (British National Grid Reference SS860486), with reference to the outliers on either side of Helwell Bay, Watchet (ST078436) and Swill Point, Doniford (ST095436) (Crothers, 1994). All three shores are extensive wave-cut platforms with shallow rock pools; that on The Gore being composed of Old Red Sandstone boulders and the others of Triassic marls (Blue Anchor Series) or interbedded Lower Jurassic limestones and shales. The Watchet site is the muddiest (Crothers, 1992).

The Gore was visited at least once in most months (except December and January) between 1987 and 1991, and again from 1994 to 1997, either by the author alone or when accompanied by students attending field courses. The other sites were visited occasionally.

Samples were taken from an area of shore where the 0+ age-class was known to occur. Within that area, collections of living topshells were made without conscious bias for size or age. Searching in and out of pools, under stones and in crevices, I took every individual I could reach without moving my knees; repeating the procedure as often as necessary. Students were instructed to select an individual, collect it and the 99 others closest to it.

Each shell was aged by counting its 'conspicuous' growth checks across the shell parallel to the shell lip, which are assumed to represent winters (Williamson & Kendall, 1981). However, the growth check representing the first winter was always ignored. It varied greatly in its conspicuousness (depending on the severity of the winter) and was never visible in old shells. Separation of 0+ and 1+ cohorts was not difficult because of their disparate sizes.

Shell length (the maximum dimension involving the apex) was measured to the nearest millimetre using callipers. All samples were returned to the shore after measurement, usually immediately.

Data collected by students are incorporated in the results. It can be argued that errors have been introduced as everyone was learning the techniques as they went along. But other errors are avoided, because students are free of the temptation to make the data conform to the earlier results! Alone, the author measured between 100 and 250 topshells per visit; classes regularly measured in excess of 1 000. Only on the rare occasions when the data collected by the students differed notably from my own obtained at the same time, did I rely on mine alone.

Results

Gore Point, Porlock Weir

The data collected on 2nd June 1996 are presented in Table 1 and those for 24 May 1997 in Table 2. In an attempt to evaluate the success of the 1995 cohort, I re-examined the data for this population collected over the last ten years. They fall into two discrete periods; 1987–1991 and 1994–1997. They were analysed to display the month-on-month increase in size (i.e. growth rate) of each cohort and also its preponderance within the population.

Growth rates

The mean size of each cohort was tabulated monthly. Sampling intensity was uneven so some entries represented single samples whilst others were averages of two or more. Few samples were taken in December or January but, as these animals do not increase the size of their shells between November and May (Crothers, 1994), it was reasonable to interpolate figures for those months. Some individuals commence growth in early May, whilst others delay until late June

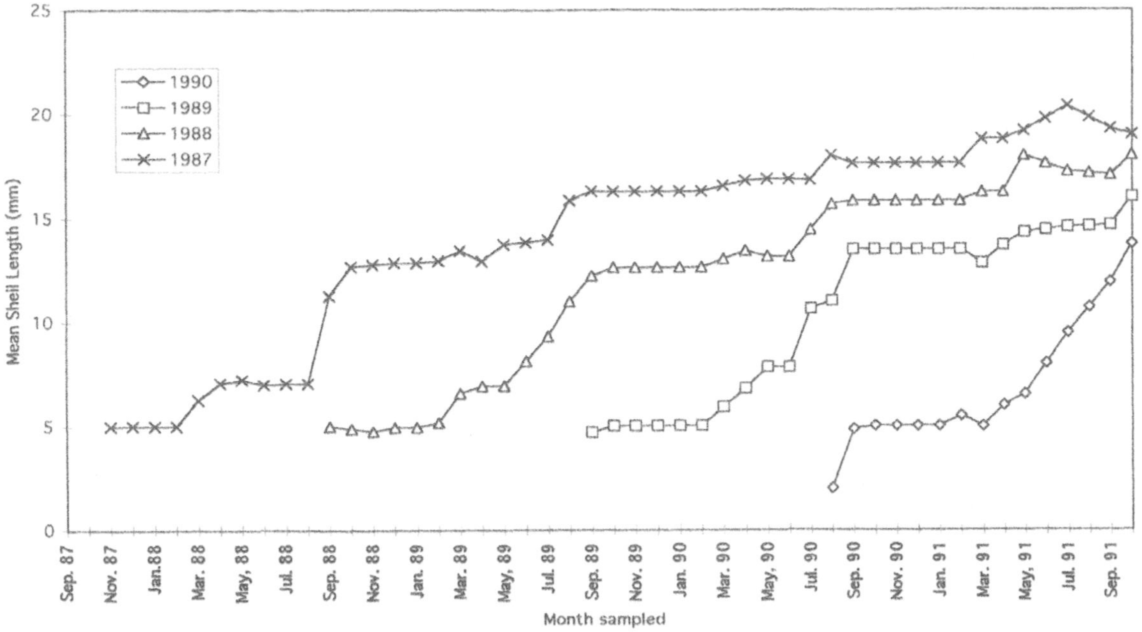

Figure 1. *Trochocochlea lineata* on Gore Point 1987–1991. Monthly mean size (shell length, mm) of the 1988, 1989, 1990 and 1991 cohorts.

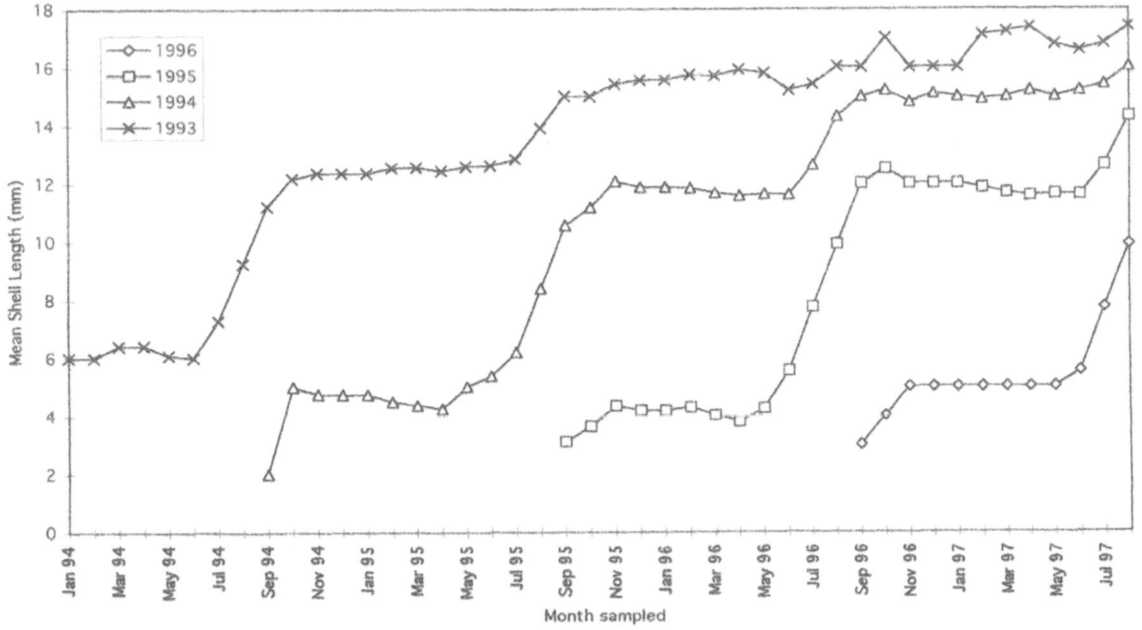

Figure 2. *Trochocochlea lineata* on Gore Point 1994–1997. Monthly mean size (shell length, mm) of the 1993, 1994, 1995 and 1996 cohorts.

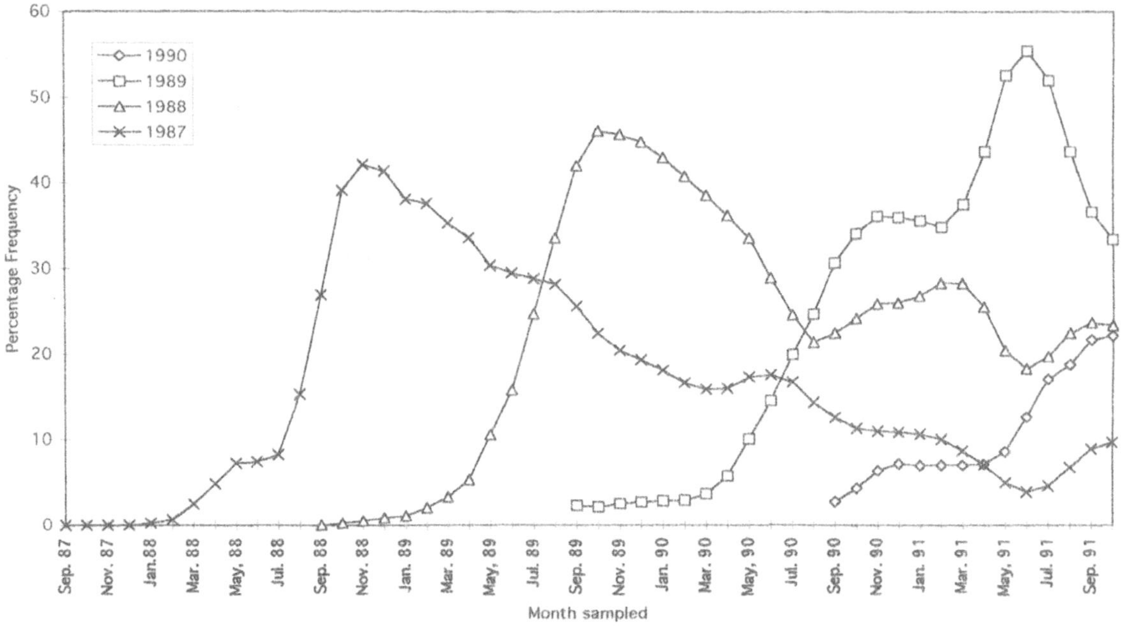

Figure 3. Trochocochlea lineata on Gore Point 1987–1991. Monthly percentage abundance of the 1988, 1989, 1990 and 1991 cohorts.

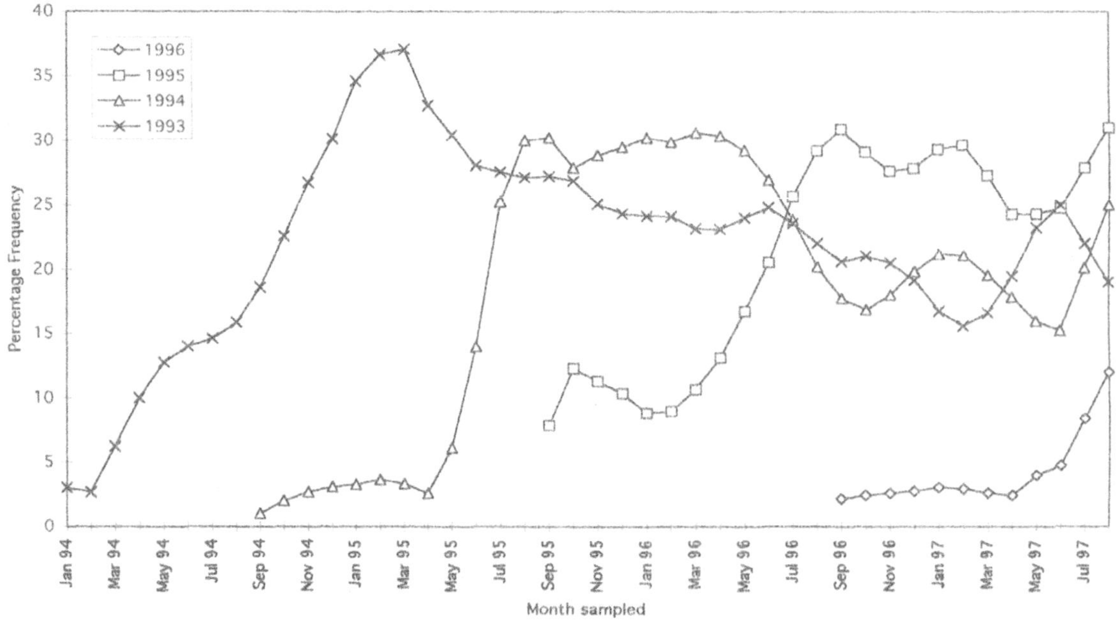

Figure 4. Trochocochlea lineata on Gore Point 1994–1997. Monthly percentage abundance of the 1993, 1994, 1995 and 1996 cohorts.

Table 1. Trochocochlea lineata on Gore Point, 2nd June 1996. Shell length is the maximum measurement that can be made across the shell from the apex.

Shell length	Presumed year of settlement									
(mm)	'95	'94	'93	'92	'91	'90	'89	'88	'87	Total
1										
2										
3	1									1
4	4									4
5	19									19
6	13									13
7	6									6
8	5									5
9										0
10		11								11
11		44								44
12		66								66
13		54								54
14		23	29							52
15			82	39	2					123
16			83	103	8					194
17			40	115	24	2				181
18			6	38	38	4	3			89
19				13	36	3	4			56
20				7	23	7	6			43
21					17	6	5	2		30
22					3	2	4	1	1	11
23					1		2	1		4
24								1		1
Totals	48	198	240	315	152	24	24	5	1	1007
Mean shell length (mm)	5.6	12.2	15.6	16.7	18.7	19.7	20.4	22.2	22.0	
% of Sample	4.8	19.7	23.8	31.3	15.1	2.4	2.4	0.5	0.1	100

– so age estimations are subjective at this time of the year.

Figures 1 and 2, which have been simplified by omitting all but the four youngest cohorts, suggest a stable pattern of growth that varied little from one year to the next. Each cohort appeared on the shore in late August/early September, showed a burst of growth in that first autumn before the winter check, grew rapidly during the following summer before experiencing a second winter's check, grew rather less vigorously during their second summer before they entered the adult population and growth rate diminished to a millimetre (or less) per annum.

There is no suggestion that the growth rate of the 1995 cohort was any different from its predecessors.

The preponderance of particular cohorts

Obviously, each cohort of snails is most abundant immediately after settlement (i.e. in its first September) and can only decrease with time. The data, however (Figures 3 and 4), show the cohorts gradually recruiting into the measured population over one or two years, usually reaching their maximum importance in the year in which they first breed (Crothers, 1994). Very few of the 0+ animals were recorded in their first winter – either because we were unable to recognise them or because they were occupying a different habitat. Considerable searching has failed to reveal a nursery area, such as was described in Milford Haven by Little et al. (1986).

Table 2. Trochocochlea lineata on Gore Point, 24th May 1997.

Shell length (mm)	Presumed year of settlement '96	'95	'94	'93	'92	'91	'90	'89	'88	Total
4										0
5	7									7
6	11									11
7	8									8
8										0
9										0
10		2								2
11		18								18
12		76	1							77
13		68	5							73
14		17	9							26
15		3	54		1	1				59
16		1	38	29	15	4	1			88
17			19	73	45	12		2		151
18			3	63	49	25	11	2		153
19			1	48	34	21		1		105
20				13	9	11	13	1		47
21				5	3	4	1			13
22					1	2	1	1		5
23						1				1
24								1		1
25							1			
Totals	26	185	130	231	157	81	28	8	0	846
Mean shell length (mm)	6.0	12.5	15.5	17.8	17.9	18.6	19.4	19.4		
% of Sample	3.1	21.9	15.4	27.3	18.6	9.6	3.3	0.9		100

However, it is clear from Figure 4 that the 1995 cohort was proportionally more abundant during its first autumn/winter than was usual for the 0+ age class.

Lovers Point, Watchet; The western side of Helwell Bay

Table 3 shows the data collected in autumn 1995. These animals had grown more rapidly and had reached a larger size that those at Gore Point. No more than 5 individuals were found on any visit in summer 1996, and none at all have been seen alive in 1997. Only one (7 mm) individual of the 1995 cohort was ever recorded.

Swill Point, Doniford; the eastern side of Helwell Bay

The data from a sample collected on 26th August 1996 are shown in Table 4. Only three cohorts were represented with the 1995 cohort predominating. No living animals have been found at this site in 1997.

Discussion

The winter of 1962/63 demonstrated the effects of an exceptionally cold winter on many species (Crisp, 1964) but *T. lineata* was affected more severely than most. On the Dorset coast, all populations east of Portland were eliminated and densities elsewhere were severely reduced, with local extinctions, as far west as the Exe Estuary (Crisp, 1964; Hawthorne, 1965). Thirty years later, *T. lineata* had recolonised its former sites only as far east as Lyme Regis, which is now considered to be the most easterly 'permanent population' along the south coast of England. A few individuals can usually be found on the Isle of Portland, but not in sufficient numbers to constitute a breeding population (Preece, 1993; personal observations). [In the distant past it probably reached the Isle of Wight for shells have been found there in archaeological sites (Preece, 1993) and old worn shells are occasionally found on beaches (Preece, 1993, and personal observations)].

Table 3. Trochocochlea lineata at Watchet. Samples collected 10th September and 13th September 1995 combined.

Shell length (mm)	Presumed year of settlement '95	'94	'93	'92	'91	'90	'89	'88	'87	'86	Total
6											0
7	1										1
8											0
9											0
10											0
11											0
12											0
13											0
14											0
15											0
16		3									3
17		4									4
18		5									5
19		2	5								7
20		1	5								6
21			5	1	1						7
22				1							1
23				1	3	1					5
24				3	3	1	2				9
25				1	2	7	2				12
26					1		3	1		1	6
27					1	1	5	1	2		10
28						1	1	1	1		4
29										2	2
Totals	1	15	15	7	11	11	13	3	3	3	82
Mean shell length	7.0	17.6	20.0	23.3	24.1	25.2	26.1	27.0	27.3	28.0	
% of Sample	1.2	18.3	18.3	8.5	13.4	13.4	15.9	3.7	3.7	3.7	100

On the Welsh coast of the Bristol Channel, the species was common around Swansea in the last century (Jeffreys, 1865) and occurred in the Porthcawl area during the 1950s but disappeared from all shores east of the Gower after 1963 (Crisp, 1964). It remains rare on the Gower and has not recolonised any of its former habitats further east (Boyden et al., 1977, and personal observations).

Bassindale's (1941, 1943) surveys of the Somerset coast recorded *T. lineata* (as *Osilinus lineatus*) only at Porlock. After the 1963 winter, no living individuals were observed along the English coast east of Woolacombe (Crisp, 1964). However, by 1995, the species had not only recolonised all of its (reported) former territory along this coast but appeared to have colonised new sites to the east.

Gore Point, west of Porlock Weir, is the only Somerset site at which all of the six or seven most recent cohorts are invariably present and is the easternmost point in the Bristol Channel at which *T. lineata* can truly be called 'common',[3] although it spreads across Porlock Bay at a lower density. In this population (Crothers, 1994), the 15–18 mm sized animals are collected most frequently. As in all sampling, the number recorded is a reflection both of the number present and the ease with which they can be found; we have never been very successful at finding the youngest age class at this site.

[3] Direct counts in 1 m^2 quadrats have ranged from 3 to 88. A mark release recapture carried out in June 1996, based on 2274 marked individuals released in a 20 m × 20 m square, suggested densities of 12 m^{-2} and 25 m^{-2} after 2 and 5 d respectively.

Table 4. Trochocochlea lineata on Swill Point, Doniford. Collected 26th August 1996.

Shell length (mm)	Presumed year of settlement					Total
	1996	1995	1994	1993	1992	
12						
13		1				1
14		2				2
15		16				16
16		32				32
17		20	3			23
18		5	9			14
19			12	1		13
20			4	1		5
21						0
Totals	0	76	28	2	0	106
Mean shell length (mm)		16.1	18.6	19.5		
% of each cohort		71.7	26.4	1.89		100

The 1995 cohort was well-established by June 1996 (Table 1) but there is no evidence (e.g. Figure 2) that these snails grew any faster (i.e. reached a larger size at a younger age) than other cohorts. Comparison of the 1995 cohort with those of 1987–1990 (Figure 3) and 1993, 1994 and 1996 (Figure 4) does suggest that they were unusually abundant in their first autumn although by 1996 they were no longer outstanding. A comparison of Table 2 with Table 1 (collected by the equivalent course in the following year) suggests that the 1995 cohort is more successful than those of 1994 and 1996, but it is too early to see whether this will prove to be a particularly successful cohort in the long term.

Boyden et al. (1977) noted isolated enclaves east of Porlock Bay on Greenaleigh Point (SS955485), at Blue Anchor (ST040438) and on the western side of Helwell Bay, Watchet (ST078436). In 1990, another enclave was found on Warren Point (ST985468), east of Minehead, but that at Blue Anchor had disappeared. Also in 1990, an outlying enclave was discovered on Swill Point, Doniford (ST095436), on the eastern side of Helwell Bay, 2 km east of Watchet.

Since 1967, at least, *T. lineata* has been much less abundant at Watchet, but individuals have grown to a much larger size. The same was true for both *Gibbula umbilicalis* and *Littorina littorea* (Crothers, 1992), which suggests that the high density of herbivorous gastropods had stunted the growth, and reduced the life expectancy, of the Gore Point populations; whilst food has been plentiful for the outlying enclaves. Some component of the breeding system, perhaps the distance from a breeding population or the mud on the rock surface inhibiting settlement, had restricted recruitment at these sites. The outlying enclaves never showed the whole range of age-classes.

At Doniford, on the eastern side of Helwell Bay, *T. lineata* was apparently confined to a single rock pool system on the top of the ridge that forms Swill Point. Most individuals were found within a circle of no more than 50 m radius. The site is almost free of fucoid algae and is much less muddy than the surrounding area. Initial investigations of this population (Crothers, 1994) assumed that the lack of older age-classes simply reflected its recent foundation. However, both succeeding years gave the same profile, so it appears that some factor was seriously constraining longevity at this site. Immatures always predominated. Samples taken in autumn 1995 indicated a very positive recruitment. 1996 samples showed this cohort to have grown strongly (and much faster than their compatriots on Gore Point) but no elderly individuals were found in a two hour search. No living *T. lineata* have been recorded at this site in 1997.

A hot summer, two cold winters and the Somerset populations of Trochocochlea lineata

Lewis et al. (1982) and Lewis (1986) concluded that some feature of the breeding system was the primary cause of a northern limit, probably something to do with larval settlement and survival of juvenile snails on the shore. The data presented here reinforce the impression left by Crisp et al. (1964) that winter cold is the main restraint to the eastern advance of *Trochocochlea lineata* in Somerset.

The hot summer of 1995 encouraged (or coincided with) a cohort of *T. lineata* that was unusually abundant in its first autumn at both Gore Point and notably abundant at Swill Point, although not at Lovers Point on the other side of Helwell Bay. At Swill Point, the 1995 cohort accounted for 85% of a sample measured on 13th November 1995 and still represented 72% the following August (Table 4). They also grew fast; reaching a mean length of 8 mm before their first winter and of 16 mm at one year old – the size that the Gore Point snails will probably not reach until 1998.

The winter of 1995/1996 was colder for longer than many of its predecessors and both then and in 1996/1997 there was a very cold period around New Year, coinciding with strong NE winds in the latter case. There is no suggestion these winters had any lasting effect on the Gore Point population but there appear to have been no survivors east of Minehead. The enclaves either side of Helwell Bay have disappeared; the first cold winter weakened them and the second eliminated them.

At Swill Point, the 1992 and 1993 cohorts represented 8% of a sample taken in November 1995 but only 2% of the August 1996 sample (Table 4). On the Watchet side of Helwell Bay (Table 3), the 1990 and older cohorts represented over 40% of the samples in autumn 1995 but only 21% in the spring of 1996.

The conclusion from these observations is that a hot summer can encourage recruitment but that the geographical limit in Somerset may be maintained by winter cold.

Acknowledgements

The assistance of students and their accompanying staff, attending field courses at Nettlecombe Court, in collecting and measuring thousands of topshells is gratefully acknowledged. None of this work would been achieved, or communicated at this Symposium, without the support of Field Studies Council.

When the first draft of this paper was written, in the autumn of 1996, no firm conclusion could be drawn from the available data. A second cold period in the winter of 1996/1997 clarified the situation and I am grateful to the editors for permission to revise the paper accordingly.

References

Bamber, R. N. & J. Coughlan, 1987. An ecological survey of the foreshore adjacent to Hinkley Point. Report TPRD/L/3127/R87 to the Central Electricity Generating Board.

Bassindale, R., 1941. Studies on the biology of the Bristol Channel IV. The invertebrate fauna of the southern shores of the Bristol Channel and Severn Estuary. Proc. Bristol Nat. Soc. 9: 143–201.

Bassindale, R., 1943. Studies on the biology of the Bristol Channel XIII. The intertidal fauna of Porlock Bay. Proc. Bristol Nat. Soc. 9: 386–399.

Boyden, C. R., J. H. Crothers, C. Little & C. Mettam, 1977. The Intertidal Invertebrate Fauna of the Severn Estuary. Field Stud. 4: 477–554.

Cesari, P., 1986. La sistematica del gen. *Monodonta* Lamck., 1799 (s.l.). 1 – Osservazione preliminari (Mollusca, Trochidae). Boll. Mus. Civ. Stor. Nat. Venezia 37: 191–212.

Crisp, D. J. (ed.) 1964. The effects of the severe winter 1962–1963 on marine life in Britain. J. anim Ecol. 33: 165–210.

Crothers, J. H., 1976. On the distribution of some common animals and plants on rocky shores of West Somerset. Field Stud. 4: 369–389.

Crothers, J. H., 1992. Shell size and shape variation in edible winkles, *Littorina littorea* (L.), from West Somerset. In J. Grahame, P. Mill & D. G. Reid (eds), Proceedings of the Third International Symposium on Littorinid Biology. The Malacological Society of London: 91–97.

Crothers, J. H., 1994. Student investigations on the population structure of the common topshell, *Monodonta lineata*, on The Gore, Somerset. Field Stud. 8: 337–355.

da Costa, E. M., 1778. Historia Naturalis Testaceorum Britanniae. London.

Fretter, V. & A. Graham, 1977. The Prosobranch Molluscs of Britain and Denmark. Part 2 – Trochacea. J. moll. Stud., Suppl. No. 3: 39–100.

Fretter, V. & A. Graham, 1994. British Prosobranch Molluscs (Second Edition). The Ray Society, London, 820 pp.

Garwood, P. R. & M. A. Kendall, 1985. The reproductive cycles of *Monodonta lineata* and *Gibbula umbilicalis* on the coast of Mid-Wales. J. mar. biol. Ass. U.K. 65: 993–1008.

Gofas, S. & A. Jabaud, 1997. The relationships of the Mediterranean trochid '*Monodonta*' *mutabilis* (Philippi, 1846) and '*Gibbula*' *richardi* (Payraudeau, 1826). J. moll. Stud. 63: 57–64.

Hawthorne, J. B., 1965. The eastern limit of distribution of *Monodonta lineata* (da Costa) in the English Channel. J. Conch. 25: 384–352.

Jeffreys, J. G., 1865. British Conchology, or an account of the Mollusca which now inhabit the British Isles and the surrounding seas. Volume III Marine Shells. Van Voorst, London, 393 pp.

Kendall, M. A., 1987. The age and size structure of some northern populations of the trochid gastropod *Monodonta lineata*. J. moll. Stud. 53: 213–222.

Lewis, J. R., 1964. The Ecology of Rocky Shores. Hodder & Stoughton, London, 323 pp.

Lewis, J. R., 1986. Latitudinal trends in reproduction, recruitment and population characteristics of some rocky littoral molluscs and cirripedes. Hydrobiologia 142: 1–13.

Lewis, J. R., R. S. Bowman, M. A. Kendall & P. Williamson, 1982. Some geographical components in population dynamics: possibilities and realities in some littoral species. Neth. J. Sea Res. 16: 18–28.

Little A. E., B. Dicks & J. H. Crothers, 1986. Studies of barnacles, limpets and topshells in Milford Haven. Field Stud. 6: 459–492.

Martin, K., 1994. An ecological survey of species with cooling water system fouling potential at Hinkley Point. Unpublished Honours Project report, University of Plymouth, 94 pp.

McMillan, N. F., 1944. The distribution of *Monodonta* (*Trochus*) *lineata* (da Costa) in Britain. Northw. Naturalist 19: 290–292.

Preece R. C., 1993. *Monodonta lineata* (da Costa) from archaeological sites in Dorset and the Isle of Wight. J. Conch. 34: 339–341.

Stanbury, D. J., 1974. Some marking experiments on a population of *Monodonta lineata* (da Costa), South Haven, Skokholm, Pembrokeshire. Skokholm Bird Observatory Report for 1974: 27–34.

Williams, E. E., 1965. The growth and distribution of *Monodonta lineata* (da Costa) on a rocky shore in Wales. Field Stud. 2: 189–198.

Williamson, P. & M. A. Kendall, 1981. Population age structure and growth of the trochid *Monodonta lineata* determined from shell rings. J. mar. biol. Ass. U.K. 61: 1011–1026.

Hydrobiologia **378:** 143–148, 1998.
R. M. O'Riordan, G. M. Burnell, M. S. Davies & N. F. Ramsay (eds), *Aspects of Littorinid Biology.*

Density and shell-size variation of *Nodilittorina lineolata* (Orbigny, 1840) in the intertidal region in southeastern Brazil

C. A. De Magalhães
Depto. de Zoologia, Instituto de Biologia, Universidade Estadual de Campinas, Campinas, SP, Brazil

Key words: Nodilittorina lineolata, shell-size gradient, density, supralittoral, Brazil

Abstract

Nodilittorina lineolata is a small littorinid characteristic of the supralittoral of rocky shores but extends its vertical range to the lower midlittoral. Shell-size and density of *Nodilittorina* were measured at two different levels, the supralittoral and barnacle zones, in six sites along the south-eastern Brazilian coast. The sites studied varied in the degree of wave exposure. Densities were lower in the supralittoral, and higher overall values occurred on the most exposed shores. Shell size increased significantly in an upshore direction at five of the six sites. However, mean-size and the magnitude of the difference between levels varied according to the site studied. The shell-size gradient can be explained by larval recruitment in the barnacle zone and subsequent ascending migration to the supralittoral zone, where the larger individuals were found. Reciprocal transfer experiments revealed that differential mortality of large individuals in the lower levels, possibly due to lack of refuges, may also influence the size-distribution. It was also shown that some individuals can be passively transported to upper levels of the shore by wave action during storms.

Introduction

Vertical shell-size gradients along the intertidal region are frequent in many species of marine gastropods. Vermeij (1972) proposed that shell size tends to increase in an upshore direction in species characteristic of the littoral fringe. Within the Littorinidae size gradients have been related to many environmental factors, such as temperature and desiccation stress, wave action and crevice availability, rock surface slope and topography, surface complexity and microhabitat refuges, and recruitment site (e.g. Raffaelli & Hughes, 1978; Chen & Richardson, 1987; Reid, 1993; Chapman, 1994; Kostylev et al., 1997). In a supralittoral species of periwinkle, shell-size increases and density decreases with increased height on the shore (Chen & Richardson, 1987). In another study the reverse was found, mostly because of the influence of wave exposure (McQuaid, 1981).

The coast of Brazil is poor in littorinid diversity, with only three species occurring on the continental part of the shore – *Nodilittorina lineolata*, *Littoraria flava* and *Littoraria angulifera* (Reid, 1990; Rios, 1994). The species of the genus *Littoraria* have more restricted distributions than *N. lineolata* and are linked to areas with mixohaline conditions. *Nodilittorina lineolata* belongs to the Caribbean *ziczac* species complex (Borkowski & Borkowski, 1969; Bandel & Kadolsky, 1982). It is the dominant, and sometimes exclusive, component of the upper levels of the shore, being abundant along the Brazilian coast. This species is distributed from the supralittoral to the upper barnacle zone, where it is found among spaces between the cirripedes and within their empty carapaces; it has a planktonic egg stage (Borkowski & Borkowski, 1969). In this work I studied the vertical size and density distributions of *N. lineolata* at two distinct levels of the shore, the supralittoral and the lower shore *Chthamalus bisinuatus* zones.

Study sites

Two different areas were selected for sampling. The first is located in the Saquarema district of Rio de

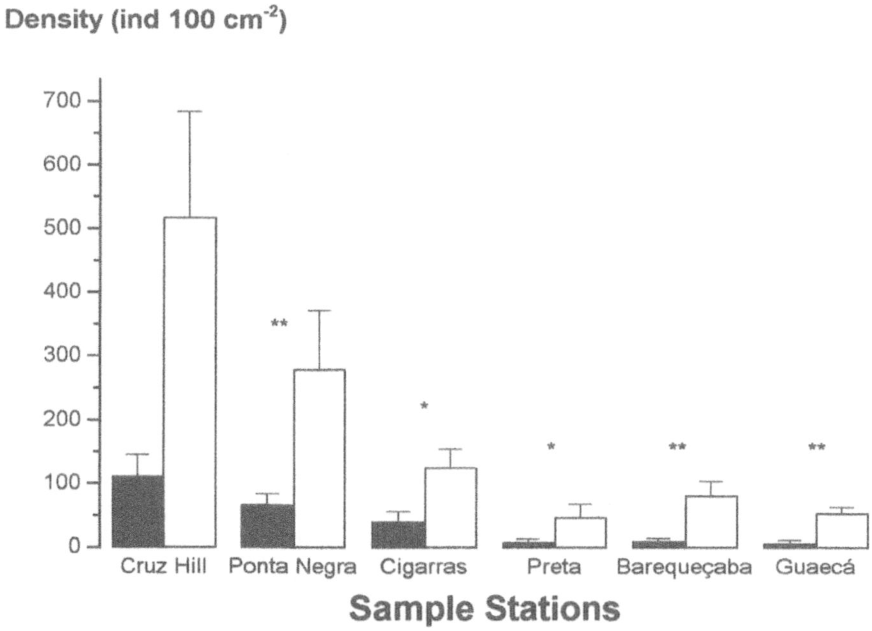

Figure 1. Nodilittorina lineolata. Mean density of individuals (SE) on the southeastern coast of Brazil. Tukey comparisons: $* = p < 0.01$; $** = p < 0.001$. Dark bars = supralittoral zone; White bars = barnacle zone.

Janeiro state and is characterised by rocky shores which are extremely exposed to wave action.

Samples in this area were taken at Ponta Negra and Cruz Hill (22°55′ S, 42°31′ W and 23°00′ S, 42°25′ W). The second region, on the landward side of São Sebastião Channel, São Paulo state, is composed of narrow strips of rocky shore protected or moderately exposed to wave action. The four samples taken in this area were from rocky surfaces on the beaches of Cigarras, Preta, Barequeçaba and Guaecá (between 23°43′ S, 45°20′ W and 23°53′ S, 45°27′ W). Both geographical areas are located in the southeastern part of Brazil and are subtropical.

Methods

Densities of *N. lineolata* were sampled at two levels on each shore by means of five randomly placed 100 cm^2 quadrats at each level. All snails within the quadrat frame were collected. Gastropod size was measured as shell length, to the nearest 0.1 mm, with vernier callipers. For the two most exposed and densely populated sites (Cruz Hill and Ponta Negra), size frequencies were constructed.

Reciprocal transfer experiments were conducted between the supralittoral and barnacle zone snails in Saquarema. Snails from control and treatment groups were collected, marked with porcelain paint using different colour codes, and replaced in the field at the same source locations as the control groups. In each of three trials, (Sept. ’94, Oct. ’94 and Apr. ’95), 200 individuals were marked from each zone and 100 released at each level of the shore (i.e. an experimental group and a control group). At regular intervals of 2, 15 and 28 d, the sites were thoroughly inspected, and all marked snails found were counted.

The variables studied were analysed by means of two-way ANOVA. Assumptions of normality and homogeneity of variances were tested through residual analysis. Tukey tests were used for *post-hoc* differentiation between pairs of means. Frequency differences between control and treatment groups of the transfer experiments were detected by the log-likelihood G statistic.

Results

Densities were lower in the supralittoral than in the barnacle zone at all shores and higher overall values occurred in the most exposed shores i.e. Cruz Hill and Ponta Negra, Rio de Janeiro (Fig. 1).

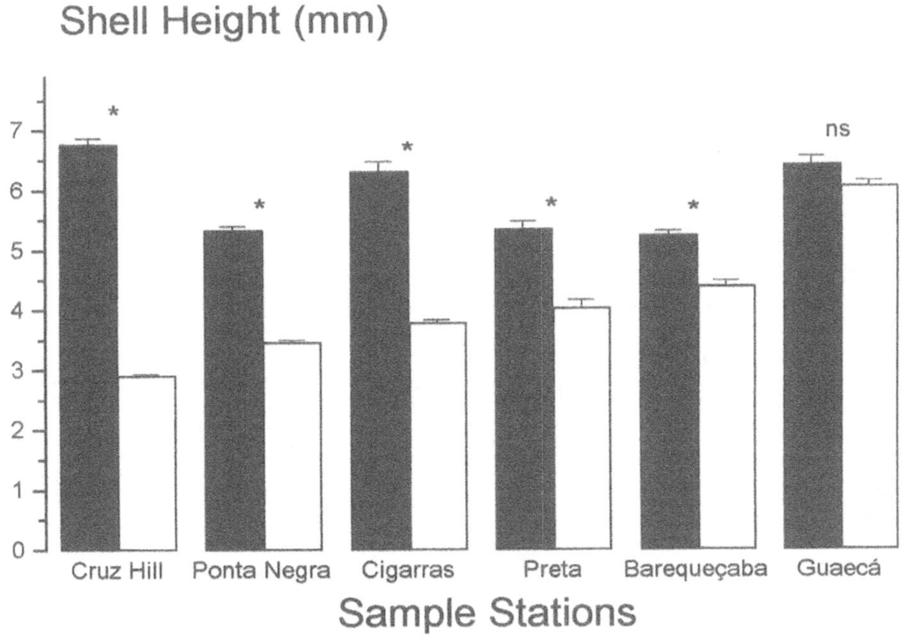

Figure 2. Nodilittorina lineolata. Mean sizes (± SE) on the southeastern coast of Brazil. Tukey comparisons: * $= p < 0.01$; ** $= p < 0.001$; ns = not significant. Dark bars = supralittoral zone; White bars = barnacle zone.

Table 1. Nodilittorina lineolata. Two-way ANOVA on shell-size data presented in Fig. 2, showing the effects of shore level and site on shell-size. df = Degrees of freedom, SS = Sum of squares, MS = Mean square, F = F-ratio, p = Probability.

Source	df	SS	MS	*F*	*p*
Level	1	1239.231	1239.231	1371.978	< 0.001
Site	5	474.403	94.881	105.044	< 0.001
Level X Site	5	764.709	152.942	169.325	< 0.0.01

Shell-size increased significantly in an upshore direction at 5 sites (Fig. 2). Also, there was a significant interaction between shore level and sample site, with average size varying differently between levels on each shore (Table 1). Guaecá, the site where snails from both levels exhibited similar sizes, is the most horizontal of the rocky profiles, being moderately exposed to wave action. All barnacle zone samples were significantly different, possibly indicating distinct recruitment periods. The supralittoral samples can be divided into two size groups, i.e. Cruz Hill, Cigarras, and Guaecá stations, and Ponta Negra, Preta and Barequeçaba stations.

The size-structures of *N. lineolata* from Rio de Janeiro (Fig. 3) revealed that mean shell length was greater in the supralittoral because of the absence of recruits and young individuals at this level of the shore. Thus, the *Chthamalus* zone appears to be the site of larval recruitment.

Snails from the barnacle zone survived as well as the controls in the supralittoral. The number of larger individuals recovered from the supralittoral zone was very low at the lower level of the shore (Fig. 4). Overall shapes of the control curves were similar at both releasing levels, indicating homogeneity of manipulative effects upon the marked individuals. In exposed sites such as the one where the mark-release-recaptures were done, the smaller individuals would find shelter more easily between barnacles at the lower levels of the shore.

The mark-release-recapture trial of September 1994 did not succeed because of rough weather shortly after the beginning of the experiment, but served to show that passive transport of snails by waves can happen. Five days after the storm, 21.5% of the marked individuals replaced in the supralittoral and 6% of the ones from the barnacle zone were found in tide pools as far as 20 linear metres above their release point.

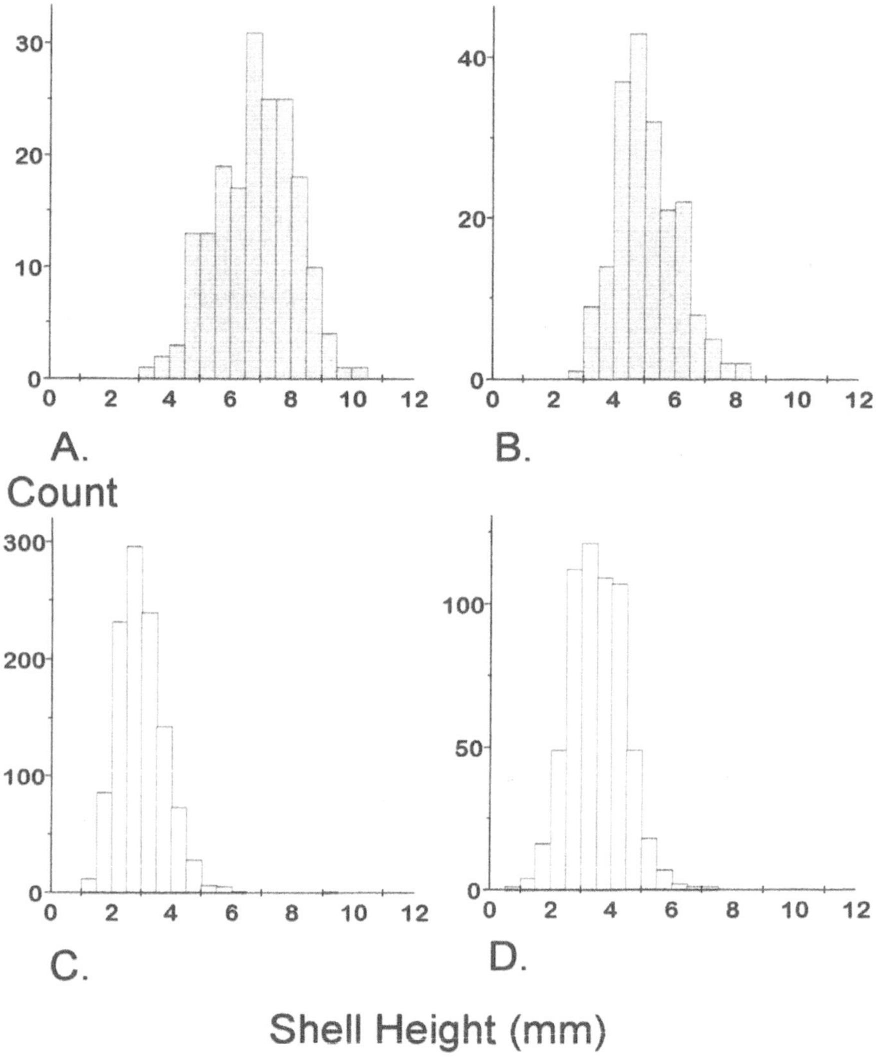

Figure 3. Nodilittorina lineolata. Size structure on two levels of exposed shores in Rio de Janeiro, Brazil. A, C, Cruz Hill; B, D, Ponta Negra. Grey bars = supralittoral, White bars = barnacle zone.

Discussion

The decreasing densities and larger sizes of *N. lineolata* in an upshore direction found in this study indicate that this species conforms to the type 1 shore-level size gradient proposed by Vermeij (1972). Gastropods exhibiting this type of gradient are characteristic of the supralittoral and high intertidal zones. According to Vermeij (1972) their juveniles occupy that part of the vertical range where mortality is expected to be lowest. In the upper edge of the intertidal region mortality generally results from physical extremes such as temperature and desiccation (Chow, 1975; Chen & Richardson, 1987).

In the very exposed shore of Saquarema, the small individuals of the barnacle zone survived as well as the large ones of the supralittoral in the upper level of the shore. Thus, it seems that both groups resist the high surface temperatures and desiccation common to this level. It is also feasible that the constant wave spray, characteristic of this area, counteracts these physical extremes.

Large snails, on the other hand, had significantly lower recovery rates as compared to the small ones in the barnacle zone. Reid (1993) noted the existence

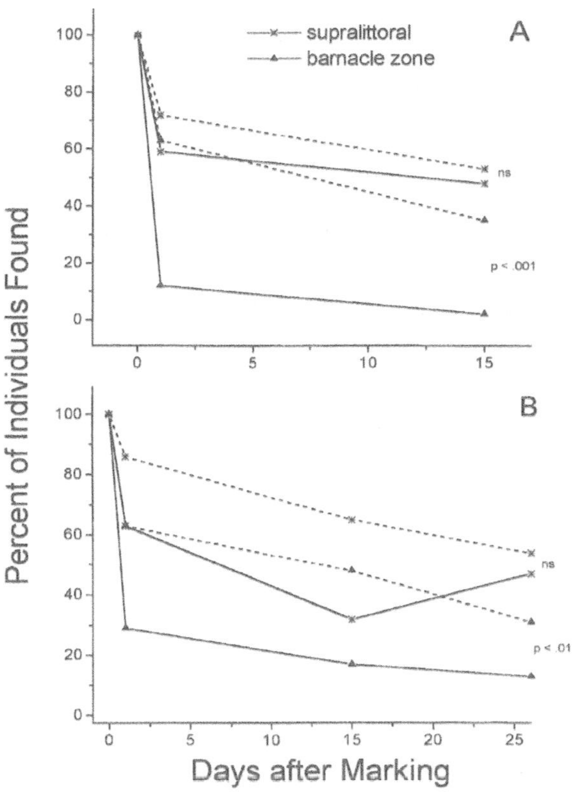

Figure 4. Nodilittorina lineolata. Reciprocal transfer experiments. Percentage recovery in Cruz Hill, Saquarema, Brazil. Solid lines = treatment, Dashed lines = control. A = October '94; B = April '95.

of small barnacle-dwelling forms in three species of *Littorina* on British shores. He suggested that these dwarf forms may be responding to the same set of environmental conditions in the barnacle zone, thereby producing analogous ecotypes. Small size is advantageous for finding shelter among and within empty barnacle shells on exposed rocky shores (Raffaelli & Hughes, 1978), and on exposed Galician mid-shores a small-sized ecotype of *L. saxatilis* is found in higher densities in barnacles than in mussel patches in contrast to a large-sized ecotype (Kostylev et al., 1997).

An inverse relationship between snail density and size was found also in *Nodilittorina unifasciata* by Chapman (1994) but it was not linked with a type 1 shell-size gradient. The correlation was a consequence of density-dependent growth, with food being limiting at high densities. Although the activity pattern of *N. lineolata* has not been studied here, foraging has been observed to occur over moist rock surfaces, beginning soon after the tide falls. On very exposed shores there is a great risk of dislodgement by waves at the lower levels which limits the foraging periods and possibly upper shell-size. Wave spray in the supralittoral could enable extended foraging excursions on the wet surface without the risk of being dislodged, allowing the attainment of larger sizes and greater densities at those sites.

In the barnacle zone, densities varied on a hundred fold scale between the two most exposed and the protected shores. However, in the supralittoral the differences are less striking. McQuaid (1996) emphasised that littorinid densities are strongly influenced by recruitment success and positively correlated with wave action, but we need also to understand the factors that affect settlers and control their recruitment into the adult population. It seems that in shores subject to severe wave action effective recruitment is more difficult to achieve than in protected shores with comparatively lower settlement rates.

From the results obtained it may be concluded that in *N. lineolata* settlement occurs in the zone dominated by the small barnacle *Chthamalus bisinuatus* with later ascending migration to the supralittoral where the larger individuals can be found. Density decreases in an upshore direction and there is a type 1 vertical-size gradient. It was also incidentally shown that some individuals can be passively transported to upper levels of the shore by wave action during storms.

Acknowledgements

I am very grateful to Dr David Reid for helping in species identification, to M. O. Tanaka for the layout of the figures, and the referees for improving and correcting the manuscript.

References

Bandel, K. & D. Kadolsky, 1982. Western Atlantic species of *Nodilittorina* (Gastropoda: Prosobranchia): comparative morphology and its functional, ecological, phylogenetic and taxonomic implications. Veliger 25: 1–42.

Borkowski, T. V. & M. R. Borkowski, 1969. The *Littorina ziczac* species complex. Veliger 11: 408–414.

Chapman, M. G., 1994. Small-scale patterns of distribution and size-structure of the intertidal littorinid *Littorina unifasciata* (Gastropoda: Littorinidae) in New South Wales. Aust. J. mar. Freshwat. Res. 45: 635–52.

Chen, Y. S. & A. M. M. Richardson, 1987. Factors affecting the size structure of two populations of the intertidal periwinkle, *Nodilittorina unifasciata* (Gray, 1839), in the Derwent River, Tasmania. J. moll. Stud. 53: 69–78.

Chow, V., 1975. The importance of size in the intertidal distribution of *Littorina scutulata*. Veliger 18: 69–78.

Kostylev, V., J. Erlandsson & K. Johannesson, 1997. Microdistribution of the polymorphic snail *Littorina saxatilis* (Olivi) in a patchy rocky shore habitat. Ophelia 47: 1–12.

McQuaid, C. D., 1981. The establishment and maintenance of vertical size gradients in populations of *Littorina africana knysnaensis* (Philippi) on an exposed rocky shore. J. exp. mar. Biol. Ecol. 54: 77–89.

McQuaid, C. D., 1996. Biology of the gastropod family Littorinidae. II. Role in the ecology of intertidal and shallow marine ecosystems. Oceanog. Mar. Biol. annu. Rev. 34: 263–302.

Raffaelli, D. G. & R. N. Hughes, 1978. The effects of crevice size and availability on populations of *Littorina rudis* and *Littorina neritoides*. J. anim. Ecol. 47: 71–83.

Reid, D. G., 1990. Trans-Arctic migration and speciation induced by climatic change: the biogeography of *Littorina* (Mollusca: Gastropoda). Bull. mar. Sci. 47: 35–49.

Reid, D. G., 1993. Barnacle-dwelling ecotypes of three British *Littorina* species and the status of *Littorina neglecta* Bean. J. moll. Stud. 59: 51–62.

Rios, E. C., 1994. Seashells of Brazil. Editora da Fundação Universidade do Rio Grande, Rio Grande, 368 pp.

Vermeij, G. J., 1972. Intraspecific shore-level size gradients in intertidal molluscs. Ecology 53: 693–700.

Hydrobiologia **378:** 149–160, 1998.
R. M. O'Riordan, G. M. Burnell, M. S. Davies & N. F. Ramsay (eds), Aspects of Littorinid Biology.

Spawning periodicity of three species of *Nodilittorina* in Hong Kong

Y. M. Mak
Department of Ecology and Biodiversity and The Swire Institute of Marine Science, The University of Hong Kong, Hong Kong
Present address: Agriculture and Fisheries Department, 13/F., Canton Road Government Offices, 393 Canton Road, Kowloon, Hong Kong

Key words: Nodilittorina, Littorinidae, spawning periodicity, fecundity, Hong Kong

Abstract

The spawning periodicity of *Nodilittorina trochoides*, *N. radiata* and *N. vidua* from two rocky shores of different wave exposure, Big Wave Bay (more exposed site) and Cape d'Aguilar, in Hong Kong was monitored for one year (November 1994–November 1995). The breeding season was similar for the three species, mainly during the summer (June–September). The spawning duration was 9 months for *N. vidua* and 7 months for the other two littorinids. Spawning was initiated at different times for the two sites. At Big Wave Bay, spawning occurred between May–November for *N. trochoides*; March–September for *N. radiata* and March–November for *N. vidua*. At Cape d'Aguilar, the spawning period of *N. trochoides* and *N. radiata* was similar (April–October), whilst for *N. vidua*, spawning extended up to December. A decrease in monthly fecundity was recorded during July and August at Big Wave Bay and Cape d'Aguilar respectively. A weak relationship was found between the size of reproductive females and egg production. The seasonal fecundity of *N. trochoides* and *N. vidua* was similar at both sites and estimated as 10 000 and 7500 eggs per female respectively. Fecundity of *N. radiata* varied between sites, with 15 000 eggs and 9000 eggs per female being recorded from Cape d'Aguilar and Big Wave Bay respectively.

Introduction

The family Littorinidae is an excellent group to study reproductive strategies, mode of reproduction and pattern of larval growth, since representative species occur in all intertidal habitats around the world (see reviews by Mileikovsky, 1971 and Underwood, 1979). The mode of reproduction in this family ranges from ovoviviparity (e.g. *Littorina saxatilis*: Berry, 1961; Ross & Berry, 1991); laying benthic egg capsules (e.g. *Littorina obtusata* and *Littorina mariae*: Goodwin, 1978; *Littorina nigrolineata*: Hughes, 1980, *Lacuna marmorata* and *Lacuna unifasciata*: Langan-Cranford & Pearse, 1995) to, in the majority of species, producing pelagic eggs with planktotrophic larvae (e.g. *Littorina littorea*: Fish, 1972; Grahame, 1975; Hughes & Answer, 1982; Chase & Thomas, 1995; *Littorina scutulata*: Murray, 1979; *Littorina pintado* and *Littorina picta*: Struhsaker, 1966; *Melarapha* (=*Melarhaphe*) *cincta* and *Melarapha oliveri*: Pilkington, 1971 and *Nodilittorina trochoides*, *N. radiata* and *N. vidua*: Kojima, 1960; Ohgaki, 1981; Berry, 1986; Mak, 1995).

Several criteria have been used to define the reproductive cycles of littorinids. Temporal changes in the incidence of copulation activities, as indicated by the onset of mating behaviour and the percentage of pairing snails (Berry, 1961; Gibson, 1964; Schmitt, 1979) and spawning (Fretter & Graham, 1962; Struhsaker, 1966; Borkowski, 1971; Pilkington, 1971; Berry and Chew, 1973; Grahame, 1975; Goodwin, 1978; Murray, 1979; Schmitt, 1979; Hughes, 1980; Hughes & Roberts, 1980; Ohgaki, 1981; Hughes & Answer, 1982; Underwood & McFadyen, 1983; Berry, 1986) have frequently been described. Seasonal changes in the morphology and histology of the reproductive tract have also been outlined (Berry, 1961; Palant & Fishelson, 1968; Borkowski, 1971; Fish, 1972; Goodwin, 1978; Hannaford Ellis, 1983; Underwood & McFadyen, 1983; Lasiak, 1987; Ross & Berry, 1991).

Spawning periods have also been inferred from data on the recruitment of juveniles (Underwood, 1974; McQuaid, 1981). These methods are able to determine the breeding period of the littorinids, but only by monitoring egg production can one estimate the reproductive effort in each spawning season.

In Hong Kong, 13 species of the family Littorinidae have been recorded, inhabiting both rocky shores and mangals (Reid, 1992). Although the three rocky shore littorinids, *Nodilittorina trochoides*, *N. radiata* and *N. vidua* are relatively common in Hong Kong, little is known about the breeding and spawning periodicity of these species. No comprehensive research with spatial variation has been conducted on the reproduction of these species, apart from studies on spawning periodicity with only one single study site in Japan (Kojima, 1960; Ohgaki, 1981) and Malaysia (Berry, 1986). Berry (1986) has also studied the lunar cycle of spawning in *N. trochoides* and *N. vidua* in Malaysia.

The aims of the present work were to follow the spawning periodicity of three *Nodilittorina* species on a monthly basis by using a new laboratory 'drainage' system and to investigate whether there is any spatial differences in spawning, from two rocky shores in Hong Kong for a year. The number of eggs spawned per female per spawning season was enumerated and mark-recapture methods were used to examine how many times a female was potentially able to spawn and, therefore, maximum individual fecundity.

Materials and methods

Samples of littorinids were obtained monthly, for approximately one year (Big Wave Bay: November 1994–November 1995, Cape d'Aguilar: December 1994–October 1995) from Big Wave Bay and Cape d'Aguilar on Hong Kong Island (22°10′ N, 114°10′ E). The exposure of the shores was defined by the dominant intertidal flora and fauna as proposed by Ballantine (1961) and also by subjective wave exposure judgement. Under this classification, Big Wave Bay is more exposed than Cape d'Aguilar. The shore at Big Wave Bay is composed of a flat rock platform of $\sim 60°$ inclination and that of Cape d'Aguilar, is a nearly vertical rock platform. A monthly sampling period was selected as previous studies have shown that female littorinids can renew their supply of mature eggs in approximately one month (Struhsaker, 1966). Collections were undertaken during lowest spring tide, to control for the influence of the lunar cycle on spawning. On the shore, ~ 80 of the largest individuals of each species were collected from the lower limit of their zonation, as littorinids migrate down shore to spawn (Mak, 1996).

Littorinids were sexed by looking for a penis behind the right cephalic tentacles and separated in the laboratory. The shell length of the 30 largest females was measured under a stereo-microscope using an ocular micrometer. Trematode infection in *Nodilittorina trochoides* and *N. radiata* from various sites in Hong Kong was found to be very low (maximum 0.6%, Tang, 1995), hence the effect of parasitic castration and gigantism was not considered in the study. These 30 females were then maintained separately in plastic vials with 75 μm plankton netting on both ends. A 'drainage' system was established to stimulate spawning activity (Figure 1). The enclosed vials, with the female littorinids, were arranged in a row on an inclined PVC trough and irrigated by filtered sea water. No food or tidal stimulation were provided. Any eggs released by the females were collected by the plankton net. Vials were checked every one or two days and the experiment terminated when there was no further egg production. Egg capsules were preserved in 10% formalin and counted in a petri dish under a stereo-microscope. The entire sample was counted when the density was < 200, otherwise only a quarter of the petri dish was counted to estimate density. The percentage of females which spawned (= reproductively active) and the amount of eggs produced per fecund female were recorded for each species.

A second, more established method, the 'container' method (e.g. Struhsaker, 1966; Underwood & McFadyen, 1983), was also used to investigate spawning for littorinids from Big Wave Bay (August 1995) and Cape d'Aguilar (August–September 1995) during the maximum spawning period. The apparatus was the same as used to obtain egg capsules in a previous study (Mak, 1995), with 30 female littorinids kept in separate containers with static, filtered sea water. Spawning was checked at the same time intervals as the 'drainage' experiment.

A mark and recapture method was initiated in March 1995 to examine whether individual females can spawn more than once during the breeding season and therefore the maximum fecundity of each species can be calculated. Snails which had spawned were marked with a Pilot marker covered with a layer of Supa Glue. The littorinids were released back to the site from which they were originally col-

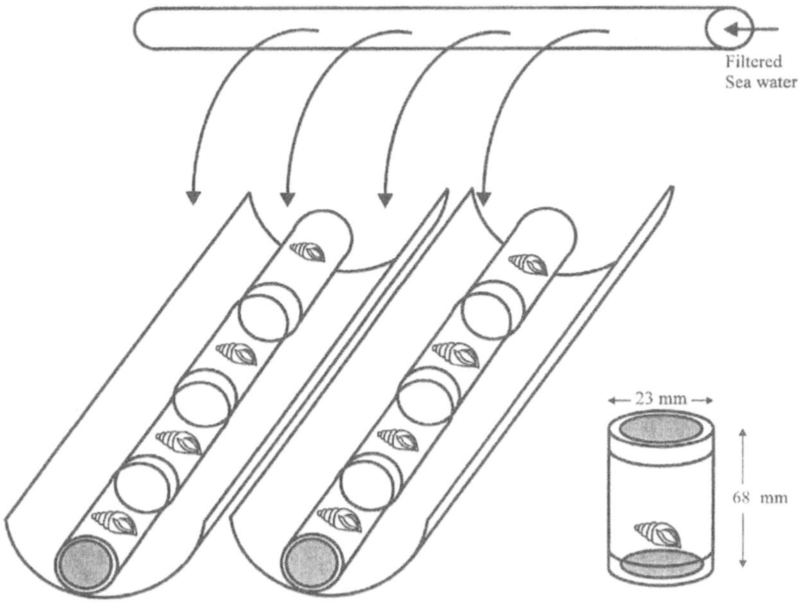

Figure 1. The 'drainage' spawning system. Experimental set up to maintain female littorinids in a running water system to stimulate egg release. The dimension of the individual vials with 75 μm plankton netting on both ends is shown.

lected. Marked littorinids were searched for at the next monthly sampling trip. In each spawning experiment after March 1995, therefore, both individuals which had previously spawned (recaptured individuals) and new individuals were used.

To investigate the relationship between littorinid size and fecundity, linear regressions were used (when $n > 5$) for each species and month. For samples from the peak spawning period (June–September 1995), two way ANOVAs were used to firstly investigate the monthly relationship between the number of eggs spawned between the different species at the two sites and secondly the effect of shell size on spawning intensity for the different sites (Big Wave Bay and Cape d'Aguilar) and species (*N. trochoides*, *N. radiata* and *N. vidua*). Egg counts and shell length measurements were $\log_{10}$ transformed prior to analysis in order to pass the assumptions of normality and homogeneity. Significant results were further separated using Student-Newman-Keuls (SNK) tests (Zar, 1996).

Results

The three littorinid species spawned in summer at both sites (Figures 2 and 3). At Big Wave Bay, spawning was between May–November (7 months) for *N. trochoides,* March–September (7 months) for *N. radiata*, whilst that of *N. vidua* was the longest between March–November (9 months). At Cape d'Aguilar, the spawning period of *N. trochoides* and *N. radiata* was shorter than at Big Wave Bay (7 months, April–October), although for *N. vidua* spawning extended up to December (9 months). The percentage of reproductively active females was lower for *N. trochoides* than *N. radiata* and *N. vidua* for both sites during the spawning season (Figure 2). The monthly average number of eggs spawned per female was, however, lowest for *Nodilittorina vidua* (Figure 3). All species showed a decrease in the percentage of spawning females and average fecundity during July for Big Wave Bay and August for Cape d'Aguilar, although peak spawning for *N. trochoides* was in August. Two peaks were recorded for *N. radiata* and *N. vidua* in June and August at Big Wave Bay (Figure 3). At Cape d'Aguilar, two peaks were also recorded for *N. trochoides* and *N. radiata,* during July and September, with a single peak for *N. vidua* in September. Reproductive output, however, varied enormously (< 200–10 000 eggs spawned per female per day).

Fecundity of *N. trochoides* did not vary with size at the two sites (Table 1) and only during August, at Big Wave Bay, did large *N. radiata* produce more eggs than smaller individuals. Large *N. vidua*, how-

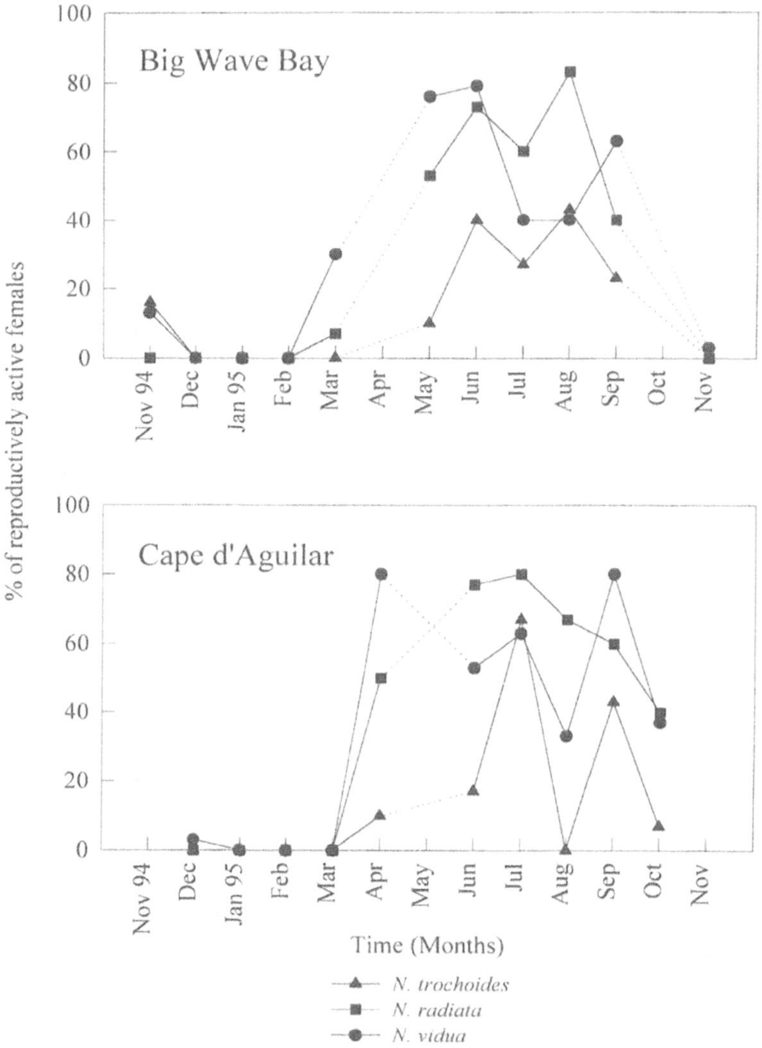

Figure 2. Nodilittorina trochoides, N. radiata and *N. vidua*. The percentage of reproductively active females per month at Big Wave Bay and Cape d'Aguilar. Dotted lines represent periods between missing monthly samples.

ever, produced more eggs at Big Wave Bay during September, and at Cape d'Aguilar in June and September, than smaller animals. Regressions on pooled samples showed a positive relationship between egg production and size for *N. vidua* at both Big Wave Bay ($r^2 = 0.125$, $p < 0.001$) and Cape d'Aguilar ($r^2 = 0.087$, $p = 0.002$) and for *N. radiata* at Cape d'Aguilar ($r^2 = 0.1$, $p < 0.001$). The relationship between the monthly fecundity and shell size was, however, not strongly positively correlated, as indicated by the low r^2 values. During the breeding season, littorinids from Cape d'Aguilar produced more egg capsules in July and September than those at Big Wave Bay (Table 2). There was no difference, however, in the monthly reproductive output between the three species, and only in September, did *N. radiata* from Big Wave Bay produce fewer egg capsules than the other species (Table 2). Reproductively active females from Cape d'Aguilar were always larger than those from Big Wave Bay and female *N. trochoides* were consistently larger than *N. radiata* and *N. vidua* from both sites (Table 3).

The yearly fecundity per female was estimated from the cumulative egg counts for individuals of each

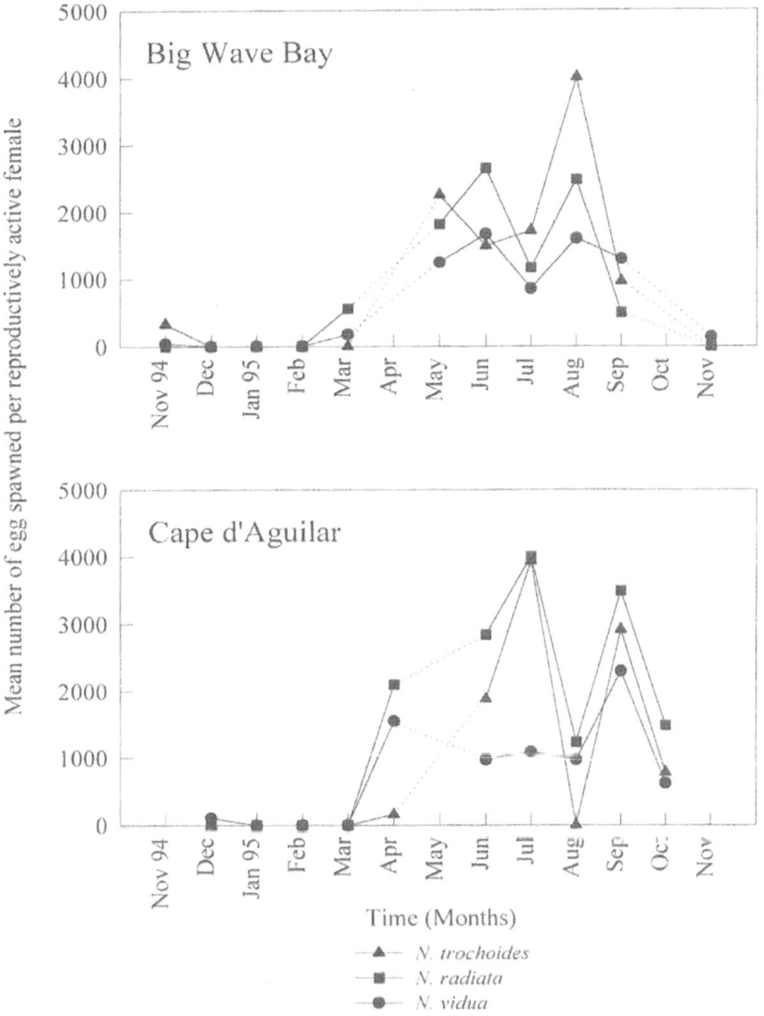

Figure 3. Nodilittorina trochoides, N. radiata and *N. vidua*. Mean number of eggs spawned per reproductively active female per month at Big Wave Bay and Cape d'Aguilar. Dotted lines represent periods between missing monthly samples. Error bars are omitted for clarity.

species at the two sites (Figure 4). The sequence for the three littorinids, from most to least fecund, was *N. trochoides* > *N. radiata* > *N. vidua* from Big Wave Bay whilst at Cape d'Aguilar the sequence was *N. radiata* > *N. trochoides* > *N. vidua*. The fecundity of *N. trochoides* and *N. vidua* was similar for Big Wave Bay and Cape d'Aguilar at ~ 10 000 eggs and 7500 eggs per female per year respectively. However, there was a pronounced difference between Big Wave Bay and Cape d'Aguilar for *N. radiata*, where an average of 9000 eggs spawned per female were recorded for Big Wave Bay and 15 000 eggs per female at Cape d'Aguilar. Spawning was initiated earlier at Big Wave Bay than Cape d'Aguilar and as a result the number of eggs spawned started to level off by September in Big Wave Bay (Figure 4).

Mark-recapture revealed that littorinids could spawn more than once within the spawning period. However, the recapture success was very low (< 50%), and only a few littorinids were able to be followed for more than two consecutive months. An individual of *N. trochoides* (shell length 9.45 mm) was able to spawn up to 8696 eggs in two consecutive samples in June and July. Even within one monthly spawning, 18 038 and 13 696 eggs were produced in August and July for Big Wave Bay and Cape d'Aguilar respec-

Table 1. N. trochoides, N. radiata and *N. vidua*. Monthly linear regressions of shell length against the number of eggs spawned for reproductively active females from Big Wave Bay and Cape d'Aguilar. Significant results ($p < 0.05$) are indicated in bold. (Abbreviation: / no fecund females recorded, < number of fecund females was less than 5 and not tested, shaded area: data not collected).

	Big Wave Bay					
	N. trochoides		*N. radiata*		*N. vidua*	
	r^2	*P*	r^2	*P*	r^2	*P*
Mar	/		<	0.002	0.9	
Apr						
May	<		0.041	0.454	0.074	0.209
Jun	0.215	0.128	0.052	0.307	0.132	0.126
Jul	0.156	0.333	0.030	0.460	0.223	0.121
Aug	0.078	0.354	**0.17**	**0.035**	0.183	0.166
Sep	0.370	0.147	0.003	0.860	**0.273**	**0.022**
Oct	/		/		<	

	Cape d'Aguilar					
	N. trochoides		*N. radiata*		*N. vidua*	
	r^2	*P*	r^2	*P*	r^2	*P*
Mar	/			/		
Apr	<		0.145	0.131	0.110	0.106
May						
Jun	<		0.128	0.094	**0.433**	**0.006**
Jul	>0.001	0.926	0.028	0.437	0.040	0.810
Aug	/		0.099	0.176	0.031	0.648
Sep	0.274	0.067	0.070	0.290	**0.224**	**0.019**
Oct	<		0.177	0.173	0.018	0.691

tively. *N. radiata* was able to spawn up to 7292 eggs (July) and 12 936 eggs (June) at Big Wave Bay and Cape d'Aguilar respectively. One individual followed from April–October produced ≥ 18 182 eggs within these 7 months. The maximum monthly reproductive output for *N. vidua* was less at 6124 eggs (Big Wave Bay, individual of 7.34 mm in shell length) and 8 271 eggs (Cape d'Aguilar individual of 8.28 mm in shell length) than the other species. A total of 10 371 eggs were recorded for one *N. vidua* individual (8.59 mm in shell length) for two consecutive sample periods in April and June.

There was a considerable difference in the number of littorinids which spawned as determined using the two methods (Table 4). The 'drainage' method was found to promote the spawning process more than the isolated 'container' method as only 13.3% of *Nodilittorina radiata* from Cape d'Aguilar were found to spawn in the container methods as compared to a maximum of 84% under the drainage system.

Discussion

In this study, two laboratory spawning systems were used to quantitatively enumerate egg capsule production. Previous methods have involved separating just female snails, or females with a male, in either a tube without water flow (Struhsaker, 1966; Borkowski, 1971; Berry & Chew, 1973; Grahame, 1975; Schmitt, 1979; Ohgaki, 1981; Underwood & McFadyen, 1983; Chase & Thomas, 1995) or a container with slight water movement (Grahame, 1975; Hughes & Roberts, 1980; Hughes & Answer, 1982). The 'container' method in this study underestimated egg production and egg release was enhanced with slight water movement (Pilkington, 1971), as seen in the 'drainage' system. Schmitt (1979) noted that agitating the spawning container, or changing the water, could induce spawning. An experimental system to record maximum egg capsule release should, therefore, be provided with a flowing water system.

There have been many laboratory studies investigating the lunar, tidal and diel cycle of littorinid spawning (Struhsaker, 1966; Berry & Chew, 1973; Grahame, 1975; Fish, 1972; Schmitt, 1979; Hughes & Roberts, 1980; Berry, 1986). The lunar cycle of *N. trochoides* and *N. vidua* has been shown to vary; *N. trochoides* spawned strictly during spring tides whilst *N. vidua* showed a less precise spawning synchronism with the tidal cycle (Berry, 1986). No lunar cycle was obtained by Pilkington (1971) and Borkowski (1971), however, in *Melarapha* species and Floridian littorinids respectively. In the present study, to account for the possible influence of the lunar cycle, spawning

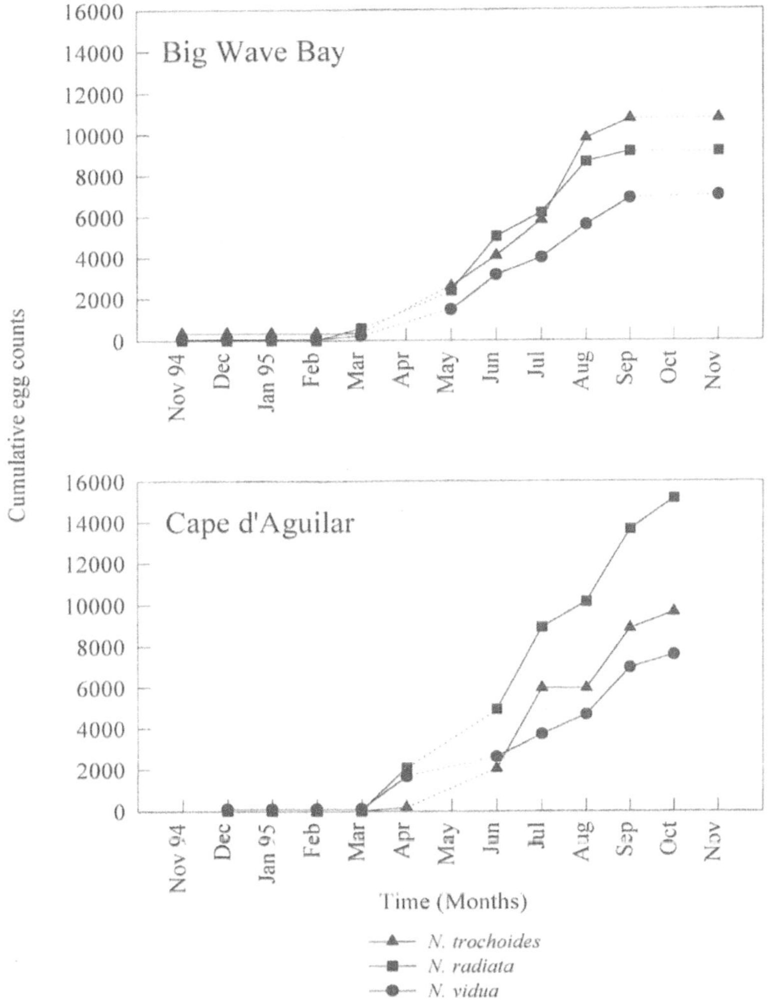

Figure 4. Nodilittorina trochoides, N. radiata and *N. vidua*. Cumulative monthly average egg capsule production per female for Big Wave Bay and Cape d'Aguilar. Dotted lines represent months not sampled.

tests were performed near to the full moon period and extended until no further eggs were produced by all the reproductive females. Tides were not reproduced nor was food provided during the whole process. This may underestimate egg production as reproductive activities are known to be affected by food supply (Berry, 1961; Spight & Emlen, 1979; Ross & Berry, 1991), although, no mortality was recorded.

The three *Nodilittorina* species spawned during summer, starting from March/May and ending by October/December. Peak egg counts were recorded between June–September. Previous studies recorded the spawning season for *N. vidua* as during the summer between June–September (Ohgaki, 1981, at Shirahama, Japan); July–August (Kojima, 1960, at Asamushi, Japan). For *N. radiata*, a similar summer spawning period was recorded from June–September (Ohgaki, 1981, at Shirahama, Japan) and between March–June (Berry, 1986, Malaysia). The spawning season of *N. trochoides* has been described between March–July, with no spawning during July–October, in Malaysia (Berry, 1986). The breeding season obtained in the present study was consistent with previous workers, despite the large longitudinal difference.

The timing of spawning has been related to seasonal environmental variables, such as critical temperatures (Fretter & Graham, 1962; Borkowski, 1971; Fish, 1972; Ohgaki, 1981) and salinity (Pilkington,

Table 2. N. trochoides, N. radiata and *N. vidua*. Monthly, two way ANOVAs and SNK tests to compare egg production against sites (Big Wave Bay, BWB; Cape d'Aguilar, CAG) and species (*N. trochoides*, NT; *N. radiata*, NR; *N. vidua*, NV). For the August sample, only two species (*N. radiata* and *N. vidua*) were tested, as no *N. trochoides* from Cape d'Aguilar spawned. Egg counts were $\log_{10}$ transformed. Significant results: $p < 0.05$.

Source	df	SS	MS	*F*	*p*
June					
Species	2	1.1635	0.5817	1.600	0.208
Site	1	0.6053	0.6053	1.665	0.200
Species*Site	2	0.2568	0.1284	0.353	0.703
Error	91	33.0883	0.3636		
July					
Species	2	2.0641	1.0320	2.438	0.093
Site	1	1.8412	1.8412	4.350	0.040
Species* Site	2	2.1046	1.0523	2.486	0.089
Error	95	40.2141	0.4233		
SNK					
Site	CAG > BWB				
August					
Species	1	0.3224	0.3224	0.696	0.407
Site	1	1.1858	1.1858	2.561	0.115
Species* Site	1	0.1121	0.1121	0.242	0.624
Error	62	28.8723	0.4630		
September					
Species	2	0.5823	0.2912	0.956	0.389
Site	1	8.5387	8.5387	28.030	< 0.001
Species* Site	2	2.7257	1.3628	4.474	0.014
Error	87	26.5024	0.3046		
SNK					
Site	CAG > BWB				
Species* Site	NR*CAG = NT*CAG = NV*CAG = NV*BWB = NT*BWB > NR*BWB				

1971). Floridian rocky shore littorinids, for example, reproduced in summer, when water temperature was > 25 °C (Borkowski, 1971). Low temperatures can restrict gonad development and copulating in *Littorina saxatilis* (Berry, 1961). In Hong Kong, water temperatures will reach 28 °C in summer which may trigger the littorinids to spawn. There is also an increase in the amount of phytoplankton in the water column during July and August in Hong Kong waters (Chan, 1995). As the three littorinids develop into planktotrophic veligers before settlement, the synchronised pattern of phytoplankton production and spawning may enhance larval growth as proposed for *Littorina acutispira* (Underwood & McFadyen, 1983).

There were, however, differences in breeding strategy between the three littorinids. *N. trochoides* spawned a greater number of eggs over a shorter period (7 months) although the number of reproductively active females was less as compared to the other species. This was a result of the low proportion of reproductively active females but the high average egg count per female, from both sites. In contrast, individuals of the lower shore *N. vidua*, spawned less per month, but more of the population were reproductively active and spawned for a longer duration (9 months).

Table 3. N. trochoides, N. radiata and *N. vidua.* Monthly, two way ANOVAs and SNK tests to compare shell length of reproductively active female littorinids against sites (Big Wave Bay, BWB; Cape d'Aguilar, CAG) and species (*N. trochoides*, NT; *N. radiata*, NR; *N. vidua*, NV). For the August sample, only two species (*N. radiata* and *N. vidua*) were tested, as no *N. trochoides* from Cape d'Aguilar spawned. Shell length measurements were $\log_{10}$ transformed. Significant results: $p < 0.05$.

Source	df	SS	MS	*F*	*p*
June					
Species	2	0.1071	0.0536	22.251	< 0.001
Site	1	0.1497	0.1497	62.180	< 0.001
Species*Site	2	0.0306	0.0153	6.348	0.003
Error	91	0.2191	0.0024		
SNK					
Species	NT > NR = NV				
Site	CAG > BWB				
Species*Site	NT*CAG = NT*BWB = NV*CAG = NR*CAG > NR*BWB = NV*BWB				
July					
Species	2	0.1613	0.0807	44.176	< 0.001
Site	1	0.2137	0.2137	117.10	< 0.001
Species*Site	2	0.0125	0.0063	3.427	0.037
Error	95	0.1734	0.0018		
SNK					
Species	NT > NR = NV				
Site	CAG > BWB				
Species*Site	NT*CAG > NT*BWB = NR*CAG = NV*CAG > NR*BWB = NV*BWB				
August					
Species	1	0.0111	0.0111	3.585	0.063
Site	1	0.0429	0.0429	13.80	< 0.001
Species*Site	1	7.997×10^{-8}	7.997×10^{-8}	2.574×10^{-5}	0.996
SNK					
Sites	CAG > BWB				
September					
Species	2	0.1094	0.0547	17.387	< 0.001
Site	1	0.1389	0.1389	44.180	< 0.001
Species*Site	2	0.0017	0.0009	0.272	0.762
Error	87	0.2736	0.0031		
SNK					
Species	NT > NV = NR				
Site	CAG > BWB				

Table 4. N. trochoides, N. radiata and *N. vidua*. Comparison of the percentage of reproductively active females (n = 30) for the two experimental methods to record spawning activity. NT = *N. trochoides*, NR = *N. radiata*, NV = *N. vidua*.

	Cape d'Aguilar				Big Wave Bay	
	August		September		August	
	Drainage	Container	Drainage	Container	Drainage	Container
NT	0 (0%)	1 (3.3%)	13 (43.3%)	0 (0%)	13 (43.3%)	1 (3.3%)
NR	20 (66.7%)	4 (13.3%)	18 (60%)	0 (0%)	25 (83.8%)	1 (3.3%)
NV	10 (33.3%)	1 (3.3%)	24 (80%)	1 (3.3%)	12 (40%)	0 (0%)

Spawning in *N. radiata* was intermediate between the other two species, with a high number of eggs per female but a short spawning period. These differences may be a result of the different tidal heights which these species inhabit. *N. trochoides*, living higher on the shore, may have a decreased opportunity for pairing and spawning, as these events occur during high tide or rough seas when the shore and littorinids will be wet by waves (Mak, 1996). To compensate for this, reproductively active females may spawn under any available conditions, but overall fewer females will spawn. The lower shore *N. vidua* is unlikely to be limited by opportunities to breed, and spawning activity, therefore, spawned less per individual at each high tide period. Overall fecundity was, however, higher for *N. trochoides* than *N. vidua* (Figure 4).

Species which live in more harsh conditions produce more, smaller, offspring (*r*-selected) as compared to species in more stable environments, which produce few, large, offspring (*K*-selected, Stearns, 1976). Hence, one may expect *N. trochoides*, living higher on the shore, to produce more eggs in one single breeding season. Since the egg size of the three species was identical (Mak, 1995), only the quantity of spawn can be varied to achieve an optimal strategy. The sequence of cumulative egg production (*N. trochoides* > *N. radiata* > *N. vidua*) in Big Wave Bay was consistent with this strategy but not at Cape d'Aguilar (*N. radiata* > *N. trochoides* > *N. vidua*, Figure 4). Differences in the overall annual fecundity between sites may be related to microalgal food supply; reproductive effort of individual *Thais* in different populations, for example, was found to vary with food supply (Spight & Emlen, 1976). Littorinids from the more sheltered site (Cape d'Aguilar), have a higher food supply (Mak, 1996), and may therefore, increase their fecundity as indicated by the cumulative number of eggs spawned by *N. radiata*.

There was a weak relationship between fecundity and shell size. A general trend has been proposed for an increase in reproductive effort with an increase in littorinid size (*Littorina nigrolineata*, Hughes, 1980; *Littorina* (= *Melarhaphe*) *neritoides*, Hughes & Roberts, 1980; *Littorina littorea*, Hughes & Answer, 1982). However, no r^2 values were stated in these studies and so it is difficult to judge the strength of the relationship with only the p value of the regression coefficient. When the data from this study were pooled low r^2 values suggest only a weak relationship between shell size and fecundity. No relationship between shell size and egg production was reported for the Hawaiian littorinid species, *Littorina pintado* and *Littorina picta* (Struhsaker, 1966).

Few studies have estimated the seasonal fecundity of littorinids and comparison between studies is confounded by different spawning estimation methods (Struhsaker, 1966; Borkowski, 1971; Grahame, 1973; Hughes & Roberts, 1980; Chase & Thomas, 1995). The fecundity of seven species of Floridian littorinids was estimated to range from 12 000 to 44 000 egg capsules per female (Borkowski, 1971). This estimation, however, was obtained by relating the biomass of egg capsules to the monthly weight loss of the female littorinids. The fecundity of the Hawaiian littorinids, *Littorina picta* and *L. pintado*, was estimated as 82 000 and 25 000 eggs respectively (Struhsaker, 1966), for *Littorina littorea* 109 000 (Grahame, 1973) and 113 600 (Chase & Thomas, 1995) eggs per season. In the present study, seasonal fecundity was obtained by the summation of the average monthly fecundity and by using mark-recapture techniques. Similar methods have been used by Hughes & Roberts (1980) in estimating the seasonal fecundity of *Littorina* (= *Melarhaphe*) *neritoides*. Female littorinids were kept in laboratory conditions for a year, and every fortnight, allowed to copulate with males. In the present experiment, females were released back to their original sites and recaptured after a month, which is a more natural situation than under laboratory conditions. Recapture success was, however, quite low, only one *N. radiata*, for example, was followed within the spawning period. The results were, however, comparable with 15 000 eggs being estimated by the laboratory experiment and 18 182 eggs from the mark-recapture method. Spawning was not recorded during April and May at Big Wave Bay and Cape d'Aguilar, and as a result, fecundity may be underestimated.

In Hong Kong, the three littorinids spawned in summer which may be triggered by warm temperatures or increased food availability in the plankton for the veligers. Species which inhabit different tidal heights may have different spawning strategies. There was no variation, however, between the sites of different exposure in spawning season, although littorinids from Cape d'Aguilar produced more eggs than Big Wave Bay.

Acknowledgements

I would like to thank Drs Gray A. Williams and Robin Kennish for improving the manuscript. This study represents part of a dissertation submitted for a Ph.D. degree at the University of Hong Kong. Attendance at the conference was funded by the Croucher Foundation, Hong Kong.

References

Ballantine, W. J., 1961. A biologically-defined exposure scale for the comparative description of rocky shores. Field Stud. 1: 1–19.

Berry, A. J., 1961. Some factors affecting the distribution of *Littorina saxatilis* (Olivi). J. anim. Ecol. 30: 27–45.

Berry, A. J., 1986. Semi-lunar and lunar spawning periodicity in some tropical littorinid gastropods. J. moll. Stud. 52: 144–149.

Berry, A. J. & E. Chew, 1973. Reproductive systems and cyclic release of eggs in *Littorina melanostoma* from Malayan mangrove swamps (Mollusca: Gastropoda). J. Zool., Lond. 171: 333–344.

Borkowski, T. V., 1971. Reproduction and reproductive periodicities of South Floridian Littorinidae (Gastropoda: Prosobranchia). Bull. mar. Sci. 21: 826–840.

Chan, L. C., 1995. The ecology of marine plankton in Tai Tam Bay, Hong Kong, with special reference to barnacle (Crustacea: Cirripedia) larvae. Ph.D. Thesis, Hong Kong, University of Hong Kong.

Chase, M. E. & M. L. H. Thomas, 1995. The effect of the rate and onset of temperature increase on spawning of the periwinkle, *Littorina littorea* (L.). J. exp. mar. Biol. Ecol. 186: 277–287.

Fish, D. J., 1972. The breeding cycle and growth of open coast and estuarine populations of *Littorina littorea*. J. mar. biol. Ass. U.K. 52: 1011–1019.

Fretter, V. & A. Graham, 1962. British Prosobranch Molluscs. Ray Society, London, 755 pp.

Gibson, D. G., 1964. Mating behavior in *Littorina planaxis*. Veliger 7: 134–139.

Goodwin, B. J., 1978. The growth and breeding cycle of *Littorina obtusata* (Gastropoda: Prosobranchiata) from Cardigan Bay. J. moll. Stud. 44: 231–242.

Grahame, J., 1973. Breeding energetics of *Littorina littorea* (L.) (Gastropoda: Prosobranchiata). J. anim. Ecol. 42: 391–404.

Grahame, J., 1975. Spawning in *Littorina littorea* (L.) (Gastropoda: Prosobranchiata). J. exp. mar. Biol. Ecol. 18: 185–196.

Hannaford Ellis, C. J., 1983. Patterns of reproduction in four *Littorina* species. J. moll. Stud. 49: 98–106.

Hughes, R. N. 1980. Population dynamics, growth and reproductive rates of *Littorina nigrolineata* Gray from a moderately sheltered locality in North Wales. J. exp. mar. Biol. Ecol. 44: 211–228.

Hughes, R. N. & P. Answer, 1982. Growth, spawning and trematode infection of *Littorina littorea* (L.) from an exposed shore in North Wales. J. moll. Stud. 48: 321–330.

Hughes, R. N. & D. J. Roberts, 1980. Growth and reproductive rates of *Littorina neritoides* (L.) in North Wales. J. mar. biol. Ass. U.K. 60: 591–599.

Kojima, Y., 1960. On the reproduction of periwinkles, Littorinidae, Gastropoda. Bull. mar. biol. station Asamushi 10: 117–120.

Langan-Cranford, K. M. & J. S. Pearse, 1995. Breeding experiments confirm species status of two morphologically similar gastropods (*Lacuna* spp.) in central California. J. exp. mar. Biol. Ecol. 186: 17–31.

Lasiak, T., 1987. Reproductive cycle of the supralittoral gastropod *Littorina kraussi* (Rosewater, 1970). S. Afr. J. Zool. 22: 170–172.

Mak, Y. M., 1995. Egg capsule morphology of five Hong Kong rocky shore littorinids. Hydrobiologia 309: 53–59.

Mak, Y. M., 1996. The ecology of the high-zoned littorinids, *Nodilittorina trochoides*, *N. radiata* and *N. vidua* on rocky shores in Hong Kong. Ph.D. Thesis, University of Hong Kong, Hong Kong.

McQuaid, C. D., 1981. Population dynamics of *Littorina africana knysnaensis* (Philippi) on an exposed rocky shore. J. exp. mar. Biol. Ecol. 54: 65–76.

Mileikovsky, S. A., 1971. Types of larval development in marine bottom invertebrates, their distribution and ecological significance: a re-evaluation. Mar. Biol. 10: 193–213.

Murray, T., 1979. Evidence for an additional *Littorina* species and a summary of the reproductive biology of *Littorina* from California. Veliger 21: 469–474.

Ohgaki, S., 1981. Spawning activity in *Nodilittorina exigua* and *Peasiella roepstorffiana* (Littorinidae, Gastropoda). Publ. Seto Mar. Biol. Lab. 26: 437–446.

Palant, B. & L. Fishelson, 1968. *Littorina punctata* (Gmelin) and *Littorina neritoides* (L.) (Mollusca, Gastropoda) from Israel: ecology and annual cycle of genital system. Israel J. Zool. 17: 145–160.

Pilkington, M. C., 1971. Eggs, larvae, and spawning in *Melarapha cincta* (Quoy & Gaimard) and *M. oliveri* Finlay (Littorinidae, Gastropoda). Aust. J. mar. Freshwat. Res. 22: 79–90.

Reid, D. G., 1992. The gastropod family Littorinidae in Hong Kong. In B. Morton (ed), The Marine Flora and Fauna of Hong Kong and Southern China III. Proceedings of the Fourth International Marine Biological Workshop: The Marine Flora and Fauna of Hong Kong and Southern China, Hong Kong. Hong Kong University Press, Hong Kong: 187–210.

Ross, B. & A. J. Berry, 1991. Annual and lunar reproductive cycles in *Littorina saxatilis* (Olivi) and differences between breeding in the marine Firth of Forth and the Forth estuary. J. moll. Stud. 57: 347–358.

Schmitt, R. J., 1979. Mechanics and timing of egg capsule release by the littoral fringe periwinkle *Littorina planaxis* (Gastropoda: Prosobranchia). Mar. Biol. 50: 359–366.

Spight, T. M. & J. Emlen, 1979. Clutch sizes of two marine snails with changing food supply. Ecology 57: 1162–1178.

Stearns, S. C., 1976. Life history tactics: a review of the ideas. Q. Rev. Biol. 57: 3–47.

Struhsaker, J. W., 1966. Breeding, spawning, spawning periodicity and early development in the Hawaiian *Littorina*: *L. pintado* (Wood), *L. picta* Philippi and *L. scabra* (Linne). Proc. malac. Soc. Lond. 37: 137–166.

Tang, C. T., 1995. Spatial variation in larval trematode infections of populations of *Nodilittorina trochoides* and *Nodilittorina radiata* (Gastropoda: Littorinidae) from Hong Kong. Asian mar. Biol. 12: 19–26.

Underwood, A. J., 1974. The reproductive cycles and geographical distribution of some common Eastern Australian prosobranchs (Mollusca: Gastropoda). Aust. J. mar. Freshwat. Res. 25: 63–88.

Underwood, A. J., 1979. The ecology of intertidal gastropods. Adv. Mar. Biol. 16: 111–210.

Underwood, A. J. & K. E. McFadyen, 1983. Ecology of the intertidal snail *Littorina acutispira* Smith. J. exp. mar. Biol. Ecol. 66: 169–197.

Zar, J. H., 1996. Biostatistical Analysis. 3rd edition, Prentice Hall, New Jersey, 662 pp.

Hydrobiologia **378:** 161–185, 1998.
R. M. O'Riordan, G. M. Burnell, M. S. Davies & N. F. Ramsay (eds), Aspects of Littorinid Biology.

What to do when there is nothing to do: the ecology of Jamaican intertidal Littorinidae (Gastropoda: Prosobranchia) in repose

R. C. Lang, J. C. Britton* & T. Metz
Department of Biology, Texas Christian University, Fort Worth, TX 76129, U.S.A.
(author for correspondence)*

Key words: Littorinidae, *Nodilittorina*, *Tectarius*, *Cenchritis*, ecology, supralittoral fringe, intertidal, thermoregulation, Jamaica

Abstract

The littorine gastropods *Cenchritis muricatus*, *Nodilittorina mespillum*, *N. angustior*, *N. dilatata*, *N. riisei*, *N. ziczac* and *Tectarius antonii* from a natural limestone platform on the north coast of Jamaica were evaluated according to several ecological parameters in an effort to define their respective niches on this shore. Distributions along transects perpendicular to the shore demonstrated that distance from the sea is one, albeit not always consistent, ecological criterion which serves to delineate these species. *N. riisei* generally occupies that region of the shore frequently wetted by wave splash, *N. mespillum* is usually found within frequently refreshed tidepools and *C. muricatus* always occupies the highest regions of the shore most distant from the sea. There is, however, considerable overlap between these and other species and, along much of the rock platform, *N. angustior*, *N. dilatata*, *T. antonii* and *C. muricatus* are sympatric. *N. dilatata* has the broadest range on this low profile limestone platform. A comparison of body temperatures and substratum temperatures taken from adjacent rock surfaces revealed that the body temperatures of *T. antonii* generally reflect that of the ambient environment. On the other hand, *C. muricatus*, *N. dilatata*, *N. riisei* and *N. angustior* seem to thermoregulate, maintaining body temperatures above ambient at temperatures below 29 °C and below ambient at temperatures above 33 °C. Data were also compiled for most species with respect to resting posture (whether hanging suspended or resting aperture-down), resting site (whether within the shelter of a crevice or upon an exposed rock surface), shell orientation and the relative frequency (presence or absence) of a mucous holdfast. Size-frequency analyses indicate that *N. riisei* and *N. angustior* re-populate the shore from the meroplankton at least sometime between January and March.

Introduction

The supralittoral fringe is distinguished from both the eulittoral and sublittoral as being an especially harsh environment, physically rather than biologically regulated, and usually occupied by only a few hardy species. High tide marks the boundary between not only the eulittoral and supralittoral fringe, but also between two primary groups of intertidal molluscs delineated on the basis of morphological, physiological and behavioural responses to water, either its availability or absence, and thermal stress (Newell, 1979; McMahon, 1990; McMahon & Britton, 1991). The shape of a shell or the size of an aperture may impact how effectively a species deals with thermal stress or desiccation (Vermeij, 1973; Britton, 1995). Thus, eulittoral and supralittoral fringe species rely on different strategies to deal with thermal stress and emersion in their respective environments (McMahon, 1990). Eulittoral molluscs employ evaporative cooling to avoid thermal stress, since they can afford modest body water loss as predictable rehydration will occur with each tidal cycle. The supralittoral fringe molluscs, which are subjected to prolonged emersion, can neither rely on nor afford evaporative cooling. They must conserve water, so evaporative cooling must be avoided to reduce water loss. These species must also be able to tolerate greater temperature fluctuations than eulittoral species. In order to cope with these constraints, all supralittoral fringe molluscs share a common, albeit frequently ignored, behavioural characteristic. They pass a significant portion of their life, and perhaps the majority of their being, in repose.

Britton (1992) described evaporative water loss, behaviour during emersion, and upper thermal tolerance limits in Jamaica in an effort to establish physiological and behavioural differences between these species. This study is a continuation of these efforts, focussing upon ecological parameters which characterise the Jamaican supralittoral fringe littorinids in repose. We endeavour to determine which, if any, of the several ecological parameters studied might be useful in differentiating the ecological niches occupied by these species.

Methods

We determined the density, distribution and some aspects of the resting ecology of the Littorinidae occupying a natural limestone platform on the northern coast of Jamaica near the Hofstra University Marine Laboratory at Priory, St. Ann's Parish. This unvegetated, micro-karsted, north-facing limestone shore rises as much as 2 m above sea level and ranges from 18 to 30 m wide for a distance of more than 200 m. High winds, especially in winter, sometimes submerge parts of the platform for a few hours or days, but normally most of it lies beyond the influence of either tidal flux or wave splash. Species density and distribution were determined along five randomly selected transects lying approximately perpendicular to the shore and extending from near mean sea level, indicated by the upper limit of dense macrophyte cover, to the maritime fringe, indicated especially by the presence of buttonwood trees, *Conocarpus erectus*. One transect was made in January 1994 and four in March 1996. The January 1994 transect data were collected several days following a winter storm with seawater incursion upon the platform and likely reflect its impact. There was no indication of recent platform flooding prior to collecting the March 1996 transect data. For each transect, distance from the sea was determined using a metric tape and elevation was estimated by eye, a string level and a metre stick. Single 0.0625 m^2 quadrats were sampled at 0.5 m intervals for shore-dwelling gastropods and 0.25 m intervals for surface irregularity and moisture along the January 1994 transect and approximately every 2 m for all data along the March 1996 transects. All gastropods within each quadrat were identified to species, counted and measured for shell height. The surface irregularity of each quadrat was estimated visually on a scale from 1 to 10, the former being uniformly smooth and the latter being extremely pitted, grooved or otherwise highly irregular. Relative moisture was estimated for each quadrat and ranked as either dry, mostly dry, mostly wet, or wet (e.g. a tidepool). These categories were converted to the rank numbers 1, 3, 5, and 7, respectively, for statistical analysis.

In addition to density and distribution data, several ecological observations were determined for 400 *C. muricatus*, 158 *N. angustior*, 312 *N. dilatata*, 100 *N. riisei*, and 257 *T. antonii* in either January or July 1994 and March 1996. Size-frequency histograms for each species were constructed representing each observation period. Additional individuals of *C. muricatus*, *N. dilatata* and *T. antonii* were examined during a 12-h temperature profile of substratum and littorinid body temperature. Not all individuals of each species were used in all analyses resulting from these data, usually because a particular measurement was not or could not be made for an individual and sometimes for other reasons. The number of individuals examined for all comparisons is always given with the results.

Ecological observations included resting site (whether within the shelter of a crevice or upon an exposed rock surface), resting posture (whether hanging suspended from a ledge or resting upon the substratum aperture-down), presence or absence of a holdfast, orientation (the compass direction toward which most of the body is exposed) and body tissue temperatures and substratum temperatures. Most observations were made during the day between 08:00 and 17:45. A few observations, taken during the 12-hour temperature profile, 5 July 1994 (see below), were made as early as 06:00 and as late as 18:00, but these were used only with respect to the temperature analysis. Resting site, resting posture, shell orientation and substratum temperature observations were made *in situ* prior to disturbing each specimen. Then the specimen was removed from the substratum and quickly assessed for the presence or absence of a mucous holdfast and body temperature. Temperature measurements were made using tele-thermometers (YSI Model 43TD or Omega Model HH21) with appropriate probes fitted to determine either rock surface temperatures or tissue temperatures. The surface temperatures were taken using a surface thermocouple placed adjacent to the snail and shielded from the sun. Body temperatures were

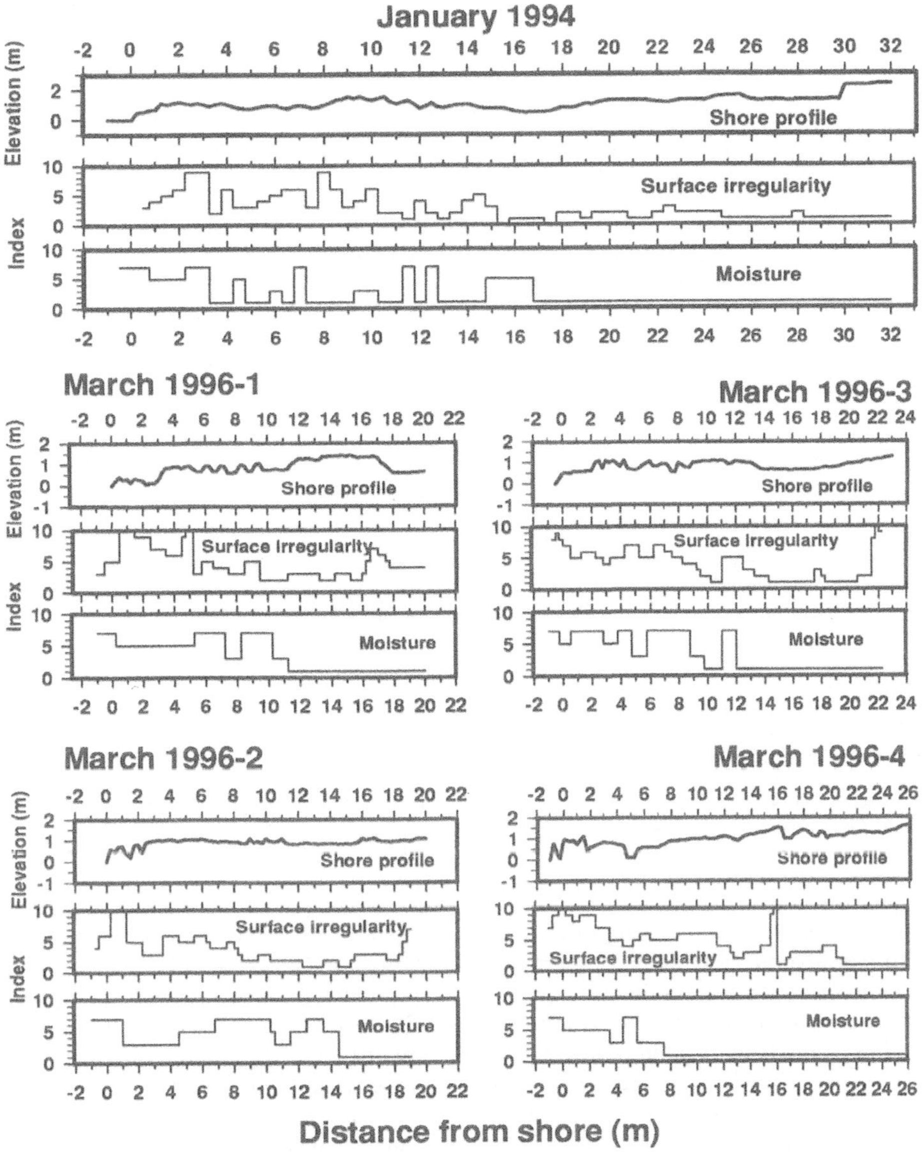

Figure 1. Physical characteristics of five transects along a coastal limestone platform near the Hofstra Marine Laboratory, Priory, Jamaica.

taken with a probe inserted between operculum and the soft body parts.

In July 1994, a substratum and body temperature profile was compiled, consisting of observations taken at 15 min intervals from 06:00 to 18:00 h at four locations on the shore. Site 1 was a small, shallow (< 1 cm depth) tidepool approximately 3.4 m from the sea. Site 2 was a large micro-karsted boulder 5.5 m from the sea. Site 3 was a very flat smooth bedrock surface 5.4 m from the sea. Site 4 was also a large micro-karsted boulder 15.3 meters from the sea. Substratum temperatures were determined for both an exposed rock surface and within a crevice at each 15 min interval. Tissue temperatures of individuals of either *C. muricatus*, *N. dilatata* or *T. antonii* were determined at each site at each sampling interval. Different individuals were used for every body temperature measurement, i.e. no individual was measured more than once during the day. Whenever possible, one individual was selected from within a crevice the other

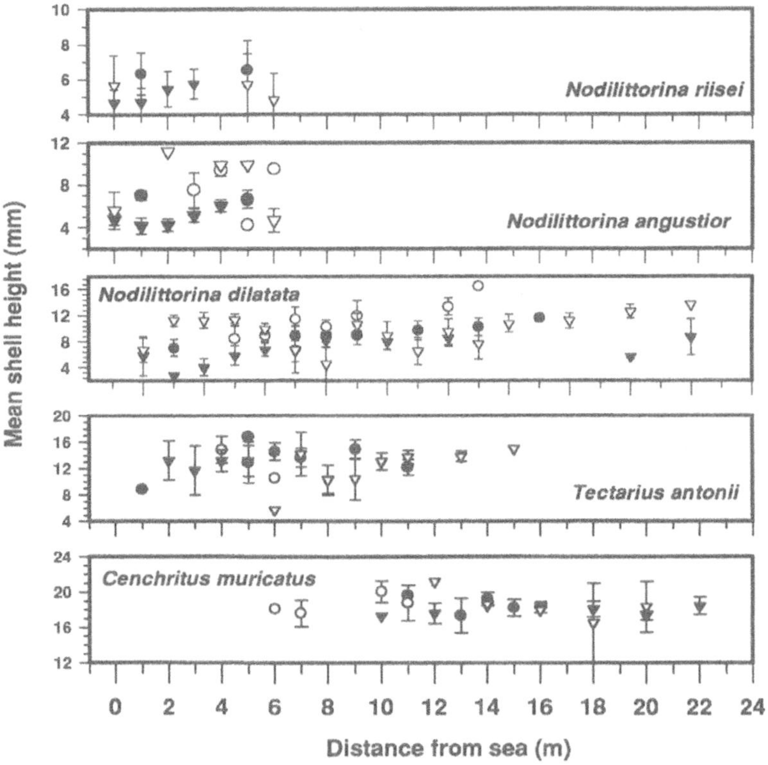

Figure 2. Mean size (shell height) for five species of Littorinidae collected along four (1996) transects from the north coast of Jamaica according to distance from shore. Error bars indicate standard deviations. Transect 1, filled circles; Transect 2, open circles; Transect 3, filled triangles; Transect 4, open triangles.

from an unsheltered site. To complete the profile, it was necessary to use several individuals of as many species as occurred within a 0.5 m radius of the site. Results of the 12 h temperature profile were examined as a contiguous record and then body temperature data collected during this part of the study were combined with those determined in either January or July 1994 and March 1996 to examine the relationship between littorinid body temperature and that of the substratum upon which they reside.

Results

Transects

Physical characteristics of the five transects examined during this study are presented in Figure 1. Generally, the limestone platform rises abruptly to an elevation of approximately 1 m above sea level within 2–3 m of the sea. The one exception to this pattern was the 1996–1 transect, where a small islet was separated from the main portion of the platform by almost 3 m, with the region between the two being constantly awash. The surface irregularity of all transects was generally greater nearest the sea, with surfaces tending to become smoother and less pitted along the landward half of platform. Similarly, surface moisture, including the standing water of tidepools, was greatest along the seaward half of all transects. Platform rocks along the landward third to half of the platform were mostly found to be dry and noticeably warm during midday.

Thirteen species of Mollusca were found within transect quadrats. The Littorinidae were well represented, including *Cenchritis muricatus* (Linnaeus, 1958), *Nodilittorina angustior* (Mörch, 1876), *N. dilatata* (d'Orbigny, 1842), *N. mespillum* (Mühlfeld, 1824), *N. riisei* (Mörch, 1876), *N. ziczac* (Gmelin, 1791) and *Tectarius antonii* (Philippi, 1846). Other molluscs included the gastropods *Planaxis lineatus* (da Costa, 1778), *Batillaria minima* (Gmelin, 1791), *Nerita versicolor* Gmelin, 1791, *N. peloronta* Lin-

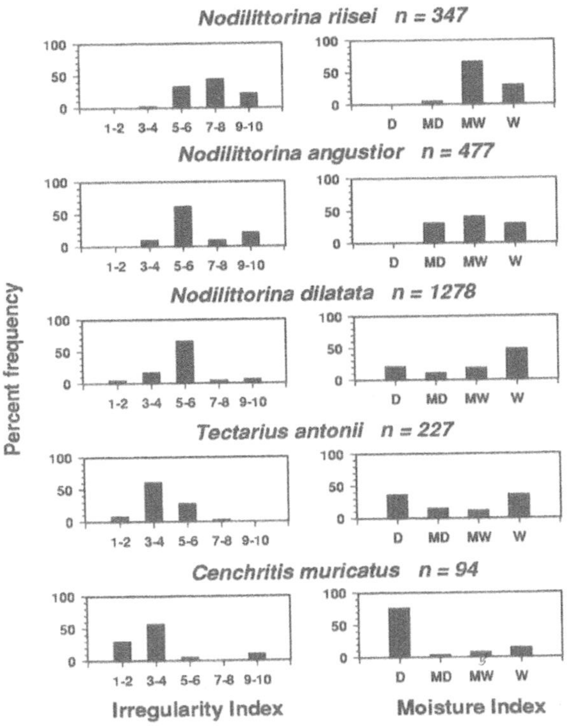

Figure 3. Percent-frequency of five species of Littorinidae from the north coast of Jamaica according to surface irregularity and substratum moisture. The surface irregularity index is a subjective assessment of surface texture, ranging from smooth (1) to highly irregular (10). The moisture index is another subjective assessment with four criteria: dry (D), mostly dry (MD), mostly wet (MW) and wet (W), each defined in Table 2.

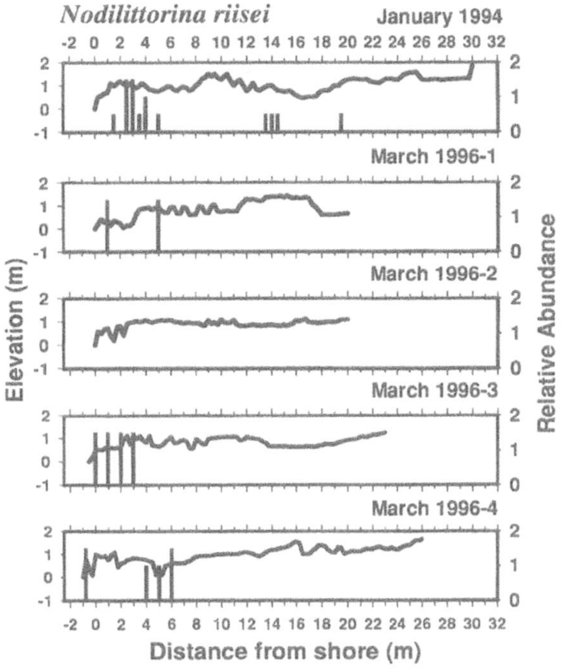

Figure 4. The distribution of *N. riisei* from five transects along a coastal limestone platform near the Hofstra Marine Laboratory, Priory, Jamaica. The elevation profile of the transect is represented by a continuous line, with the elevation scale indicated on the left side of each profile. The relative abundance of *N. riisei* is indicated by vertical bars as follows: 0.5, fewer than 3 individuals within a 0.0625 m^2 quadrat; 1.0, between 3 and 12 individuals within a 0.0625 m^2 quadrat; 1.5, more than 12 individuals within a 0.0625 m^2 quadrat. The absence of vertical bars indicates the species was not present in the transect. The relative abundance scale is on the right side of each transect.

naeus, 1758 and *Puperita pupa* (Linnaeus, 1767) and the polyplacophoran *Acanthopleura granulata* (Gmelin, 1791). Density and spatial distribution were determined only for the littorinids. The mean sizes of the five most abundant littorinid species according to their distance from the sea are presented in Figure 2 and their distribution with respect to surface irregularity and moisture are presented in Figure 3.

Three littorinids occurred most often on the seaward face of the platform. *N. riisei* was restricted to locations generally within 6 m of the sea (Figure 4 and Table 1). It favoured moist sites frequently wetted by wave splash (Table 2 and Figure 3) with moderate to considerable surface irregularity (Table 3 and Figure 3). In January 1994 several individuals were recovered on the central portion of the platform between 13–15 m from the sea, but they were probably washed there as seawater from a winter storm flooded the platform. In March 1996, no *N. riisei* were found at distances greater than 6 m from the sea. *N. mespillum* always appeared from 1–7 m from the sea (Figure 5 and Table 1), either submerged or on the wetted perimeter of tidepools (Table 2 and Figure 3) and on relatively smooth substrata (Table 3 and Figure 3). *N. ziczac* was uncommon on the platform but, when present, generally occupied the same part of the otherwise dominated by *N. riisei* (Figure 6 and Tables 1, 2 and 3).

Three other littorinids occupied the central portion of the platform, each having a slightly different distribution. *N. angustior* occurred on the platform 0–16 m from the sea, but was most abundant within the first 8 m (Figure 7 and Table 1), preferring moderately irregular surfaces (Table 3 and Figure 3). Individuals were found on the platform as much as 16 m from the sea in both the transects and elsewhere on this shore. About 50% of the *N. angustior* observed along

Table 1. Frequencies and percent frequencies of each species collected from transect quadrats in January 1994 and March 1996, ranked according to their distance from the sea. Zeros indicate that the species was not present in any quadrat of the distance class. *N. mespillum* was not collected in the 1996 transects, nor were size classes of *N. dilatata* differentiated

	Species	Distance from the sea (m) 1–5	6–10	11–15	16–20	21–25	26–30	Total
	Nodilittorina mespillum	181	0	0	0	0	0	181
		100%						
	Nodilittorina riisei	102	0	6	1	0	0	109
		93.6%		5.5%	0.9%			
	Nodilittorina angustior	31	5	8	0	0	0	44
		70.5%	11.4%	18.2%				
	Nodilittorina ziczac	0	0	0	0	0	0	0
1994	*Nodilittorina dilatata*	123	109	64	39	1	0	336
		36.6%	32.4%	19.0%	11.6%	0.3%		
Transect	*N. d.* < 8 mm	55	22	16	0	0	0	93
		59.1%	23.7%	17.2%				
	N. d. > 8 mm	68	87	48	39	1	0	243
		28.0%	35.8%	19.8%	16.0%	0.4%		
	Tectarius antonii	30	41	3	1	4	0	79
		38.0%	51.9%	5.8%	1.2%	5.1%		
	Cenchritis muricatus	0	5	14	3	48	24	94
			5.3%	14.9%	3.2%	51.1%	25.5%	
	Nodilittorina riisei	331	18	0	0	0	0	349
		94.8%	5.2%					
	Nodilittorina angustior	540	147	0	0	0	0	687
		78.6%	21.4%					
1996	*Nodilittorina ziczac*	32	2	0	0	0	0	34
		94.1%	5.9%					
Transect	*Nodilittorina dilatata*	676	404	191	9	0	0	1280
		52.8%	31.6%	14.9%	0.7%			
	Tectarius antonii	89	88	49	1	0	0	227
		39.2%	38.8%	21.6%	0.4%			
	Cenchritis muricatus	0	14	32	32	16	0	94

the transects in 1994 were from relatively smooth, dry substrata, but more than 75% of those encountered in 1996 were from quadrats with at least moderate surface moisture and irregularity (Tables 2 and 3 and Figure 3).

N. dilatata was more broadly distributed than any other littorinid, extending from the sea to as much as 24 m landward (Figure 8 and Table 1). This species seems to have a broad tolerance for different substratum moisture conditions (Tables 2 and 3 and Figure 3). In January 1994, individuals were distributed further across the platform than those sampled in March 1996 (Table 1). As with *N. angustior*, there were also more individuals of *N. dilatata* found on smoother, drier substrata in January 1994 than in March 1996, apparently a consequence following platform flooding (see discussion). In January 1994, individuals of *N. dilatata* smaller than 8 mm occupied the shore from sea level to 7 m from the sea while those larger than 8 mm ranged from approximately 1.5 to 24 m from the sea (Table 1). The highest density of the larger *N. dilatata* occurred between about 4 and 15 m from the sea. In March 1996, Transects 1, 3 and 4 also displayed a tendency for body size of *N. dilatata* to increase with increasing distance from the sea (Figure 2).

T. antonii also inhabited the central platform, generally between 2 to 16 m from the sea but a few individuals were found up to 24 m in January 1994

Table 2. Frequencies and percent-frequencies of each species collected from transect quadrats in January 1994 and March 1996, ranked according to the moisture condition of the quadrat. Dry indicates the entire surface of the quadrat was devoid of water. Mostly dry (M. dry) indicates less than 25% of the surface of the quadrat was damp or wet. Mostly wet (M. wet) indicates that at least 75% of the quadrat was wet but with little or no standing or pooled water. Wet indicates that at least 50% of the quadrat contained standing water, as in a tidepool. Zeros indicate that the species was not represented in the moisture class, dashes indicate that the species was not evaluated or the categories were not differentiated.

	1994 Transect					1996 Transect				
Species	Dry	M. Dry	M. Wet	Wet	Tot.	Dry	M Dry	M. Wet	Wet	Tot.
N.	0	0	6	175	181	–	–	–	–	–
mespillum			3.3%	96.7%						
N.	16	0	30	63	109	1	17	205	126	349
riisei	14.7%		27.5%	57.8%		0.3%	4.9%	58.7%	36.1%	
N.	21	0	11	12	44	1	145	159	382	687
angustior	47.7%		25.0%	27.3%		0.1%	21.1%	23.1%	55.6%	
N.	0	0	0	0	0	1	2	17	14	34
ziczac						2.9%	5.9%	50.0%	41.2%	
N.	180	37	71	48	336	268	109	234	669	128
dilatata	53.6%	11.0%	21.1%	14.3%		20.9%	8.5%	18.3%	52.3%	0
N. d.	35	14	25	19	93	–	–	–	–	–
< 8 mm	37.6%	15.1%	26.9%	20.4%						
N. d.	145	23	46	29	243	–	–	–	–	–
> 8 mm	60.0%	9.5%	18.9%	11.9%						
T.	58	5	4	12	79	81	26	37	83	227
antonii	73.4%	6.3%	5.1%	15.2%		35.7%	11.4%	16.3%	36.6%	
C.	83	3	0	8	94	72	8	0	14	94
muricatus	88.3%	3.2%		8.5%		76.6%	8.5%		14.9%	

(Figure 9 and Table 1). This species tolerates a broad range of substratum moisture conditions, similar to that of *N. dilatata* (Table 2 and Figure 3), and also prefers moderately irregular surfaces (Table 3 and Figure 3).

C. muricatus always occupied the higher, drier and smoother parts of the platform (Figures 3 and 10 and Tables 1, 2 and 3). It was reliably found both along the landward margin and elevated central portions of the rock platform. The latter is most evident within the January 1994 transect, where *C. muricatus* occurred in two patches from about 9 to 15 m and from 20 to 28 m from the sea.

Size

Shell height was determined during the course of this study for 3674 Littorinidae belonging to five species, *N. riisei*, *N. angustior*, *N. dilatata*, *T. antonii* and *C. muricatus*. Size-frequency histograms for each are presented in Figure 11.

N. mespillum is the smallest littorine on this shore, but its size was not quantified, largely because, in its favoured habitat (tidepools), it is rarely in repose during the day. Among the remaining littorines, *N. riisei*, with a mean shell height of 6.41 mm ($n = 447$; $SD = 2.46$), is the next smallest species. The mean size of *N. angustior* is only slightly larger, 6.55 mm ($n = 634$; $SD = 2.93$), although many individuals attain a shell height over 11 mm. In contrast, although only 39 *N. ziczac* were measured during this study, it was decidedly larger than the other littorinids occupying rocks near the front of the shore, with a mean shell height of 10.4 mm ($SD = 1.84$). The January 1994 and March 1996 size-frequency histograms for *N. riisei* and *N. angustior* (Figure 11) demonstrated a bimodal distribution of sizes. In January, almost 70% of *N. riisei* were near maximum size, 12–13 mm, and about 60% of *N. angustior* were between 11–16 mm. In March, more than 70% of the population of *N. riisei* was between 6–8 mm and about 70% of the population of *N. angustior* was between 5–8 mm. By July,

Table 3. Frequencies and percent frequencies of each species collected from transect quadrats in January 1994 and March 1996, ranked according to the surface irregularity of the quadrat. Rock surfaces were ranked on a scale of 1 to 10 with a larger number indicating increasing irregularity. Those ranked 1–3 were relatively smooth, those ranked 4–7 were moderately pitted, grooved, elevated and/or depressed, especially by micro-karsting. Zeros indicate that the species was not represented in the quadrats of specified irregularity classes; dashes indicate that the species was not evaluated or the categories were not differentiated

	1994 Transect				1996 Transect			
Species	(1–3)	(4–7)	(8–10)	Total	(1–3)	(4–7)	(8–10)	Total
Nodilittorina mespillum	115	56	0	181	–	–	–	–
	63.5%	36.5%						
Nodilittorina riisei	12	34	63	109	0	119	230	349
	11.0%	31.2%	57.8%			34.1%	65.9%	
Nodilittorina angustior	22	12	10	44	5	539	143	687
	50.0%	27.3%	22.7%		0.7%	78.5%	20.8%	
Nodilittorina ziczac	0	0	0	0	2	10	22	34
					5.9%	29.4%	64.7%	
Nodilittorina dilatata	155	149	32	336	105	1046	129	1280
	46.1%	44.3%	9.5%		8.2%	81.7%	10.1%	
N. d. < 8 mm	22	57	14	93	–	–	–	–
	23.6%	61.3%	15.1%					
N. d. > 8 mm	133	92	18	243	–	–	–	–
	54.7%	37.9%	7.4%					
Tectarius antonii	23	39	17	79	18	202	7	227
	29.1%	49.4%	21.5%		7.9%	89.0%	3.1%	
Cenchritis muricatus	89	5	0	94	58	27	9	94
	94.7%	5.3%			61.7%	28.7%	9.6%	

about 50% of the population of *N. riisei* was between 9–10 mm, but there remained numerous smaller individuals from 5–7 mm. In contrast, there was only a modest number of *N. angustior* between 4–6 mm in July, but two size-frequency peaks at 7–8 mm and 10–11 mm.

N. dilatata was represented by individuals ranging in size from 3–19 mm in both 1994 and 1996, with the mean size for all being 8.54 mm ($n=1584$; SD = 3.06). Individuals of *N. dilatata* less than 8 mm are well represented in all transects, especially near the sea. The relative frequency of smaller individuals (< 8 mm) seems larger in 1996 than in either January or July 1994 (Figure 11), but this is likely the result of more intense quantitative sampling in 1996, for photographic records from 1994 show small *N. dilatata* to be at least locally abundant. Expressed another way, smaller *N. dilatata* were well represented in each of the sampled months (January, March or July), in contrast to the relative paucity of smaller *T. antonii* and, especially, *C. muricatus* at the same time (see below).

The mean size for all *T. antonii* measured from this shore was 13.5 mm. In January and July 1994, shell heights of *T. antonii* ranged mostly from 11–18 mm. No individuals less than 10 mm of this species were observed in 1994. In March 1996, the size-frequencies of *T. antonii* were similar to those observed in 1994, except there were more smaller individuals, with 8.4% of the 1996 population ranging in height from 3–10 mm.

C. muricatus was the largest littorinid on this shore (Figure 11). The mean size for all *C. muricatus* measured during this study was 18.3 mm ($n=490$; SD = 1.99). Shell heights less than 14 mm were uncommon for *C. muricatus* in both 1994 and 1996 and individuals less than 9 mm were never observed or recorded. The mean size of *C. muricatus* was 18.1 mm ($n=275$; SD = 1.96) in January 1994, 18.3 mm ($n=94$; SD = 2.05) in March 1996 and 18.6 mm ($n=121$; SD = 2.00) in July 1994.

Resting site

In January and July 1994, a total of 100 *N. riisei*, 157 *N. angustior*, 300 *N. dilatata*, 251 *T. antonii* and

Table 4. Comparisons of individuals in repose either resting on exposed rock surfaces or found within rock crevices in January and July 1994. MSL = mean shell length, N = number, SD = standard deviation. Non-parametric comparisons were made when normality tests failed. Comparisons with statistically significant differences ($\alpha \leqslant 0.05$) are indicated by probability (p) values in bold

Species	Variable	January 1994		July 1994	
		Exposed	Crevice	Exposed	Crevice
	MSL (mm)	17.9	18.6	18.4	18.8
	N	109	154	65	56
C. muricatus	SD	2.13	1.86	2.17	1.79
	Size comparison	Mann Whitney Rank Sums Test $U = 13812.5$ $p = 0.34$		Mann Whitney Rank Sums Test $U = 3527.5$ $p = 0.56$	
	Location comparison	z-test $z = 2.624$ **p=0.009***		z-test $z = 0.629$ $p = 0.529$	
	MSL (mm)	11.5	11.6	8.6	9.6
	N	49	39	40	29
N. angustior	SD	2.31	2.19	3.02	2.22
	Size comparison	Mann Whitney Rank Sums Test Test $U = 1756.5$ $p = 0.86$		Mann Whitney Rank Sums Test $U = 1179.0$ **p=0.047***	
	Location comparison	z-test $z = 0.876$ $p = 0.107$		z-test $z = 1.069$ $p = 0.28$	
	MSL (mm)	12.2	12.2	11.5	12.0
	N	65	133	48	54
N. dilatata	SD	2.02	2.06	2.35	2.33
	Size comparison	Mann Whitney Rank Sums		t-test $t = 1.15$ $p = 0.25$	
	Location comparison	z-test $z = 4.43$ **p=0.01***		z-test $z = 0.386$ $p = 0.69$	
	MSL (mm)	11.7	12.2	9.1	8.2
	N	19	19	34	28
N. riisei	SD	1.26	0.87	1.94	2.24
	Size comparison	Mann Whitney Rank Sums Test $U = 323$ $p = 0.170$		t-test $t = 1.62$ $p = 0.111$	
	Location comparison	z-test $z = -0.324$ $p = 0.75$		z-test $z = 0.497$ $p = 0.62$	
	MSL (mm)	14.2	14.0	14.8	14.5
	N	88	100	34	29
T. antonii	SD	1.83	1.44	1.45	1.49
	Size comparison	t-test $t = 1.62$ $p = 0.111$		t-test $t = 0.913$ $p = 0.36$	
	Location comparison	z-test $z = 0.730$ $p = 0.21$		z-test $z = 0.380$ $p = 0.70$	

384 *C. muricatus* were examined with respect to resting site, that is, whether they lie in repose within the shelter of a crevice or upon an exposed rock surface. The results of these analyses are presented in Table 4. Individuals of each species were examined to determine whether they displayed such a preference on the basis of size and the entire group was examined to determine if preferences differed according to species. Generally, no such preferences were discovered. Only *N. angustior* displayed a significant size relationship with respect to site of repose, with larger individuals appearing more commonly in crevices in July 1994 (Table 4). No such relationship was found among individuals of *N. angustior* in January 1994 nor for any other species examined (Table 4). A significantly greater number of *C. muricatus* found repose in rock crevices in January 1994, but the trend was not evident in July 1994 (Table 4). Similarly, signifi-

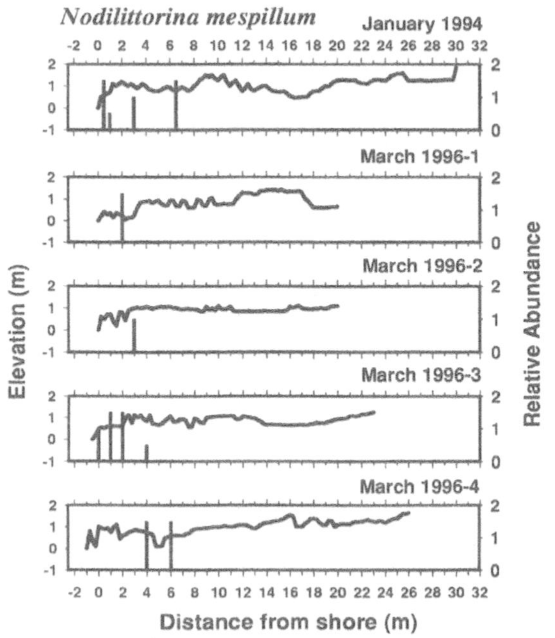

Figure 5. The distribution of *N. mespillum* from five transects along a coastal limestone platform near the Hofstra Marine Laboratory, Priory, Jamaica. The legend is described under Figure 4.

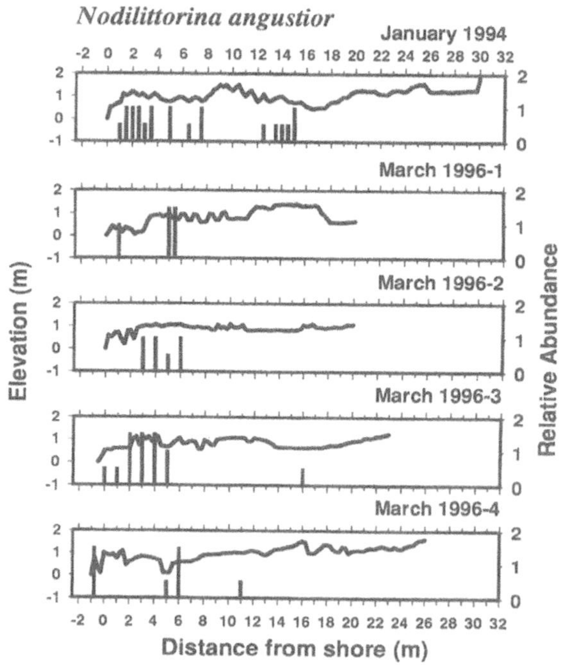

Figure 7. The distribution of *N. angustior* from five transects along a coastal limestone platform near the Hofstra Marine Laboratory, Priory, Jamaica. The legend is described under Figure 4.

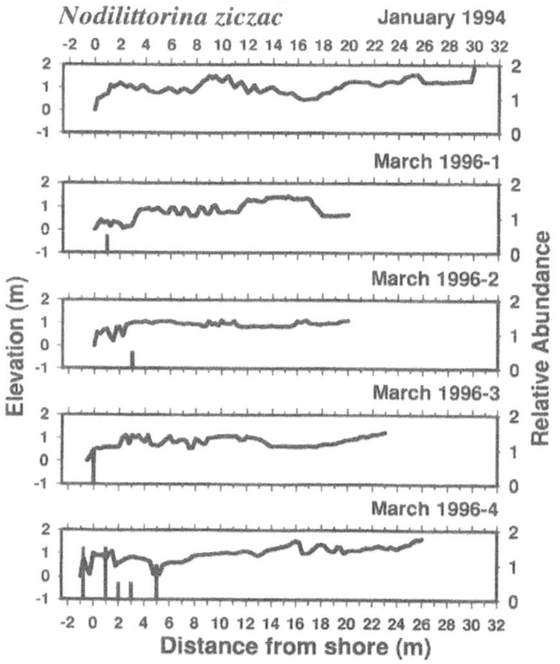

Figure 6. The distribution of *N. ziczac* from five transects along a coastal limestone platform near the Hofstra Marine Laboratory, Priory, Jamaica. The legend is described under Figure 4.

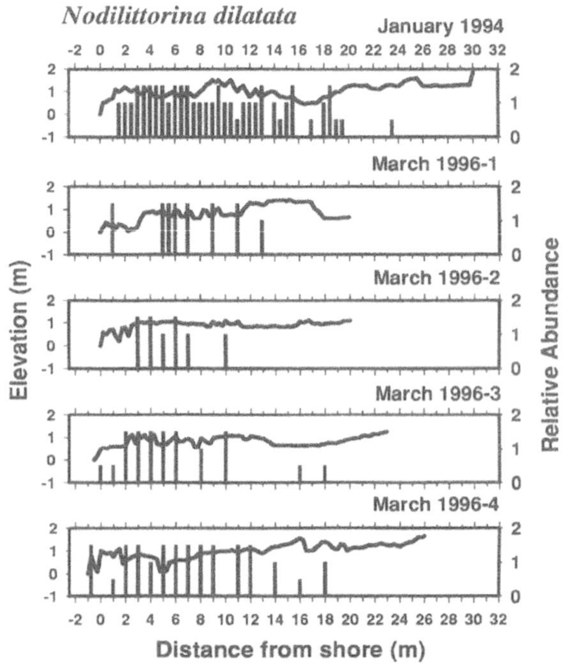

Figure 8. The distribution of *N. dilatata* from five transects along a coastal limestone platform near the Hofstra Marine Laboratory, Priory, Jamaica. The legend is described under Figure 4.

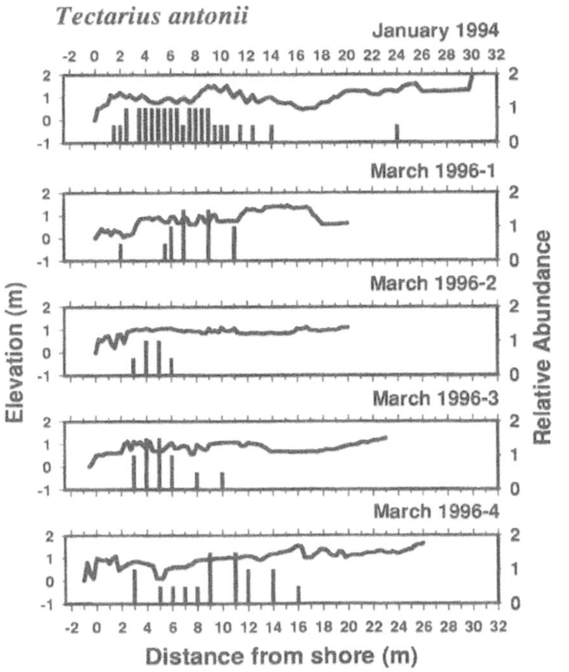

Figure 9. The distribution of *T. antonii* from five transects along a coastal limestone platform near the Hofstra Marine Laboratory, Priory, Jamaica. The legend is described under Figure 4.

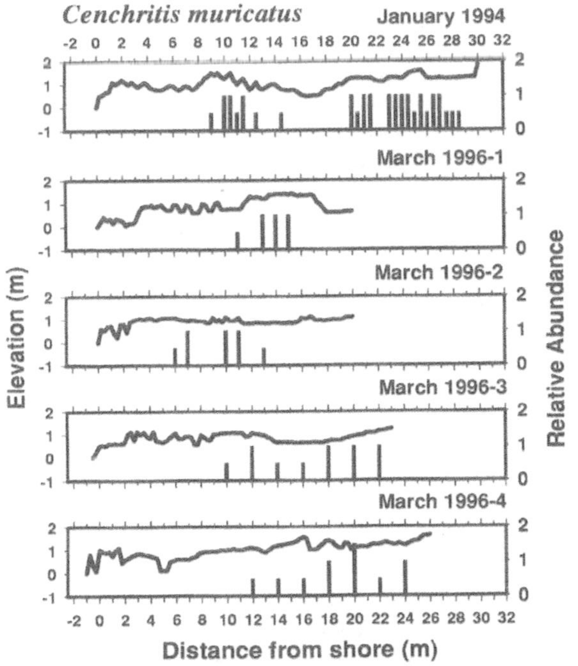

Figure 10. The distribution of *C. muricatus* from five transects along a coastal limestone platform near the Hofstra Marine Laboratory, Priory, Jamaica. The legend is described under Figure 4.

cantly more individuals of *N. dilatata* (67%) occupied crevices than exposed rock surfaces in January 1994, but not in July 1994 (Table 4). Neither *N. riisei* nor *T. antonii* showed a statistically significant preference for occupying crevices in either January or July.

Mucous holdfast

In January and July 1994, a total of 100 *N. riisei*, 155 *N. angustior*, 259 *N. dilatata*, 223 *T. antonii* and 376 *C. muricatus* were examined with respect to whether or not, they attached to the substratum during repose by means of a mucous holdfast extending from the aperture. The results of these analyses are presented in Table 5. At least a few individuals of all species were observed attached to the substratum by a mucous holdfast, but generally individuals of all species more frequently did not so attach. The relationship was significant in all cases except for *C. muricatus* observed in July 1994. The use of a mucous holdfast was more prominent in those species occurring highest onshore (Table 5). Only 6% of *N. riisei* and 8% of *N. angustior* were found attached by a mucous holdfast, but since these species often occupied moist or wet surfaces, it was difficult to determine whether or not such a structure was present. Higher on the platform, 15% of all *N. dilatata*, 20% of all *T. antonii* and 46% of all *C. muricatus* attached by a mucous holdfast. There was a significant relationship between size and the presence or absence of a holdfast for *N. dilatata* observed in January and in the combined data for January and July, with larger individuals utilising mucous holdfasts more often. None of the other species demonstrated a relationship between size and the use of a mucous holdfast.

Shell orientation

In January and July 1994, a total of 104 *N. riisei*, 138 *N. angustior*, 287 *N. dilatata*, 235 *T. antonii* and 377 *C. muricatus* were examined to determine the compass direction toward which most of the exposed body surface was facing (Table 6). All species except *N. riisei* showed significantly different patterns of shell orientation during repose between January and July samples. There was no consistent pattern with respect to shell orientation, either within a species or between species, but the secondary directions (NE, SE, SW, and NE) seemed generally under-represented, probably attributable to sampling bias.

Table 5. Comparisons of individuals in repose either attached to a rock by a mucous holdfast or not so attached in January and July 1994. MSL = mean shell length, *N* = number, SD = standard deviation. Non-parametric comparisons were made when normality tests failed. Comparisons with statistically significant differences ($\alpha \leqslant 0.05$) are indicated by probability (p) values in bold

Species	Variable	January 1994		July 1994	
		Holdfast Present	Holdfast Absent	Holdfast Present	Holdfast Absent
	MSL (mm)	18.4	17.9	18.5	18.7
	N	105	150	68	53
C. muricatus	SD	1.59	2.17	2.12	1.85
	Size comparison	Mann Whitney Rank Sums Test $U = 14437.0$		Mann Whitney Rank Sums Test $U = 3255.0$	
		$p = 0.086$		$p = 0.91$	
	Holdfast comparison	z-test $z = 2.64$		z-test $z = 1.170$	
		p=0.008*		$p = 0.24$	
	MSL (mm)	11.3	11.6	10.6	9.1
	N	10	76	2	67
N. angustior	SD	2.78	2.21	5.30	2.67
	Size comparison	Mann Whitney Rank Sums Test $U = 430.0$ $p = 0.95$		Mann Whitney Rank Sums Test $U = 78.0$ $p = 0.79$	
	Holdfast comparison	z-test $z = 5.211$		z-test $z = 4.138$	
		p=<0.001*		**p=< 0.001***	
	MSL (mm)	12.9	12.1	12.7	11.6
	N	39	128	11	91
N. dilatata	SD	1.77	2.13	1.43	2.41
	Size comparison	t-test $t = -2.30$		t-test $t = -1.46$	
		$p = 0.023$*		$p = 0.147$	
	Location comparison	z-test $z = 5.875$		z-test $z = 5.824$	
		p=<0.001*		**p=< 0.001***	
	MSL (mm)	11.3	12.0	10.4	8.5
	N	3	35	3	59
N. riisei	SD	1.08	1.11	3.58	2.06
	Size comparison	Mann Whitney Rank Sums Test $U = 36.0$ $p = 0.23$		Mann Whitney Rank Sums Test $U = 129.5$ $p = 0.26$	
	Location comparison	z-test $z = -3.11$		z-test $z = 4.257$	
		$p = 0.002$*		$p =< 0.001$*	
	MSL (mm)	14.3	14.0	14.5	14.7
	N	36	124	9	54
T. antonii	SD	1.44	1.77	1.47	1.48
	Size comparison	t-test $t = -0.875$		t-test $t = 0.327$	
		$p = 0.38$		$p = 0.75$	
	Location comparison	z-test $z = 5.90$		z-test $z = 4.192$	
		p=<0.001*		**p=< 0.001***	

Table 6. Results of the field survey assessing the compass direction toward which most of the body surface of five species of Jamaican littorinid faced in natural repose. Chi-square comparisons were made between January and July samples and performed on actual frequencies, with significant differences between January and July indicated in bold. Percent frequencies are given here for ease of comparison among and between species. In all cases df = 7

Species	1994 Month	Comparison	*n*	Percent Frequency in Compass Quadrant							
				N	NE	E	SE	S	SW	W	NW
N.	Jan.	$\chi^2 = 5.18$	38	28.9	7.9	15.8	7.9	15.8	7.9	5.3	10.5
riisei	Jul.	$p = 0.638$	66	19.7	4.5	16.7	3.0	15.2	10.6	15.2	15.2
N.	Jan.	$\chi^2 = 24.7$	69	15.9	7.2	18.8	5.8	33.3	2.9	11.6	4.3
angustior	Jul.	$\mathbf{p{=}< 0.001}$	69	20.3	24.6	10.1	10.1	10.1	10.1	4.3	10.1
N.	Jan.	$\chi^2 = 24.9$	174	22.4	2.9	15.5	5.7	30.5	6.3	13.2	3.4
dilatata	Jul.	$\mathbf{p< 0.001}$	113	15.9	9.7	25.7	10.6	11.5	6.2	14.2	6.2
T.	Jan.	$\chi^2 = 14.9$	171	26.9	7.0	17.0	4.1	17.0	8.2	18.1	1.8
antonii	Jul.	$\mathbf{p{=}0.038}$	64	15.6	7.8	12.5	12.5	20.3	14.1	10.9	6.3
C.	Jan.	$\chi^2 = 20.9$	257	37.0	10.5	9.7	2.0	9.7	4.3	18.3	8.6
muricatus	Jul.	$\mathbf{p{=}0.004}$	120	22.5	18.3	14.7	6.7	8.3	8.3	11.7	10

Table 7. Results of the field survey assessing the position of repose of five species of Jamaican Littorinidae, whether resting aperture-down on a rock surface or hanging suspended from a rock ledge with the aperture generally facing upwards. Chi-square comparisons were made between January and July frequencies, with significant differences between the respective proportions indicated in bold. Percent frequencies are given adjacent to actual frequencies. In all cases df = 1

Species	1994 Month	Comparison	Position of repose: Hanging	Resting
N. riisei	Jan.	$\chi^2 = 0.661$	9 23.7%	29 76.3%
	Jul.	$p = 0.42$	22 33.3%	44 66.7%
		Total	31 29.8%	73 70.2%
N. angustior	Jan.	$\chi^2 = 0.039$	16 16.2%	53 83.8%
	July	$p = 0.84$	18 26.1%	51 73.9%
		Total	34 24.6%	104 76.4%
N. dilatata	Jan.	$\chi^2 = 0.009$	52 29.9%	122 70.1%
	Jul.	$p = 0.92$	34 30.1%	79 69.9%
		Total	86 30.0%	201 70.0%
T. antonii	Jan.	$\chi = 6.68$	20 11.7%	151 88.3%
	July	$\mathbf{p{=}0.018}$	17 26.6%	47 73.4%
		Total	37 15.7%	198 84.3%
C. muricatus	Jan.	$\chi = 5.89$	87 33.9%	170 66.1%
	Jul.	$\mathbf{p{=}0.015}$	57 47.5%	63 52.5%
		Total	144 38.2%	233 61.8%

Posture of repose

The same individuals examined for shell orientation were also examined to determine the posture of repose, that is, whether the gastropod positioned its aperture down upon a rock substratum, here referred to as resting, or hung suspended from a ledge with the aperture generally directed upward. In the latter case, the aperture could either be exposed to air or, more frequently, applied to the rock substratum underneath a ledge. A greater proportion of individuals of all species were found resting in both January and July 1994 (Table 7). This was most strongly expressed by *T. antonii*, with 88.3% of those examined in January, 73.4% of those in July and 84.3% of all individuals examined in the resting position of repose. Similarly, 83.8% of the individuals of *N. angustior* examined in January, 73.9% of those in July and 76.4% of the combined number were found in the resting position. *C. muricatus* had the greatest number of individuals hanging (47.5% in July). Only two species, *T. antonii* and *C. muricatus*, had significantly different proportions of individuals hanging and resting from January to July, with 14.9% and 13.6% more individuals, respectively, hanging. Although all of the Jamaican littorinids preferred the resting posture, a Chi-square analysis of the totals for each species indicated that there was a significant difference between the species with respect to the proportions of individuals in one position or the other (Chi-square = 36.9; df = 4; $p =< 0.001$). There was no obvious trend, however, with position on the shore and resting position, except within the highest zoned species, *C. muricatus*, which was most likely to assume a hanging posture.

The relationship between body size and position of repose was evaluated by analyses of variance (Table 8). When size data (shell height) for a species failed either homogeneity of variance or normality tests, a Kruskal-Wallis One Way ANOVA on ranks was used instead of a parametric One Way ANOVA. When differences were detected between groups, suitable pairwise multiple comparison procedures were employed to identify which groups differed (Table 8). No significant differences in posture preference were detected for individuals of different sizes in either January or July for *N. dilatata*. The analysis of variance indicated a significant difference ($P = 0.049$; $\alpha = 0.05$) among groups of *T. antonii*, however subsequent pairwise Tukey multiple comparisons failed to detect significant differences (Table 8). Analyses of variance for the other three species detected some significant differences in mean size between individuals either at different observation periods and/or different positions of repose. The mean size of *N. riisei* observed in January was generally larger than those observed in July (Table 8). Those in the resting posture in January were significantly larger than individuals found either resting or hanging in July (Table 8). Similarly, January individuals that were hanging were significantly larger than those individuals resting in July (Table 8). Similar results were obtained for *N. angustior* and, in addition, individuals of this species found in the hanging position of repose in January were significantly larger than those found in the same position in July (Table 8). Thus, for *N. riisei* and *N. angustior* there were no significant size differences between individuals in different positions of repose within an observation period, but there were several significant size differences within and between positions of repose for individuals observed at different times. Individuals of *C. muricatus* in the resting position of repose in January were significantly smaller than those in the same position in July (Table 8).

Temperature relationships

Results of the 12-hour temperature profile are presented in Figures 12 and 13 and Table 9. Maximum rock surface temperatures were obtained between 11:00 and 12:00 (Table 9), followed by a gradual decline through the afternoon (Figure 12). There were no statistically significant differences in temperature between exposed areas and crevices within a site according to Mann-Whitney rank sums tests (Table 9) nor were there significant differences in temperature between the sites for an exposed surface (Kruskal-Wallis oneway ANOVA on ranks: H = 2.54, $P = 0.47$, df = 3). However, there were significant differences between surface temperatures taken within crevices between sites (Kruskal-Wallis oneway ANOVA on ranks: H = 10.2, $P = 0.017$, df = 3). Pairwise multiple comparisons employing the Student-Newman-Keuls method demonstrate that the crevice temperatures at Site 3 were significantly greater than those at all other crevice sites.

Mean body temperatures at exposed locations and in crevices were determined for *C. muricatus* at all sites, those for *N. dilatata* at Sites 1 and 2 and those for *T. antonii* at Site 1. Mean body temperature for *N. angustior* was determined at only the exposed location at Site 3. The mean body temperature for *C. muricatus* was greater than that for mean exposed rock surface

Table 8. Results from Analyses of Variance tests on size (shell height) relationships for Littorinidae either resting on (R) or hanging from (H) rock surfaces in January (Jan) and July (Jul) 1994

Species	Test Comparison	Pairwise Multiple Method	Significant differences Mean shell height
N. riisei	Kruskal-Wallis ANOVA on ranks $H = 48.2$ df = 3 ***p*=< 0.001**	Dunn's	Jan R – Jul R Jan R – Jul H Jan H – Jul R
N. angustior	Kruskal-Wallis ANOVA on ranks $H = 37.5$ df = 3 ***p*=< 0.001**	Dunn's	Jan R – Jul R Jan R – Jul H Jan H – Jul R Jan H – Jul H
N. dilatata	Kruskal-Wallis ANOVA on ranks H = 4.45 df = 3 ***p*=0.216**	none	none
T. antonii	One Way Parametric ANOVA F = 2.662 df = 3 ***p*=0.049**	Tukey	none
C. muricatus	Kruskal-Wallis ANOVA on ranks H = 11.4 df = 3 ***p*=0.010**	Dunn's	Jan R – Jul R

temperature at Sites 1 and 3 and for mean crevice surface temperature at Site 1, but less than that for mean exposed rock surface temperature at Sites 2 and 4 and for mean crevice surface temperature at Sites 2, 3 and 4 (Table 9). The mean body temperature of *N. dilatata* was 0.1 °C less than that of the mean exposed rock surface temperature at Site 1, equal to it at Site 2 and less than the mean crevice surface temperature at both Sites1 and 2. The mean body temperatures of *T. antonii* were less than that of mean rock temperatures at both exposed and crevice locations at Site 1. Finally, the mean body temperature of *N. angustior* was 0.1 °C less than the mean rock temperature of exposed locations at Site 3. Mean body temperatures for all sampled species of Littorinidae were consistently lower for individuals in crevices than those from exposed locations. Significant differences between the mean body temperatures of individuals from exposed sites and crevices were found for *N. dilatata* at Site 1 and for *C. muricatus* at Sites 2, 3 and 4 (Table 9).

When body temperature data for all sites are combined and the individual deviations from local rock temperatures are plotted, a slightly different picture emerges (Figure 13). Not unexpectedly, the pattern of body temperatures plotted throughout the day (Figure 13, upper cell of each set) roughly parallels the pattern observed with rock surface temperatures (Figure 12). However, when plotted as the deviation from adjacent rock temperatures (Figure 13 lower cell of each set), the body temperatures of all measured species, more often than not, are usually less than those of adjacent rock surfaces during the middle of the day. In the morning, especially from 06:00 to 08:00, and to a lesser degree, late in the afternoon, body temperatures of especially *C. muricatus* and *N. dilatata* were frequently greater than adjacent rock temperatures.

Substratum and body temperature data from the 12-hour profile were combined with other similar measurements made in January and July 1994 to pro-

Table 9. Minimum, maximum and mean temperatures recorded on exposed rock surfaces, in rock crevices and the body temperatures of four species of Littorinidae occupying these habitats during a continuous 12 h period in July 1996. Results of Mann-Whitney rank sum tests comparing exposed and crevice temperatures and the body temperature of littorinids occupying these sites are also presented. Significant differences are indicated by asterisks in the probability (P) column. The first number in the n column is the number of measurements from exposed sites, the second from crevice sites

		Exposed				Crevice						
Site and distance from sea	Temp measured	min temp (°C)	max temp (°C)	mean temp (°C)	SD	min temp (°C)	max temp (°C)	mean temp (°C)	SD	Mann – Whitney U	P	n
1. Tide-pool (3.4)	Rock	25.0	37.5 (11:00)	32.4	3.18	24.9	36.0 (11:45)	31.6	2.76	2648.5	0.114	49/49
	C. muricatus	31.8	34.4 (16:45)	32.9	1.13	32.1	32.9	32.4	0.342	26.0	0.841	5/5
	N. dilata	28.6	35.2 (10:45)	32.3	1.83	27.6	34.5	31.1	1.69	530.0	0.024*	26/28
	T. antonii	25.5	33.8 (13:00)	30.0	2.40	27.2	32.8	29.6	1.69	151.5	0.661	15/11
2. Boulder (5.4)	Rock	24.6	37.9 (12:00)	32.4	3.31	24.3	36.0 (11:00)	31.4	2.76	2696.5	0.055	49/49
	C. muricatus	26.7	37.9 (11:15)	32.1	2.46	26.7	37.9	30.7	2.39	1231.0	0.011*	38/38
	N. dilatata	30.3	35.5 (11:15)	32.4	2.05	30.2	32.0	30.7	0.638	29.0	0.073	6/6
3. Flat bedrock (5.5)	Rock	25.1	38.8 (11:15)	32.8	3.51	24.9	36.3 (11:15)	32.4	3.12	2531.0	0.456	49/49
	C. muricatus	26.1	39.1 (13:45)	33.4	2.50	25.2	37.6	31.1	3.47	506.0	0.010*	18/62
	N. angustior	29.7	34.1 (13:30)	32.7	1.68	–	–	–	–	–	–	6/0
4. Boulder 15.3	Rock	24.1	37.0 (11:15)	32.0	3.13	24.9	37.0 (12:00)	31.1	2.68	2680.0	0.062	49/49
	C. muricatus	25.8	35.8	31.5	2.14	26.5	36.4	30.8	2.03	1563.0	0.048*	39/52

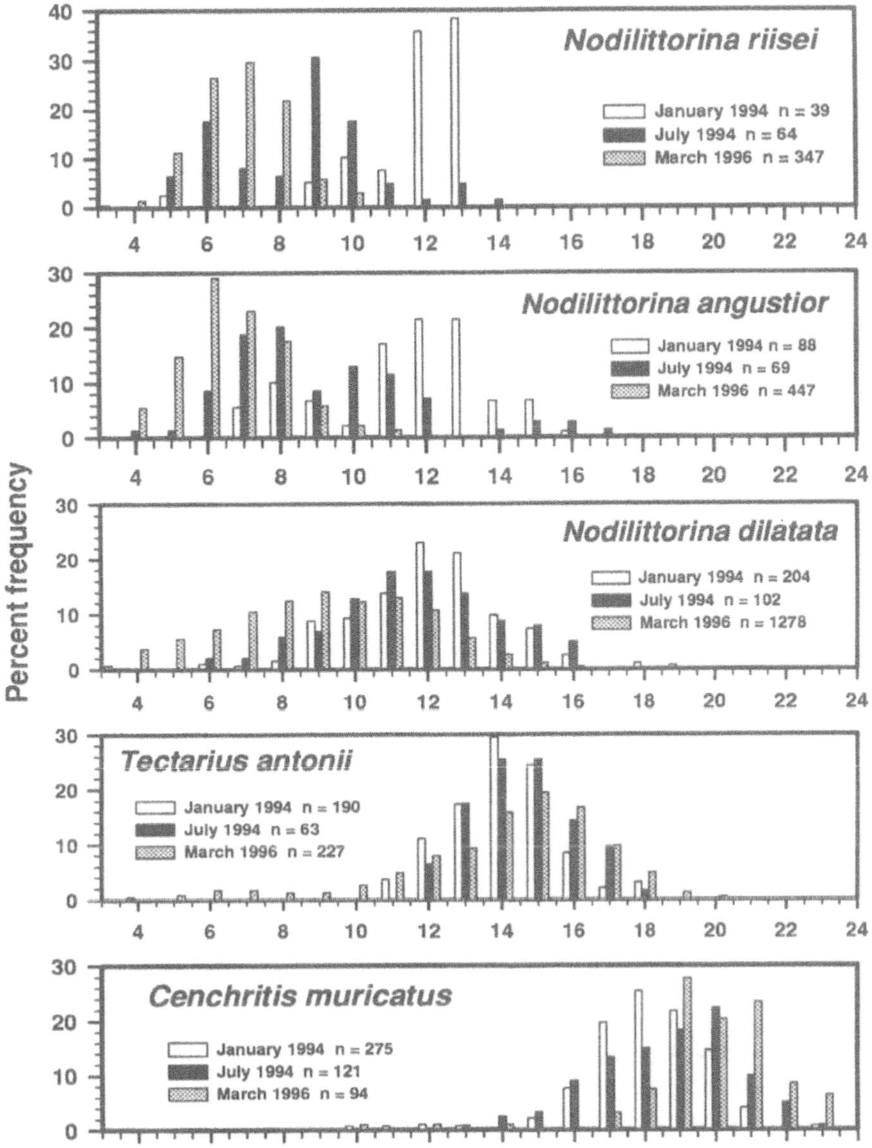

Figure 11. Size-frequency distributions of five species of Littorinidae from a coastal limestone platform in January and July 1994 and March 1996. Size frequencies have been converted to percent-frequencies of the total sample, with the total (n) indicated on each plot.

duce a general body temperature profile for 81 *N. riisei*, 82 *N. angustior*, 187 *N. dilatata*, 127 *T. antonii* and 401 *C. muricatus* (Figure 14). The pattern observed for *C. muricatus* and *N. dilatata* in the body temperature deviations from rock temperature (Figure 13) are also indicated for the same species in Figure 14. At a mean rock temperature of 27 °C, the mean body temperature of *C. muricatus* was significantly higher.

Similarly, at mean rock temperatures of 25 and 27 °C, the mean body temperature of *N. dilatata* were significantly higher. A similar pattern is also evident for *N. angustior* and *N. riseii* (Figure 14), but not for *T. antonii*. At temperatures between approximately 29 to 32 °C, but deviating from this range depending upon species, mean body temperatures of all five species are approximately the same as that of the rocks upon which they live. Then, above about 32 °C body

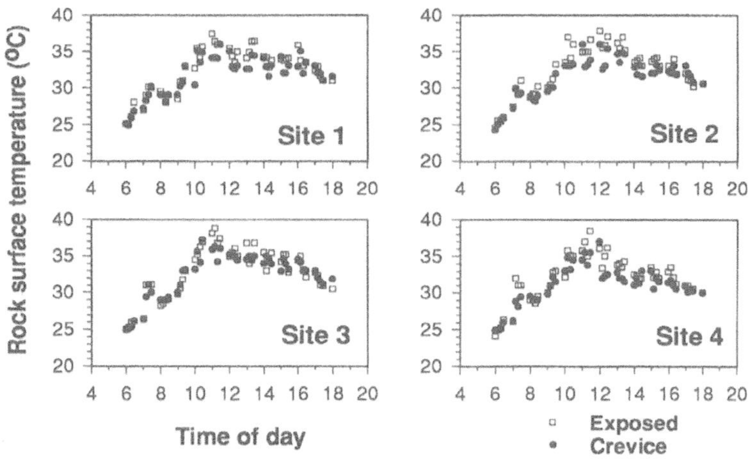

Figure 12. Rock surface temperatures recorded during a 12 h temperature profile conducted at four sites on a coastal limestone platform on the north shore of Jamaica on 5 July 1994. Open squares represent temperatures recorded on exposed rock surfaces whereas filled circles represent temperatures recorded in crevices.

temperatures of all species except *T. antonii* are generally lower than mean rock temperatures, sometimes significantly so.

Discussion

The species of Littorinidae examined during the course of this study have received considerable ecological attention throughout the tropical western Atlantic. Unfortunately, those unfamiliar with the nomenclatorial history of these species will find it difficult to access much of this information because of extensive taxonomic confusion. Prior to 1964, all the striped littorinids of the Caribbean basin and the Gulf of Mexico were collectively referred to as *Littorina ziczac* (Bequaert, 1943; Hedgpeth, 1953). Any paper published prior to the late 1960s devoted to the ecology of *L. ziczac* might concern one or more of as many as seven different species. Rodriguez (1959) recognised two forms of striped Littorinidae occupying different positions on the shore on Margarita Island, Venezuala, but did not formally distinguish them. Abbott (1964) was the first to recognise two species, *Littorina ziczac* (Gmelin, 1791) and *L. lineolata* d'Orbigny, 1840, although the latter epithet still confused at least two species as one and eventually proved to be a junior synonym of yet another nominal species. Not long thereafter, Borkowski & Borkowski (1969) extracted a third striped Caribbean species from this complex, *L. lineata* d'Orbigny, 1841, which later was also partially synonymised. Bandel (1974) recognised yet another species and, in 1982, Bandel & Kadolsky raised the number of species of striped Caribbean Littorinidae to seven, recognising them among or relegating them to the genus *Nodilittorina*. During the same time, the nodulose Caribbean littorines were also suffering nomenclatorial confusion, especially wih respect to generic assignments (Abbott, 1954; Lewis, 1960; Borkowski, 1971, 1974). Although Bandel & Kadolsky (1982) and Reid (1989) resolve some of the taxonomic confusion, the tropical Western Atlantic Littorinidae would be well served with a modern, comprehensive, taxonomic revision.

Transects

Discrepancies in the distributional patterns of the central platform littorinids between January 1994 and March 1996 transect samples probably are the result of platform flooding immediately prior to the collection of the January 1994 transect data. One consequence of seawater flooding indicative of changed distributional patterns was the striking redistribution of two eulittoral rocky shore molluscs, the chiton *Acanthopleura granulata* and the gastropod *Nerita versicolor*. During the flooding, *A. granulata* moved 2–3 m landward from its normal position along the wave-cut escarpment that usually marks the seaward edge of the platform. *N. versicolor*, normally found on the lower shore in crevices or tidepools frequently refreshed by wave splash, displayed a significantly greater displace-

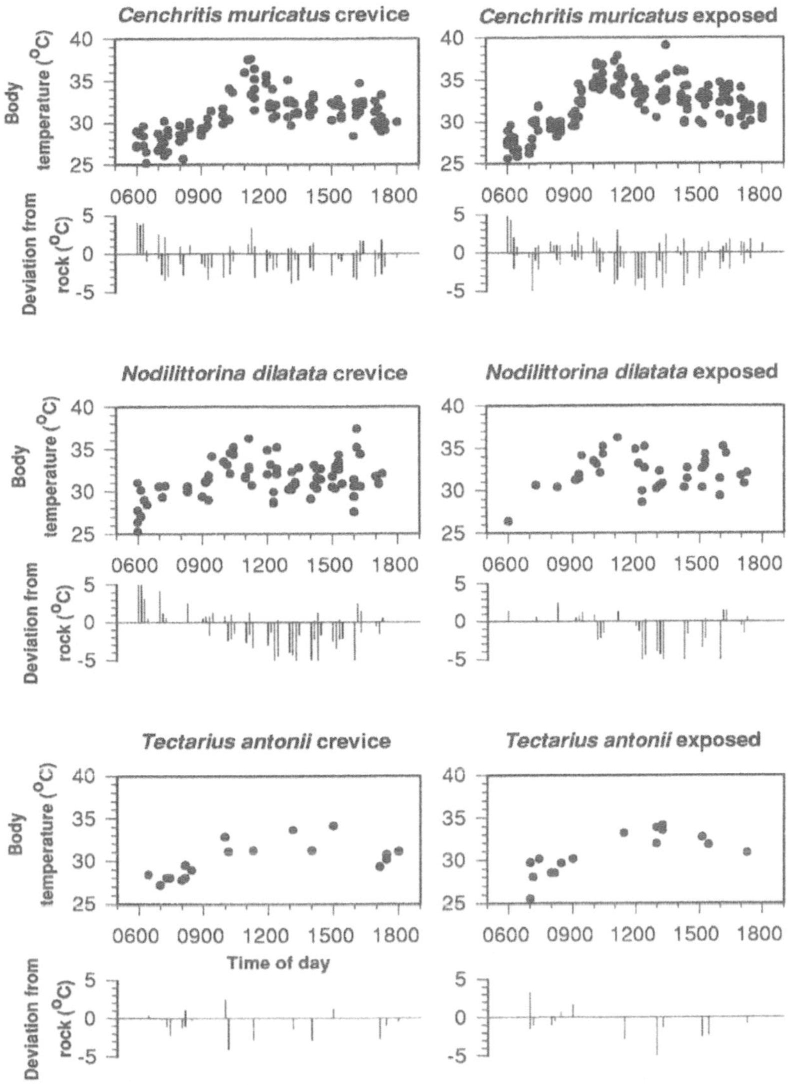

Figure 13. Body temperatures of three species of Littorinidae recorded during a 12 h temperature profile conducted at four sites on a coastal limestone platform on the north shore of Jamaica on 5 July 1994. Individuals from rock crevices are on the left whereas individuals from exposed rock surfaces are on the right. The upper plot of each set illustrates the distribution of body temperatures from 06:00 to 18:00 h. The lower plot indicates the amount each body temperature deviated from the corresponding adjacent rock temperatures.

ment. Immediately after floodwater receded from the coastal platform, numerous individuals of *N. versicolor* were 15–20 m from the sea amid large clumps of *C. muricatus*. Within three days after the water subsided, most of the displace nerites had migrated bck to their normal position on the shore (Britton, unpulished data).

When faced with abnormally high water, the central platform littorinids may have also adjusted their positions, by either migrating upshore or being washed there by waves or currents. Although the January transect samples were taken several days after floodwater had subsided, it is likely that the effects of littorinid re-distribution in the face of such flooding were recorded in the quadrat samples. Thus, transect sampling was repeated in March 1996 after the platform had stood without seawater inundation for at least several weeks. The March 1996 quadrat samples, with different and probably more typical distributions of the mid-platform littorinid species, help corroborate

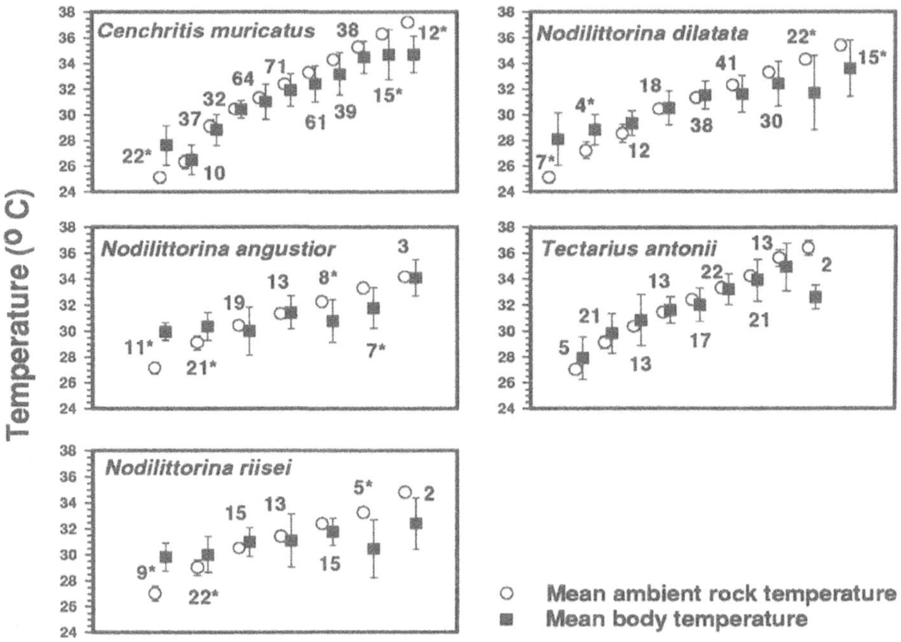

Figure 14. Mean body temperatures recorded at specific substatum temperatures, including data from both the 12 h temperature profile and additional recordings made in January and July 1994. Mean rock temperatures are indicated by open circles whereas mean body temperatures are indicated by filled squares. Error bars represent standard deviations. The number of individuals examined at each rock temperature is indicated beside each pair of data points. Significant differences between mean rock temperatures and mean body temperatures are indicated by asterisks (*).

our assertion that distributions observed in January 1994 were influenced by platform flooding. *N. angustior*, *N. dilatata* and, to a lesser degree, *T. antonii*, were distributed more widely across the platform in January 1994 (Table 1) and were found on smoother, drier rocks (Tables 2 and 3) than corresponding transect samples taken in March 1996 (Figure 3).

The pattern of distribution, especially that reflected by the March 1996 transects, conforms well with similar patterns reported for these species in other tropical western Atlantic localities. *C. muricatus* is consistently reported as the highest zoned littorinid (Lewis, 1960; Vermeij, 1973; Bandel, 1974; Bandel & Kadolsky, 1982), and in the presence of sea cliffs, it ranges considerably higher on the shore (Lewis, 1960; Britton, 1976). We found *N. riisei* nearer the sea and in a more restricted range than indicated by Bandel & Kadolsky (1982, Figure 4). *N. dilatata* was distributed more widely on the Jamaican platform than any other species, extending into the lower third of the *C. muricatus* range, which is somewhat higher than that indicated by Bandel & Kadolsky (1982). Both *N. dilatata* and *C. muricatus* are capable of withstanding prolonged emersion. *N. dilatata* experienced only 25% total body water loss and *C. muricatus* 27.3% total body water loss after 28 days in air (Britton, 1992).

T. antonii had a relatively restricted distribution on the central platform, but broader than that illustrated by Bandel & Kadolsky (1982, Figure 4). Peckol et al. (1989) showed that the distribution of Bahamian *T. antonii* partially overlapping that of *Nerita versicolor* but we never observed such an overlap, except during and shortly after platform flooding in January 1994 when *N. versicolor* was displaced upwards. *T. antonii* is as capable of prolonged emersion as either *C. muricatus* or *N. dilatata*, experiencing only slightly more than 9% total body water loss after six days emersion (Peckol et al., 1989) for individuals from the Bahamas and 26.1% total body water loss after 30 days of exposure to air for individuals collected from this Jamaican study site (Britton, 1992).

Size

The five species for which size-frequency histograms were compiled (Figure 11) recruit new generations onto the shore at different times of year. All of these species produce pelagic egg capsules (Borkowski, 1971; Mileikovsky, 1975; Bandel & Kadolsky, 1982) and thus re-occupy the shore from the meroplankton. The January populations of both *N. riisei* and *N. angustior* consisted of a greater proportion of larger individuals while the March populations consisted of a greater proportion of smaller individuals, suggesting an influx of new recruits for both species on this shore sometime between these sampling months each year. The July population of *N. riisei* consisted of a larger number of large individuals than the March population, but a significant number of small individuals still remained, perhaps indicating either continued spring recruitment or a second summer recruitment period for this species on this shore. In contrast, a small number of *N. angustior* comprised the the July population, but the peaks in July histograms (Figure 11) could be interpreted to represent two later winter or early spring recruitment episodes for this species. There was no clear evidence for the recruitment of other species on this shore during this time.

Another indication of winter recruitment of *N. riisei* and *N. angustior* appears when comparing body size asociated with the position of the repose data between January and July (Table 8). The mean size (shell height) of both *N. riisei* and *N. angustior* decreased from January to July. The mean size of *N. dilatata* did not change significantly during this period, whereas that of *T. antonii* and *C. muricatus* increased slightly. One can infer that the latter three species either did not recruit significant numbers of new individuals from the plankton from January to July or, in the case of *N. dilatata* which was always represented by a significant number of individuals less than 8 mm, may have been recruiting new individuals continually. Thus, these five species of Littorinidae colonise this shore from the meroplankton at different times.

In contrast to the *Nodilittorina* species, Borkowski (1971) found that all individuals of *C. muricatus* and *T. antonii* had spent gonads from December to April. He recovered egg capsules of *C. muricatus* from the plankton from June to August, but only twice for *T. antonii* at unspecified times. Lewis (1960) recovered planktonic egg capsules of *C. muricatus* from the waters off Barbados from mid-June to August. If juveniles of *C. muricatus* and *T. antonii* colonise the shore from the plankton in late summer or early autumn, this could explain why our size-frequency distributions, especially during March and July, consist primarily of large individuals.

Resting site

About half the littorinids on this platform, excluding *N. mespillum*, occupy the shelter of crevices. A majority of *N. dilatata* (62%) were found in crevices, whereas only 43% of *N. angustior* were found there. Smaller individuals of *N. dilatata* were more abundant in crevices than larger ones. Among the remaining species, individuals occupying the upper shore were more likely to seek a crevice shelter than those positioned closer to the sea (Table 4). Temperatures on this shore (Table 9) never exceeded upper thermal tolerance limits determined for these species (Britton, 1992), but may contribute to elevated rates of desiccation (Wolcott, 1973). Thus, individuals occupying the upper shore which seek crevices likely benefit from the shelter provided by them, as refuge in these microhabitats has been found to suppress excess body water loss (Garrity, 1984). Surface irrregularities producing crevices are less common on the upper third of this shore (Figure 1). Nevertheless, 55% of the sampled *C. muricatus* occupied the shelter of crevices. Thus, the spatial distribution of *C. muricatus* is frequently clumped, either around such sites or around one another, suggesting that the species behaviour during repose is influenced by the need to minimise desiccation.

Micro-karsted crevices and pits are considerably more abundant on the seaward half of the platform, yet only *N. dilatata*, especially the smaller individuals, demonstrates a strong preference for such microhabitats, presumably to minimise body water loss by desiccation. This may not be, however, the only factor driving littorinids into crevices. Peckol et al. (1989) reported a significant positive regression relationship ($r^2 = 0.79$, $F = 14.9$, $P < 0.02$) between the density of *T. antonii* on a Bahamian limestone shore and its presence in crevices or micro-karsted pits. They found only 17% of 1656 *T. antonii* were found on exposed sites, 1% in tidepools and 81% in crevices or pits. Prior to analysing these data from the Jamaican platform, it was also our impression that the majority of *T. antonii* on this shore also found repose within crevices. In fact only about half the population of *T. antonii* seeks such shelters on the Jamaican platform (Table 4). The different responses may reflect different conditions at the two sites. The Bahamian population studied by Peckol et al. (1989) occupied shore positions closer to the sea than those we studied in Jamaica and were apparently subjected to considerable wave splash at high tide. In Jamaica, only storm waves wetted most of *T. antonii* habitat. Thus, Bahamian *T. antonii* may have sought crevice shelters preferentially to minimise dislodgement by waves. On the Jamaican shore, where wave disturbance of *T. antonii* is rare, such shelter from physical dislodgement is usually unnecessary.

Among seven species of Jamaican Littorinidae, *T. antonii*, was most capable of conserving body water, losing only about 17% of total body water after 12 days emersion, with *N. dilatata* the next most capable (Britton, 1992). *C. muricatus* was less capable, losing about 23% of total body water after 12 days emersion. It is more selectively advantageous for *C. muricatus* to minimise desiccation body water loss in crevice shelters than for either *N. dilatata* or *T. antonii.* There is little danger of desiccation for *N. riisei* or *N. angustior* frequently wetted by waves in the lower portions of the limestone platform. The latter apparently also experience minmial physical stress from waves, as indicated by the majority of both species found on exposed rock surfaces (Table 4).

The mucous holdfast and position of repose

Some Littorinidae utilise a mucous thread to attach the shell to the substratum and thereby assume the hanging position of repose for thermoregulation. This posture minimises conductive transfer of heat from the substratum to the shell by reducing the amount of shell surface in contact with the substratum (Vermeij, 1971). Further, this posture exposes more of the shell surface to air currents, allowing heat to be transferred from the shell to the air through convective cooling (Vermeij, 1971). Species which often hang suspended from a mucous thread, e.g. *Littorina irrorata* (Bingham, 1972), *Littorina scabra* (McMahon & Britton, 1985), and *Littorina striata* (Britton, 1995), maximise the benefits of both thermoregulatory strategies. Although all species of Jamaican Littorinidae examined may produce a mucous holdfast and assume a hanging posture, none favour either strategy, even during the warmest months of the year. This could imply that the limestone habitat upon which these species is infrequently heated to excessive levels. The temperature studies conducted during this study support this view (see below).

Shell orientation

There was no discernible pattern with respect to which direction any of the Jamaican littorinids orient the shell during repose. This parameter was especially difficult to quantify, as evidenced by the prevalence of individuals assigned to primary compass directions (N, S, E and W) and the paucity of individuals reported for the secondary compass directions (NE, SE, SW and NW). The discrepancy is more likely the result of sampling bias than reflective of the actual pattern on the shore. Nevertheless, even if frequencies obtained for the primary compass directions are considered by themselves or in various combinations with frequencies in adjacent secondary quadrants, no pattern is apparent with these data. Shell orientation seems to be more related to matters of thermoregulation, desiccation avoidance or physical stability than to compass direction. *L. irrorata* preferentially orients the spire downward when it attaches to marsh grasses by means of a mucous holdfast, presumably because this is the most secure or stable posture that it can assume (Bingham, 1972, 1973). Garrity (1984) showed the importance of shell orientation to thermoregulation in tropical littorinids. Individuals placed on a substratum with the longest dimension of the shell perpendicular to the midday sun and relatively more shell in contact with the substratum experienced hotter body temperatures than those allowed to orient spires upward with little more than apertural contact to the substratum. We rarely observed any Jamaican littorinid in repose with its long axis perpendicular to the midday sun. Usually the spire was held noticeably above or, in the case of the hanging posture, beneath the substratum. With the exception of this kind of posturing, none of the Jamaican littorinids positioned themselves in any particular direction during repose.

Temperature relationships

Rock surface temperatures rarely vary significantly with respect to height on shore or microenvironment. The 12-hour temperature profiles of microenvironments at four sites on the Jamaican platform revealed no significant differences between surface temperatures on exposed rocks and those within crevices. Furthermore, a comparison of temperature measurements taken from exposed rock surfaces at four sites representing different heights on shore indicated that the average temperature for the day did not differ significantly between the sites. Similar observations were reported by Wolcott (1973).

In July 1994, none of the Jamaican platform littorinids experienced ambient substratum temperatures in excess of 37.9 °C, a value considerably below the Heat Coma Temperatures (HCT) determined for these species. Britton (1992) found the mean HCT for *C.*

muricatus to be 42.1 °C, that for *T. antonii* 43.4 °C and that for *N. dilatata* 45.4 °C. Recently, McMahon (personal communication) determined the mean HCT for *N. riisei* to be 44.1 °C. It is possible that the Jamaican platform littorinids could occasionally experience exceptionally warm temperatures which either approach or even briefly exceed their respective HCT limits, but even if this occurs, brief incursions into HCT ranges are usually not lethal. Thus, normal rock surface temperature fluctuations experienced by the littorinid inhabitants of this shore are generally unthreatening.

C. muricatus, *T. antonii*, much of the population of *N. dilatata* and some individuals of *N. angustior* on this shore may pass many days or sometimes weeks emerged without being wetted. When they are rehydrated, it is often from rainfall rather than seawater. Therefore, the effect of elevated ambient temperatures accelerating desiccation seems to be a more serious problem for these species than lethal tissue temperatures, especially considering the length of time they lie emerged. Our data do not address this problem directly, but provide a few insights. At least one species, *N. dilatata*, uses water in tidepools for rehydration. Many *N. dilatata* were found either within high-perched pools or aggregated at their margins. These pools are infrequently refreshed with seawater, filling mostly after rains. Thus, salinity can fluctuate from nearly fresh water to hypersaline brine. We found *N. dilatata* in pools with water at both extremes. Less commonly, *N. angustior* and *T. antonii* also repose in the high-perched tidepools. Many individuals of these species and most *C. muricatus* are uncommon in standing water. This, coupled with the absence of mucous holdfasts or aestivation membranes among at least half of the observed individuals leaves the question of how these species avoid excessive desiccation in the face of prolonged emersion.

Comparison of the body temperatures of snails from exposed areas and crevices showed crevice dwellers were significantly cooler in four of seven cases, including *N. dilatata* at Site 1 and *C. muricatus* at Sites 2, 3 and 4. Body temperatures of organisms within a limestone crevice have previously been reported lower than those on exposed limestone rocks (Garrity, 1984; McQuaid & Scherman, 1988), and our observations generally follow this trend. In those cases where crevice dwellers did not have significantly different temperatures than their exposed counterparts, the rocks may conduct heat from the exposed surfaces to surfaces within the crevice masking the effects of shade or the crevice may open directly toward the warming sun. Also, the mineralogy of the substratum can produce large variations in microenvironment temperatures (Britton et al., 1991).

Perhaps the most interesting result from comparisons between rock and body temperatures was the relative relationship between these two variables. At lower ambient temperatures, most individuals had a body temperature slightly above ambient (Figure 13). During the day, as rock temperatures attained temperatures between about 29 °C and 32 °C, the littorinid body temperatures generally approximate that of the ambient rock temperatures. Then, as rock temperatures rise above about 33 °C, the mean body temperatures of these littorinids somehow are maintained below that of the ambient rock temperatures. The first thought as to how this might be achieved is by evaporative cooling, but mean total body water loss values determined for these species (Britton, 1992) make this unlikely. The mechanism by which such observed thermoregulation among these littorinids occurs requires more study.

Summary relationships among the platform littorinids

This study has uncovered several striking similarities and differences among the Jamaican rocky shore littorinids. Of the five species receiving primary treatment, *N. riisei*, *N. angustior*, *N. dilatata*, *T. antonii* and *C. muricatus*, the first two are the most similar ecologically. Although *N. riisei* is shorter, more globose and with a relatively larger aperture than *N. angustior* (Bandel & Kadolsky, 1982) and apparently has a more restricted distribution toward the front of the shore, these species nevertheless share several features which distinguish them from other Jamaican littorinids. Both have darkly pigmented 'zebra-striped' shells, a feature they share with only *N. ziczac*. Both are usually found on the front third of the rock platform (Figure 2), but *N. angustior* also occurs on the central portion of the platform (Figure 7). Both prefer moist or wet and moderately to highly irregular rock surfaces (Figure 3). Both apparently recruit juveniles on the shore during spring (Figure 11). Both prefer to repose on relatively more exposed rock surfaces than within the shelter of crevices (Table 4) and few individuals of either species produce a mucous holdfast. Except for those *N. angustior* which occupy the central part of the platform, these two species prefer shore sites which experience periodic wetting, either by tides, waves or the standing water of tidepools. In this respect, they are similar to

N. ziczac, but because of its rarity on this shore, no additional comparisons can be made.

N. dilatata was the most widely distributed species on the Jamaican platform (Figures 2 and 8) and displayed the broadest tolerances for several ecological conditions. Although it favoured moderately irregular rock surfaces, it was found in repose from very smooth to highly irregular substrata (Figure 3). It showed a preference for wet substratum moisture conditions (Figure 3), but almost 30% of the individuals occupied a dry substratum. It was consistently represented by both large (> 11 mm) and small (< 8 mm) individuals, with the latter preferentially distributed toward the front half of the rock platform (Table 2). It is unclear whether the presence of small individuals of *N. dilatata* from January to July represent a prolonged winter to summer recruitment period, whether individuals are being recruited from the plankton onto this shore throughout the year or whether a portion of the *N. dilatata* has a persistent prolonged neotenous stage. The latter seems unlikely for at least two reasons. First, prolonged neoteny has been observed most often when individuals occupy especially stressful environments (Britton, 1995; Reid, 1996; De Wolf et al., 1994). The environment occupied by small *N. dilatata* in Jamaica is not especially stressful. Second, the smaller individuals of species which may express prolonged neoteny, such as *L. striata* of the eastern Atlantic Macaronesian islands (Britton, 1995) or *N. australis* of Australia (Reid, 1996) are also noticeably more nodulose than their larger conspecifics. Again, this is not the case with *N. dilatata* (Metz, unpublished data). *N. dilatata* may be recruited on this shore over a prolonged period, but the details of such recruitment is unknown. *N. dilatata*, especially smaller indivduals, prefers to repose in crevices. It rarely employs a mucous holdfast to attach to the substrtaum.

The shell of *T. antonii* resembles that of *N. dilatata* but bears larger, more coarse, nodules, a distinctly different operculum (Bandel & Kadolsky, 1982) and is generally more grey and less yellowish than *N. dilatata*. *T. antonii* has a more restricted range than *N. dilatata* (Figure 2) on the central platform. Unlike populations on other shores which favour highly microkarsted substrata (Peckol et al., 1989), we found *T. antonii* mostly associated with slightly to moderately irregular rock surfaces on this rock platform. It occupies a broad range of moisture conditions from dry to wave-wetted rocks (Figure 3). Small individuals of this species, although always uncommon, were most abundant in March (Figure 11). If these individuals represented annual juvenile recruitment, the population is replaced at a low rate and the adults on shore must survive several years. Alternately, and perhaps more likely, we never witnessed the primary juveile recruitment of *T. antonii* on this shore. Unlike populations from the Bahamas where 85% of the individuals sought the shelter of crevices, only about 50% of the individuals of *T. antonii* on the Jamaican rock platform occupied them (Table 4). Only 20% of the individuals examined were found to attach to the substratum by means of a mucous holdfast (Table 5).

C. muricatus is the largest littorinid on this shore (Figure 2) and occupies its highest elevations, favouring dry, relatively smooth substrata (Figure 3). We never encountered an individual of this species with a shell height less than 9 mm and a surprisingly small number less than 15 mm (Figure 11). Other evidence (e.g. Lewis, 1960) suggests that juveniles are most likely recruited on this shore in the autumn. Because *C. muricatus* occupies the relatively smoother parts of the shore, a significant percentage of the population are unable to find shelter within a crevice or along a surface irregularity. The majority of the population (55%) do repose within a crevice (Table 4), and since such shelters are less common within a portion of the shore occupied by this species, the spatial distribution of *C. muricatus* is decidedly clumped. This species is also more likely than not to attach to the substratum by means of a mucous holdfast (Table 5).

Four of the Jamaican platform littorinids for which body temperatures were attained, that is *N. riisei*, *N. angustior*, *N. dilatata* and *C. muricatus*, seem to display a moderate thermoregulatory ability. At ambient substratum temperatures below 28 °C, the mean body tissue temperatures of these species were generally one or two degrees above the ambient temperature. Between about 29 to 32 °C, the mean body tissue temperature of these species approximated that of ambient substratum temperature. When the latter exceeded about 32 °C, the mean body tissue temperatures of these species were generally one to three degrees lower (Figures 16 and 17). The mechanism(s) by which thermoregulation is accomplished by these species is unknown. *T. antonii* did not display any clear evidence of thermoregulation (Figure 14).

Acknowledgements

We are indebted to the Hofstra University Marine laboratory for lodging, logistics and laboratory support without which this study would not have been possible. We also are appreciative of the help provided by many students of the Texas Christian University Tropical Biology classes, 1994 and 1996, including many hours of data recording and field sampling. R. C. L. is especially indebted to Dr Michael Clarke, who served as field assistant during July 1994, and shared the agony of conducting the 12-hour temperature profile. Some of the financial support for this study was provided by the Texas Christian University Department of Biology Atkins Fund.

References

Abbott, R. T., 1954. Review of the Atlantic periwinkles, *Nodilittorina, Echininus* and *Tectarius*. Proc. U.S. natn. Mus. (Nat) 103: 449–464.

Abbott, R. T., 1964. *Littorina ziczac* (Gmelin) and *L. lineolata* Orbigny. Nautilus 78: 65–66.

Bandel, K., 1974. Studies on Littorinidae from the Atlantic. Veliger 17: 92–114.

Bandel, K. & D. Kadolsky, 1982. Western Atlantic species of *Nodilittorina* (Gastropoda:Prosobranchia): Comparative morphology and its functional, ecological, phylogenetic and taxonomic implications. Veliger 25: 1–42.

Bequaert, J. C., 1943. The genus *Littorina* in the western Atlantic. Johnsonia 1: 1–27.

Bingham, F. O., 1972. The mucus holdfast of *Littorina irrorata* and its relationship to relative salinity and humidity. Veliger 15: 48–50.

Bingham, F. O., 1973. The influence of environmental stimuli on the direction of movement of the supralittoral gastropod *Littorina irrorata*. Bull. mar. Sci. 22: 309–335.

Borkowski, T. V., 1971. Reproduction and reproductive periodicities of southern Floridian Littorinidae (Gastropoda: Prosobranchia). Bull. mar. Sci. 21: 826–840.

Borkowski, T. V., 1974. Growth, mortality and productivity of southern Floridian Littorinidae (Gastropoda: Prosobranchia). Bull. mar. Sci. 24: 409–438.

Borkowski, T. V. & M. R. Borkowski, 1969. The *Littorina ziczac* species complex. Veliger 11: 408–414.

Britton, J. C., 1976. The shallow water marine mollusks of the Swan Islands, Honduras. Bull. am. Malacological Un. Inc. for 1975: 33–40.

Britton, J. C., 1992. Evaporative water loss, behaviour during emersion, and upper thermal tolerance limits in seven species of eulittoral-fringe Littorinidae (Mollusca:Gastropoda) from Jamaica. In J. Grahame, P. J. Mill & D. G. Reid (eds), Proceedings of the Third International Symposium on Littorinid Biology: 69–83.

Britton, J. C., 1995. The relationship between position on shore and shell ornamentation in two size-dependent morphotypes of *Littorina striata*, with an estimate of evaporative water loss in these morphotypes and *Melarhaphe neritoides*. In P. J. Mill & C. D. McQuaid (eds), Developments in Hydrobiology 111. Advances in Littorinid Biology, Kluwer Academic Press, Dordrecht, The Netherlands: 129–142.

Britton, J. C., R. F. McMahon & J. Hart, 1991. Relationships between topography, substratum composition and surface temperature, and the spatial distribution of intertidal fauna on rocky shores of south-western Australia. In F. E. Wells, D. I. Walker, A. H. Kirkman & R. Lethbridge (eds), Proceedings of the Third International Marine Biological Workshop: The Marine Flora and Fauna of Albany, Western Australia,Volume II. Western Australian Museum, Perth: 521–540.

DeWolf, H., S. Van Dongen, C. Brito & T. Backeljau, 1994. Population genetics of *Littorina striata* at a microgeographical scale. Cah. Biol. mar. 35: 243–244.

Garrity, S. D., 1984. Some adaptations of gastropods to physical stress on a tropical rocky shore. Ecology 65: 559–574.

Hedgpeth, J. W., 1953. An introduction to the zoogeography of the northwestern Gulf of Mexico, with references to the invertebrate fauna. Proc. Inst. Mar. Sci., University of Texas 3: 110–224.

Lewis, J. B., 1960. The fauna of rocky shores or Barbados, West Indies. Can. J. Zool. 38: 391–435.

McMahon, R. F., 1990. Thermal tolerance, evaporative water loss, air-water oxygen consumption and zonation of intertidal prosobranchs: a new synthesis. Hydrobiologia 193: 241–260.

McMahon, R. F. & J. C. Britton, 1985. The relationship between vertical distribution, thermal tolerance, evaporative water loss rate, and behaviour on emergence in six species of mangrove gastropods from Hong Kong. In B. Morton & D. Dudgeon (eds), Proceedings of the Second International Workshop on the Malacofauna of Hong Kong and Southern China, Volume 2. Hong Kong University Press, Hong Kong: 563–582.

McMahon, R. F. & J. C. Britton, 1991. The relationship between vertical distribution, rate of evaporative water loss, behavior during emergence, and morphometrics in six species of rocky shore gastropods from Princess Royal Harbour, Western Australia. In F. E. Wells, D. I. Walker, A. H. Kirkman & R. Lethbridge (eds), Proceedings of the Third International Marine Biological Workshop: The Marine Flora and Fauna of Albany, Western Australia, Volume II. Western Australian Museum, Perth: 675–692.

McQuaid, C. D. & P. A. Scherman, 1988. Thermal stress in a high shore environment: morphological and behavioural adaptations of the gastropod *Littorina africana*. In G. Chelazzi & M. Vannini (eds), Behavioural Adaptation to Intertidal Life. Plenum Press, New York: 213–224.

Mileikovsky, S. A., 1975. Types of larval development in Littorinidae (Gastropoda:Prosobranchia) of the world ocean, and ecological patterns of their distribution. Mar. Biol. 30: 129–135.

Newell, R. C., 1979. Biology of Intertidal Animals. Marine Ecological Surveys Ltd., Faversham, Kent, ix + 781 pp.

Peckol, P., S. Guarnagia & M. Fisher, 1989. Zonation and behavioral patterns of the intertidal gastropods *Nodilittorina (Tectininus) antonii* (Philippi, 1846) and *Nerita versicolor* Gmelin, 1791, in the Bahamas. Veliger 32: 8–15.

Reid, D. G., 1989. The comparative morphology, phylogeny and evolution of the gastropod family Littorinidae. Phil. Trans. r. Soc., Lond. B 324: 1–110.

Reid, D. G., 1996. Systematics and evolution of *Littorina*. The Ray Society, London. 463 pp.

Rodriguez, G., 1959. The marine communities of Margarita Island, Venezuela. Bull. mar. Sci. Gulf Caribb. 9: 237–280.

Vermeij, G. J., 1971. Temperature relationships of some tropical Pacific intertidal gastropods. Mar. Biol. 10: 308–314.

Vermeij, G. J., 1973. Morphological patterns in high intertidal gastropods: Adaptive strategies and their limitations. Mar. Biol. 20: 319–346.

Wolcott, T. G., 1973. Physiological ecology and intertidal zonation in limpets (*Acmaea*): a critical look at 'limiting factors'. Biol. Bull. 145: 389–422.

Hydrobiologia **378:** 187–191, 1998.
R. M. O'Riordan, G. M. Burnell, M. S. Davies & N. F. Ramsay (eds), Aspects of Littorinid Biology.

Seasonal variation in soluble uric acid concentration in *Littorina saxatilis* (Olivi)

D. C. Smith[1] & T. P. Smith[2]
[1] Dept. of Biological Sciences, State University of New York, Brockport, NY 14420, U.S.A.
[2] Dept. of Physics, University of New Hampshire, Durham, NH 03824, U.S.A.

Key words: *L. saxatilis*, uric acid, seasonality

Abstract

13 samples of *Littorina saxatilis* were collected from a Maine shore on an approximately monthly basis and analysed for soluble uric acid concentration. This concentration varied significantly, with summer concentrations as much as 6 times higher than winter concentrations. While it may be that the accumulation of uric acid in summer is related to desiccation, we suggest that it is more likely to be a result of increased protein catabolism.

Introduction

The role of uric acid in the conservation of water in gastropods has been recognised since the classic papers of Delaunay (1931) and Needham (1935, 1938). Uricotelism is more energy consuming than ammonotelism, but the advantage of uric acid is that, being highly insoluble in water, it can be precipitated in the excretory organ and then voided with minimal loss of water (Potts, 1967). The actual end-product of nitrogen metabolism (ammonia, urea, or uric acid) is a function of the toxicity of ammonia and the water available for its excretion. Thus aquatic gastropods are generally ammonotelic while terrestrial gastropods are ureotelic or uricotelic, though apparently all gastropods accumulate uric acid to some degree (Duerr, 1967).

Littoral gastropods may be thought of as being intermediate between aquatic and terrestrial snails with regard to their mode of nitrogen excretion; species living high in the intertidal zone contain more uric acid than those from lower levels (Potts, 1967).

The gastropod body may accumulate considerable amounts of uric acid under various conditions. Increased amounts of uric acid have been shown to occur during aestivation (Delaunay, 1931; Haggag & Fouad, 1968; Speeg & Campbell, 1968; Nayeemunnsia, 1972) and starvation (not associated with desiccation) (Emerson & Duerr, 1966). Seasonal changes in uric acid have been reported in land snails (Jezewska et al., 1963; Laziradou-Dimitriadou & Kaloyianni, 1989). *Littorina littorea*, which occupies a region lower in the intertidal zone than *L. saxatilis*, was reported by Delaunay (1931) to have a higher rate of uric acid excretion in summer than in winter. Spitzer (1937) measured more uric acid in the digestive gland of *L. littorea* in summer than in winter, but reported that the reverse is the case in the nephridium.

A previous study (Smith et al., 1995) reported that position on the shore significantly altered uric acid content of *L. saxatilis* and *L. arcana*. It was noted in that study that a possible seasonal variation in uric acid content may exist in these species. For this reason this present study was conducted to determine whether uric acid varied seasonally over an extended period, with samples of *L. saxatilis* taken on an approximately monthly basis from a single site on the Maine coast.

Materials and methods

Collections

13 collections of periwinkles (*Littorina saxatilis*) were made between July 1993 and September 1994 at approximately monthly intervals, with the average sam-

ple size being 36 snails. Animals were selected to assure that the samples included a range of body sizes; while the range therefore was large (from 10.0 to 209.6 mg), the mean weight of snails collected in the winter months (71.5 mg) was not significantly different than that of snails collected in the summer months (75.2 mg). Collections were made within a band 4 m horizontal and 1 m vertical on the surface of a rock outcrop near the upper limit of the periwinkle's range at Seapoint Beach near Kittery, Maine. During the winter, snails were very scarce on the open surface and had to be collected from shallow crevices or nearby boulders. Collections were made at the low tide occurring nearest 10:00 am; all but three (Aug. 1993, Sept. 1993, Nov. 1993) were made during the morning hours.

Uric acid determination

This method is the same as that which was reported in an earlier paper (Smith et al., 1995). The shells were cracked and the body removed under a dissecting microscope. The body was weighed and then homogenised in 100 ml of phosphate buffer (pH 7.8) at room temperature. After centrifugation the supernatant was removed and frozen for later analysis for uric acid.

The supernatant was assayed using a modification of the Sigma Diagnostics procedure no. 685. This procedure essentially follows the method of Duerr (1967) and utilises uricase to promote the oxidation of uric acid to allantoin, CO_2 and H_2O_2. The H_2O_2 then reacts, in the presence of peroxidase, with 4-aminoantipyrine and 3,5-dichloro-2-hydroxybenzenesulphonate to form a quinoneimine dye. The intensity of the colour formed is proportional to the concentration of uric acid and is read at 520 nm. Concentration is reported as μg of uric acid per gram ($\mu g\ g^{-1}$) of tissue (wet weight). This method of extracting uric acid would not have extracted insoluble, crystalline uric acid; therefore uric acid concentrations in this paper refer to the soluble fraction of the uric acid body burden.

Water content

Three collections of 31 animals each were made to determine any seasonal change in water content. The summer sample was collected on July 9, the autumn on November 8, and the winter on January 29. The animals were transported to the laboratory in sealed plastic bags containing wet seaweed. The shell was cracked and the live body was extracted, blotted briefly on filter paper, weighed, and placed in a drying oven at 56 °C for 24 hours. The animals were reweighed and the percent decrease in weight is taken as the water content. Statistical testing between sample means was by t-test following arcsin transformation.

Temperatures

Mean monthly air temperatures for the study period were obtained from the Climatological Data (NOAA, 1993, 1994) from the nearest coastal station, Kennebunkport, Maine and are shown in Table 1. Mean air temperatures were also taken from the same source for the day of collection and the two days preceding the collection; the mean of these three days' temperatures was used in Figure 2 and in determining correlation between temperature and soluble uric acid concentration.

Statistics

Testing for significance between sample means was done using Student's t-test; in the case of comparing water contents, where the measurement was a percentage, arcsin transformation preceded statistical testing. Correlation testing was done using product-moment correlation (r). In all cases, statistical significance was taken to be $p < 0.05$ unless otherwise noted.

Results

Seasonal changes in uric acid

Figure 1 shows the mean sample concentrations and statistical significance of uric acid over the 15 month (13 samples) study period. From the July 1993 concentration of $208.0 \pm 14.6\ \mu g\ g^{-1}$ (mean $\pm$ SE) the uric acid concentration declined through the autumn months to reach a low of $54.1 \pm 6.5\ \mu g\ g^{-1}$ on January 30; nearly a fourfold reduction. Concentrations increased during the spring months to reach the highest level seen during the study period, $346.1 \pm 28.4\ \mu g\ g^{-1}$ in September 1994, greater than a sixfold increase from the January 30 sample and, in fact, significantly greater than the highest concentration of the previous summer.

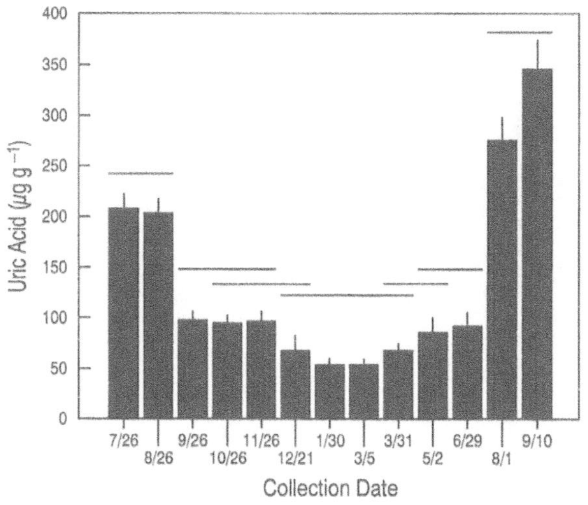

Figure 1. Mean concentration (mean + SE) of uric acid ($\mu g\ g^{-1}$ wet weight) in winkles (*Littorina saxatilis*) collected between July 26 1993 and September 10 1994. Horizontal lines above bars indicate adjacent samples or groups of samples that do not differ significantly ($p > 0.05$).

Table 1. Monthly mean maximal and minimal temperatures (°C) for the nearest NOAA reporting coastal station, Kennebunkport, Maine during the collection period

Collection	Mean Max. Temp.	Mean Min. Temp.
1993 Jul.	25.4	14.1
Aug.	24.6	12.9
Sep.	19.3	9.3
Oct.	13.1	0.4
Nov.	9.3	−2.9
Dec.	3.3	−6.5
1994 Jan.	−4.6	−16.7
Feb.	−2.0	−14.2
Mar.	4.2	−4.4
Apr.	11.7	0.1
May.	15.4	5.2
Jun.	22.5	12.0
Jul.	25.4	15.9
Aug.	23.6	12.8
Sep.	19.5	8.8

Water content

Water content was remarkably uniform within samples, and changed little between the three seasonal samples taken. The summer collection had a water content of 72.64 ± 0.18% (mean ± SE); the autumn sample was 73.31 ± 0.41%, and the winter sample 74.31 ± 0.31%. There was no statistically significant difference between any of the samples (t-test following arcsin transformation, $p>0.05$).

Temperatures

Table 1 shows the mean maximum and mean minimum air temperatures for each of the months of collecting. Additionally, daily mean air temperatures were obtained from the same source (NOAA 1993, 1994) for each day of collection as well as the two days prior to collection. These three daily temperatures were averaged and fitted to a linear regression against mean body burdens of soluble uric acid. The results are shown in Figure 2. There is a significant correlation between ambient air temperature and soluble uric acid concentration ($r = 0.669, p < 0.05$).

Discussion

Our data show remarkably large changes in uric acid content of snails over the course of the seasons, with the lowest concentrations reached in winter, and 4 to 6 fold greater concentrations achieved in mid to late summer. We have no immediate explanation for the fact that the concentrations reached during the second summer of the study were so much greater than those of the first summer. The data show transitions during the autumn and spring months, but in both cases rather sharp changes occur between late summer (Aug. 26) and early autumn (Sep. 26) and again between early summer (Jun. 29) and late summer (Aug. 1) which suggest fairly abrupt physiological adjustments to seasonal demands.

If the desiccating conditions of summer actually caused a great reduction in water content, then the high concentration of uric acid would be simply a concurrent effect of water loss. We have shown, however, that water content actually changes very little from summer to winter, and is remarkably uniform within samples; much more so than is uric acid concentration. The changes in soluble uric acid concentration therefore do not merely reflect water gain or loss, but represent real changes in uric acid production or retention. It is well established that uric acid production helps to conserve water and protect against the increased osmotic pressure that occurs during desiccation (Needham, 1935; Duerr, 1967; Heil & Eichelberg, 1983; Lazaridou-Dimitriadou & Kaloyianni, 1989). It is therefore tempting to postulate that the high uric acid concentrations seen in summer are due to desic-

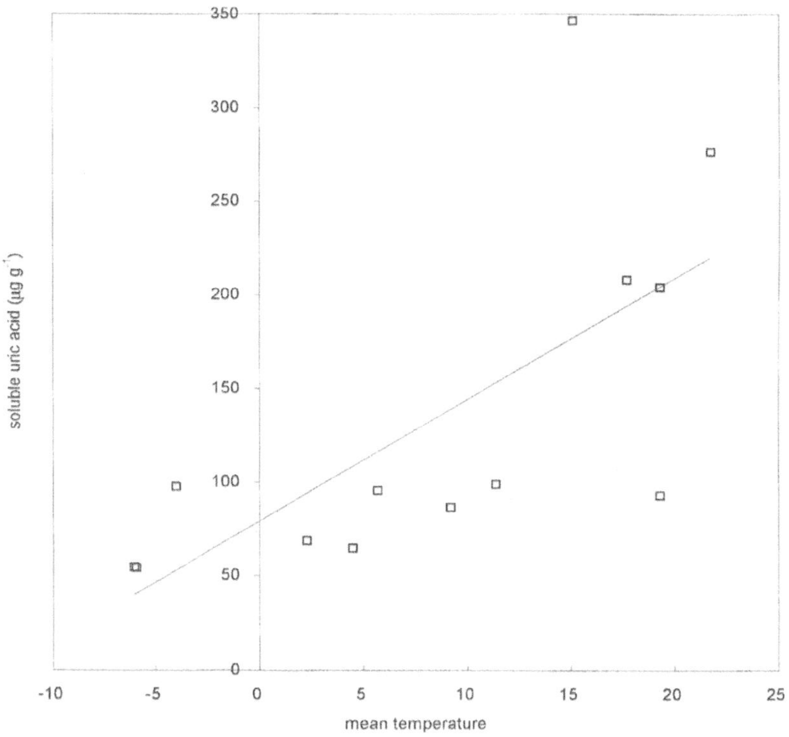

Figure 2. Soluble uric acid concentration (μg g^{-1}) as a function of ambient air temperature (mean temperature (°C) for day of collection and the two previous days). Each point represents the mean concentration for a monthly sample (at one point two samples are superimposed). The computed regression line is shown, $r = 0.669$, $p < 0.05$.

cation resulting from higher temperatures. Snails are not subjected to high temperatures in the winter; moreover, it was our observation that, in the winter, snails were not often found on the face of the rock, but often had to be sought in shallow crevices in the rock. It has previously been shown (Smith et al., 1995) that winkles occupying crevices have lower uric acid concentrations than animals on boulder faces; presumably those in the crevices are protected from desiccation effects. While the higher uric acid of snails in summer could be due to desiccation, we believe that this is unlikely because of the remarkably stable body water content. Daguzan & Razet (1972) and Daguzan (1976) have identified various physical factors (such as tide cycle, emersion period, etc.) and biotic factors (size, sex) that affect nitrogenous excretion. We have attempted to control for these factors by consistency in collecting time relative to the tidal cycle and by including a wide range of body size in each sample.

Since these animals show no evidence of desiccation stress it seems unlikely that the seasonal changes in body burden of soluble uric acid that we observed were a result of increased summer desiccation. Rather it may be a result of the effects of seasonal temperature change on metabolic rate and the protein catabolism required to support it. During winter the animals are clearly inactive in crevices at low temperature. Thus their metabolic rates are greatly reduced compared to snails in summer. Reduced metabolic rate would result in reduced soluble uric acid body burdens. In contrast, increased temperatures, active foraging and increased tissue growth and reproductive effort during the summer months lead to a greatly increased metabolic rate and, therefore, a greatly increased rate of uric acid production and the corresponding increases in soluble uric acid burdens we have described. The clear correlation between soluble uric acid content and ambient air temperature is evidence that seasonal variation in metabolic rate, and, therefore, protein catabolic rate, is the main driver of seasonal variation in soluble uric acid level. Since winkles may store all of the uric acid that they produce through life, the seasonal changes in uric acid would probably not be observed if total body burden of uric acid (soluble + insoluble) had been measured rather than the soluble fraction alone.

Acknowledgements

This research was supported by a Faculty Incentive Award from SUNY College at Brockport. The authors wish to acknowledge the able technical assistance of Mr Will Smith.

References

Daguzan, J., 1976. Variations de l'excretion azotée et de la teneur en eau en fonction de l'âge, du sexe et de divers facteurs ecologiques chez les *Littorinidae* armoricains (Mollusques Gasteropodes Prosobranches). Haliotis 6: 247–253.

Daguzan, J. & P. Razet, 1972. Rélation entre l'excretion azotée, la marée et l'âge. Arch. Sci. Physiol. 26: 401–413.

Delaunay, H., 1931. L'excretion azotée des invertebres. Biol. Rev. 6: 265–301.

Duerr, F. G., 1967. The uric acid content of several species of marine prosobranch snails. Comp. Biochem. Physiol. 22: 333–340.

Emerson, D. & F. G. Duerr. 1966. Some physiological effects of starvation in the intertidal prosobranch, *Littorina planaxis* (Philippi, 1847). Comp. Biochem. Physiol. 20: 45–53.

Haggag, G. & Y. Fouad, 1968. Comparative study of nitrogenous excretion in terrestrial and fresh-water gastropods. Z. vergl. Physiol. 57: 428–431.

Heil, H. P. & D. Eichelberg, 1983. Untersuchungen zum Harnsauremetabolismus von *Littorina littorea* (Gastropoda). Helgolander miss. Meeresunters. 35: 465–472.

Jezewska, M. M., B. Gorzkowski & J. Heller, 1963. Seasonal changes in the excretion of nitrogen wastes in *Helix pomatia*. Acta Biochim. Pol. 10: 309–314.

Laziradou-Dimitriadou, M. & M. Kaloyianni, 1989. Age, seasonal and diel variations of the uricolytic enzymatic system in the snails *Xeropicta arenosa* (Ziegler) and *Cernuella virgata* (Da Costa). J. Comp. Physiol. B 159: 115–121.

Nayeemunnsia, 1972. Aspects of nitrogen excretion during aestivation in the amphibious snail, *Pila globosa*. I. Evidence for neuroendocrine involvement in the pattern of excretion. Comp. Biochem. Physiol. 42A: 1049–1053.

Needham, J., 1935. Problems of nitrogen catabolism in invertebrates. II. Correlation between uricotelic metabolism and habitat in the Phylum Mollusca. Biochem. J. 29: 238–251.

Needham, J., 1938. Contributions of chemical physiology to the problem of reversibility in evolution. Biol. Revs. Cambridge Phil. Soc. 13: 225–251.

NOAA, 1993. Climatological Data, New England. 105, nos. 7–12.

NOAA, 1994. Climatological Data, New England. 106, nos. 1–9.

Potts, W. T., 1967. Excretion in the molluscs. Biol. Rev. 42: 1–41.

Smith, D. C., P. J. Mill & J. Grahame, 1995. Environmentally induced variation in uric acid concentration and xanthine dehydrogenase activity in *Littorina saxatilis* (Olivi) and *L. arcana* (Hannaford Ellis). Hydrobiologia 309: 111–116.

Speeg, K. V. & J. W. Campbell, 1968. Purine biosynthesis and excretion in *Otala (Helix) lactea:* an evaluation of the nitrogen excretory potential. Comp. Biochem. Physiol. 26: 579–595.

Spitzer, J. M., 1937. Physiologisch-okologische untersuchungen uber den excretstoffwechsel der mollusken. Zool. Jb. 57: 457–496.

Hydrobiologia **378:** 193–197, 1998.
R. M. O'Riordan, G. M. Burnell, M. S. Davies & N. F. Ramsay (eds), Aspects of Littorinid Biology.

Intersex and *Littorina littorea* in Cork Harbour: results of a medium-term monitoring programme

J. D. Casey, S. De Grave & G. M. Burnell
Department of Zoology and Animal Ecology, University College, Cork, Ireland

Key words: Littorina littorea, intersex, tributyltin, Cork Harbour, bio-monitoring

Abstract

Cork Harbour, located on the south coast of Ireland, is a complex, multi-user environment. In the present study *Littorina littorea* (L.) is used as a bio-indicator of TBT pollution, as it has a wide range within the harbour, unlike the more common bio-indicator, *Nucella lapillus* (L.), which is limited to rocky substrata near the mouth of the harbour. Intersex in *L. littorea* is used for the first time in Ireland as a medium-term indicator of TBT pollution and results show that there are no deviations in levels of intersex over the three year period from 1994–1996. These results are discussed.

Introduction

Cork Harbour, located on the south coast of Ireland, is a busy port with a high volume of shipping traffic. There is a large number of manufacturing plants in the area including those producing pharmaceuticals, fertilisers and steel, plus oil refineries and shipyards. In addition, the harbour contains a naval base and a large international ferryport.

TBT-based antifoulants were first used in the mid-1960s and became very popular throughout the 1970s. They were used primarily by the leisure-craft industry, and areas of high input would have been areas where there was a large number of yachts, such as the Owenboy Estuary in the south western part of Cork Harbour. The first direct evidence of TBT contamination in Irish waters was from the Owenboy Estuary in 1985, where Pacific oysters (*Crassostrea gigas*) were found to have distorted shells associated with TBT contamination (Minchin & Duggan, 1986).

A number of studies have looked at imposex in neogastropod molluscs which results in the formation of male sexual characteristics in females. This phenomenon has been described for a large number of neogastropods (Gibbs & Bryan, 1986; Fioroni et al., 1991; Stroben et al., 1992; Kohn & Almasi, 1993; Horiguchi et al., 1994). In European waters, *Nucella lapillus* (L.), which is very sensitive to imposex development, is widely used as a monitor of TBT pollution (Gibbs et al., 1987; Minchin et al., 1995). A related phenomenon, intersex, was first described by Bauer et al. (1995) for the periwinkle *Littorina littorea* (L.). Intersex is not fatal and affected populations do not exhibit a male-dominated population structure (Bauer et al., 1995) as can happen with *N. lapillus* in organotin-contaminated areas. At present, intersex appears to be limited to *L. littorea* but it should be noted that only sexually immature and juvenile periwinkles are able to develop intersex, unlike imposex in *N. lapillus* (Bauer, pers. comm). Also, owing to the reproductive strategy of *L. littorea*, which reproduces by means of pelagic larvae, populations do not necessarily become extinct as a result of intersex. The calculated threshold concentration for intersex development is 15 ng TBT as Sn l^{-1} (Bauer et al., 1995), so its reproductive strategy, coupled with its wide habitat range, makes it ideal for monitoring areas of high TBT concentrations, especially in areas in which *N. lapillus* is absent (Bauer et al., 1995).

In 1986, a survey found TBT to be a major problem in Irish waters (Minchin & Duggan, 1986). As a result of this, and other studies on a world-wide basis, a partial ban was introduced in 1987 prohibiting the use of TBT based paints on boats $\leq$ 25 m in length, on aquaculture netting and on all structures found in Irish coastal waters.

To study the effectiveness of the ban, a study was undertaken in 1993, using imposex in *N. lapillus*, to determine TBT levels in Ireland (Minchin et al., 1995). The results of this study showed that there had been a significant decrease in levels of imposex in areas where the ban had imposed restrictions on TBT usage. However, in areas where there was a high intensity of traffic of vessels exempted from the ban, there had been no such reductions, and in some cases there had been increased levels of imposex.

The present study is the first one in Irish waters using intersex in *L. littorea* as part of a medium-term monitoring programme.

Methods

The intersex index (ISI), as described by Bauer et al. (1995), was used. Intersex is defined as a change in the female pallial oviduct towards a male morphological structure. It is based on 5 stages (0–4) describing the change in the female morphological structure, and the intersex index (ISI) is the mean value of intersex stages in the total female population. In the last three stages female *L. littorea* become sterile, though reproductive capability is reduced in Stage 1 due to sperm being lost into the mantle cavity during copulation, as a result of either loss of capsular material (Stage 2) or loss of capsule formation glands (Stage 3 & 4). Unlike imposex, however, the male characteristics do not get super-imposed onto the female structure and a penis is only found in Stage 4. In order to allow a meaningful comparison with the 1994 data, the sampling protocol of Minchin et al. (1996) was closely adhered to, hence only *L. littorea* larger than 20 mm were sampled. Nineteen sites (Figure 1) around Cork Harbour were chosen to correspond with the study carried out in June 1994 by Minchin et al. (1996). Samples from 1995 were collected in August/September and samples from 1996 were collected in July/August.

At each site 45–50 snails were collected from the mid-shore level wherever possible. Snails were immediately stored at –20 °C until dissections took place. The shell height and aperture height were measured to the nearest 0.1 mm, with sliding vernier callipers. Shells were crushed using a vice and the soft tissue removed. Specimens were sexed and parasitised individuals, observed by eye under a binocular microscope, were removed from further analysis, as parasites have been known to cause changes in the female morphological structure (Lauckner, 1980), obscuring intersex changes.

Results

After removal of parasitised individuals, a total of 890 snails were included in the statistical analysis. A one-way ANOVA ($F_{2,889} = 6.38$; $p<0.05$) was used to determine if there was any difference between sampled shell height over the three years. *A posteriori* testing by means of the Tukey Honestly Significant Difference (HSD) test revealed that there was no significant difference between the 1994 and 1995 samples but that the mean height of the 1996 population was significantly smaller than either 1994 or 1995 (Table 1). As the employed intersex scale is essentially categorical in nature, no parametric statistical tests can be used, nor should any derived statistical indices, such as means and standard deviations, be calculated. On the other hand, Bauer (pers. comm.) has pointed out the extreme importance of the potential influence of shell height in determining intersex, as this condition can only be induced in juveniles/sexually immature individuals. As no non-parametric equivalent of an ANCOVA test exists, this can only be addressed by reducing the data set to the level at which this influence is no longer detected. A preliminary analysis of the relationship between shell height and intersex presence, using logit regression had demonstrated that a significant influence was exerted by shell height on ISI presence. This influence was removed from the data set by sequentially removing the largest/smallest size classes and running further logit regressions. By reducing the entire data set to only 592 snails, with shell heights between 22.0 mm and 28.9 mm, this was achieved. A one-way ANOVA test on shell heights of the reduced data set revealed that there was a significant difference in shell height over the three years studied ($F_{2,589} = 3.89$; $p<0.05$), similar to the one observed for the full data set.

A series of Kruskal-Wallis (non-parametric ANOVA) tests was run on the inter-annual differences per sampled site and it was deduced that there was no significant difference in ISI values at each of the individual sites over the three years ($p > 0.05$) (Table 2). Subsequently, the data from all sites were combined, indicating that there was no difference in ISI levels over the three year period ($H = 0.11$, $p =$ n.s).

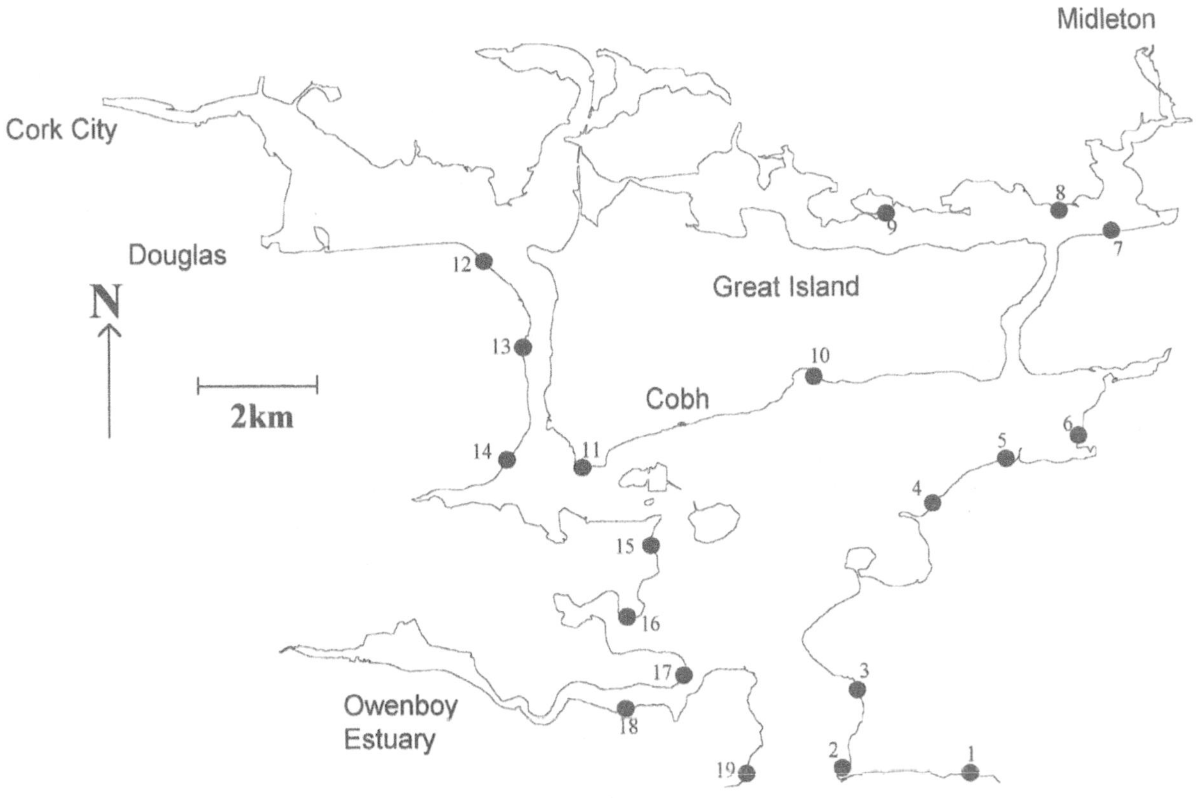

Figure 1. Map of Cork Harbour, indicating sampling sites.

Table 1. Inter annual variation in sampled shell height for full and reduced data sets. SD = Standard Deviation, *N* = Sample Size. (1994 data from Minchin et al., 1996)

	Year	Mean Height	SD	*N*
Full data set	1994	25.599	2.799	296
	1995	25.573	3.054	305
	1996	24.842	2.872	289
Reduced data set	1994	25.137	1.66	211
	1995	25.364	1.602	182
	1996	24.912	1.479	199

Discussion

Imposex in *N. lapillus* has been used extensively as a medium to long-term bio-indicator of TBT contamination. Evans et al. (1996) used imposex to determine a decrease in TBT levels over time periods ranging from 4 years to 10 years in areas of high commercial shipping traffic. Minchin et al. (1995) have also used imposex as a medium to long-term bio-indicator in Ireland, finding an increase in imposex levels in areas where there was a high frequency of commercial shipping, including Cork Harbour, from 1987 to 1993, but decreasing in areas of aquaculture and marinas over the same time period.

The possibility of size overlap, resulting from the sampling protocol followed, can be accounted for by looking at the mean shell height data from Table 1. Assuming that zero growth is not possible over the period 1994 to 1995, and similarly that negative growth is not possible for the period 1995 to 1996, then it can be assumed that in each year a different year class was sampled.

Based on statistical analysis of ISI levels from 1994–1996 (Table 2), it appears that levels of TBT contamination have remained unchanged over this period. However, as there are no recorded levels of intersex within Cork Harbour previous to 1987, it is not possible to determine if there has been any deviation from levels prior to the introduction of the ban.

Table 2. Results for Kruskall-Wallis tests per sampled site

No.	SITE	*n*	*H*	*p*
1	Gyleen	25	0.162	n.s
2	Roches Point	25	0.859	n.s
3	White Bay	26	0.423	n.s
4	Aghada	19	1.533	n.s
5	Aghada Pier	32	2.185	n.s
6	Rostellan Pier	36	2.682	n.s
7	Rathcourcey	33	2.017	n.s
8	Ahanesk Pier	36	0.401	n.s
9	Brick Island	39	0.427	n.s
10	Cuskinny	34	0.924	n.s
11	Black Point	29	0.599	n.s
12	Passage	25	1.222	n.s
13	S. of Ferry	29	0.049	n.s
14	Monkstown	34	0.048	n.s
15	Golden Rock	44	2.310	n.s
16	Lough Beg	30	1.246	n.s
17	Currabinny	30	0.535	n.s
18	RCYC	28	0.296	n.s
19	Church Bay	38	0.108	n.s

A number of other factors which are known to influence TBT levels, such as high fine sediment concentrations (Langstone & Pope, 1995) and sewage outfalls (Fent & Müller, 1991), are found within the harbour, and high levels of intersex can be related to these sources. The highest incidence of intersex is found at the Royal Cork Yacht Club (RCYC) in the Owenboy Estuary. This area was subjected to a heavy input of TBT from two yacht marinas prior to the ban in 1987. Though this source was removed following implementation of the 1987 ban, it has been shown that TBT binds to the sediment (De Mora et al., 1989), and as there are extensive mud flats in this area, the high levels of intersex may be related to this. The levels of intersex found in Cork Harbour over the three year period are comparable with values found in other areas with a high TBT input, such as Killybegs (Minchin et al., 1997). Future research should aim at elucidating the relationship between organotin source, sediment, water and intersex levels in coastal embayments, which are subjected to heavy organotin inputs.

Acknowledgements

The authors would like to thank the following: P. Randall, A. Power and J. Finn for assistance in sampling over the two years; Dr D. Minchin for his assistance with the 1994 data and B. Bauer for her help in identifying intersex stages.

References

Bauer, B., P. Fioroni, I. Ide, S. Liebe, J. Oehlmann, E. Stroben & B. Watermann, 1995. TBT effects on the female genital system of *Littorina littorea*: a possible indicator of tributyltin pollution. Hydrobiologia 309: 15–27.

De Mora, S. J., N. G. King & M. C. Miller, 1989. Tributyltin and total tin in marine sediments: Profiles and the apparent rate of TBT degradation. Envir. Technol. Lett. 10: 901–908.

Evans, S. M., P. M. Evans & T. Leksono, 1996. Widespread recovery of dogwhelks, *Nucella lapillus* (L.), from tributyltin contamination in the North Sea and Clyde Sea. Mar. Pollut. Bull. 32: 263–269.

Fent, K. & M. D. Müller, 1991. Occurrence of organotins in municipal wastewater and sewage sludge and behaviour in a treatment plant. Envir. Sci. Tech. 25: 489–493.

Fioroni, P., J. Oehlmann & E. Stroben, 1991. The pseudohermaphroditism of prosobranchs; morphological aspects. Zool. Anz. 226: 1–26.

Gibbs, P. E. & G. W. Bryan, 1986. Reproductive failure in populations of dogwhelk, *Nucella lapillus*, caused by imposex induced by tributyltin from antifouling paints. J. mar. biol. Ass. U.K. 66: 767–777.

Gibbs, P. E., G. W. Bryan, P. L. Pascoe & G. R. Burt, 1987. The use of the dogwhelk *Nucella lapillus*, as an indicator of tributyltin (TBT) contamination. J. mar. biol. Ass. U.K. 67: 507–523.

Horiguchi, T., H. Shiaishi, M. Shimizu & M. Morita, 1994. Imposex and organotin compounds in *Thais clavigera* and *T. bronni* in Japan. J. mar. biol. Ass. U.K. 74: 651–669.

Kohn, A. J. & K. N. Almasi, 1993. Imposex in Australian *Conus*. J. mar. biol. Ass. U.K. 73: 241–244.

Langstone, W. J. & N. D. Pope, 1995. Determinants of TBT adsorption and desorption in estuarine sediments. Mar. Pollut. Bull. 31: 32–43.

Lauckner, G., 1980. Diseases of Mollusca: Gastropoda. In O. Kinne (ed.), Diseases of marine animals.Wiley & Sons, Chichester: 311–424.

Minchin, D. & C. B. Duggan, 1986. Organotin contamination in Irish waters. ICES CM1986/F:48: 1–8.

Minchin, D., J. Oehlmann, C. B. Duggan, E. Stroben & M. Keatinge, 1995. Marine TBT antifouling contamination in Ireland, following legislation in 1987. Mar. Pollut. Bull. 30: 633–639.

Minchin, D., E. Stroben, J. Oehlmann, B. Bauer, C. B. Duggan & M. Keatinge, 1996. Biological indicators used to map organotin contamination in Cork Harbour, Ireland. Mar. Pollut. Bull. 32: 188–195.

Minchin. D., B. Bauer, J. Oehlmann, U. Schulte-Oehlmann & C. B. Duggan, 1997. Biological indicators used to map organotin contamination from a fishing port, Killybegs, Ireland. Mar. Pollut. Bull. 34: 235–243.

Stroben, E., J. Oehlmann & P. Fioroni, 1992. The morphological expression of imposex in *Hinia reticulata* (Gastropoda: Buccinidae): A potential indicator of tributyltin pollution. Mar. Biol. 113: 625–636.

Hydrobiologia **378:** 199–213, 1998.
R. M. O'Riordan, G. M. Burnell, M. S. Davies & N. F. Ramsay (eds), Aspects of Littorinid Biology.

Imposex in *Nucella lapillus* and intersex in *Littorina littorea*: interspecific comparison of two TBT-induced effects and their geographical uniformity

J. Oehlmann[1], B. Bauer[2], D. Minchin[3], U. Schulte-Oehlmann[2], P. Fioroni[2] & B. Markert[1]
[1] *Internationales Hochschulinstitut Zittau, Lehrstuhl für Umweltverfahrenstechnik, Markt 23, D-02763 Zittau, Germany*
[2] *Universität Münster, Institut für Spezielle Zoologie und Vergleichende Embryologie, Hüfferstr. 1, D-48149 Münster, Germany*
[3] *Marine Institute, Fisheries Research Centre, Abbotstown, Dublin 15, Ireland*

Key words: Nucella lapillus, Littorina littorea, imposex, intersex, TBT effect monitoring, geographical differences

Abstract

Two different tributyltin (TBT)-induced virilisation phenomena in prosobranch snails – intersex in *Littorina littorea* and imposex in *Nucella lapillus* – are compared in order to facilitate their simultaneous use in geographical large scale effect monitoring surveys. Imposex in dogwhelks is a more sensitive biomarker and should be used in areas that are only slightly or moderately contaminated with TBT (ambient TBT concentrations < 2.0 ng as Sn l^{-1}). The assessment of intersex intensities in periwinkle populations has considerable advantages in areas with higher TBT concentrations and should be used also wherever dogwhelks are absent irrespective of the TBT exposure level. The intersex index (ISI) and vas deferens sequence (VDS) index are proposed as the most suited parameters for effect monitoring purposes. The geographical uniformity of intersex and imposex is analysed and proven for the coasts of Ireland, France, and Germany. A relative loss of TBT sensitivity in females can be found, but to a varying extent in both species. The implications of this result for biological TBT effect monitoring programmes are discussed in light of the fact that intersex and imposex have both been found to be irreversible. Because it is the objective of these programmes to assess current TBT contaminations and resulting biological effects, only relatively young specimens should be considered in the sampling strategy.

Introduction

Butyltin compounds are used for a variety of applications, e.g. as biocides, stabilisers, catalysts and intermediates in chemical syntheses. Tributyltin (TBT) compounds exhibit the greatest toxicity of all organotins and have even been characterised as one of the most toxic groups of xenobiotics ever produced and deliberately introduced into the environment (Müller et al., 1989; Stewart et al., 1992). TBT is not only used as an active biocidal compound in antifouling paints which are designed to prevent marine and limnic biota from settlement on ships' hulls, harbour and offshore installations, but also as a biocide in wood preservatives, textiles, dispersion paints, agricultural pesticides and as a UV stabiliser in many plastics.

Among the broad variety of malformations caused by TBT in aquatic animals (for review see Bryan & Gibbs, 1991), molluscs have been found to be one of the most sensitive groups of invertebrates. TBT-induced detrimental effects in Great Britain, Ireland and France, e.g. malformations of oyster shells and the imposex phenomenon of prosobranchs, were first noticed in the 1980s. Consequently, legislative restrictions were drawn up to reduce TBT concentrations in coastal waters and the application of TBT based paints was banned, with few exceptions, for all boats less than 25 m overall length. The imposex phenomenon of prosobranchs has been successfully used as a biological effect monitoring system to determine the degree of environmental TBT pollution and also allows assessment of the possible amelioration of the exposure situation following legislative controls. In most cases the dogwhelk *Nucella lapillus* has been used as the sentinel species (e.g. Gibbs et al., 1987; Harding et al., 1992); considerations of additional species were occasional exceptions in European surveys (Stroben et al., 1992a–c; Oehlmann et al., 1993, 1996a; Ten Hallers-Tjabbes et al., 1994).

The established TBT effect monitoring species which exhibit imposex do not occur in many European coastal regions such as the southern part of the North Sea and the entire Baltic, mostly because of unsuitable habitats. In addition, they become extinct near TBT point sources, like harbours and marinas, due to the increasing level of pollution. There is a need for a further effect monitoring system which can be used in those areas where dogwhelks and other neogastropods are absent. Bauer et al. (1993, 1995) and Oehlmann et al. (1994) were the first to describe the intersex phenomenon in the periwinkle *Littorina littorea*, which is characterised by a phenotypic disturbance of sex determination between gonad and genital tract. Since the first description of intersex in periwinkles further investigations have been performed in France, Ireland (Minchin et al., 1996, 1997) and Germany (Oehlmann et al., 1996b; Bauer et al., 1997). These analyses have shown that *L. littorea* is well suited for TBT effect monitoring especially in regions where a relatively high level of contamination exists and that even here the species is very common and can be sampled in sufficient numbers because: a) it is tolerant of high TBT levels, b) it recruits from the plankton and c) it can occur in areas where dogwhelks have expired. The intersex phenomenon has also been considered by OSPARCOM (Oslo and Paris Commissions) along with the imposex response in the dogwhelk for use in biological TBT effect monitoring surveys in the entire convention area.

For such large-scale surveys two different prerequisites have to be clarified beforehand: the interspecific comparison of intersex and imposex and the geographical uniformity of these two responses, both now described in this account.

Materials and methods

Between 1993 and 1996 a total of 8 335 *Littorina littorea* from 158 stations in Ireland (4 181 specimens from 124 stations), France (286 animals from 8 stations) and Germany (3 868 specimens from 26 stations) were analysed morphometrically as reported in Bauer et al. (1995). Additionally, between 1988 and 1996 a total of 11 521 *Nucella lapillus* from 178 stations in Ireland (4 724 animals from 110 stations) and France (6 797 specimens from 68 stations) were analysed according to Oehlmann et al. (1991). All snails were narcotized in a 7% (w/v) $MgCl_2$ solution (in distilled water), shell and aperture height measured and, after cracking the shell with a vice, the external properties of the genital system were analysed with a stereo dissecting microscope. The extension of all sexual organs was measured to the nearest 0.1 mm.

For the determination of imposex and intersex intensities in the analysed populations the intersex index (ISI) as the average intersex stage in a population was determined as reported in Bauer et al. (1995) for *Littorina littorea* and the vas deferens sequence (VDS) index as the average imposex stage in a population according to Gibbs et al. (1987) and Oehlmann et al. (1991). Further intersex and imposex parameters were calculated but are not included in this paper because the ISI and the VDS index have been shown to be the most sensitive and best indices (Oehlmann, 1994; Oehlmann et al., 1996b; Stroben et al., 1996).

TBT compounds in French *Nucella lapillus* samples were determined as reported by Stroben et al. (1992b). The complete tissues of 2–5 animals were homogenised in stoppered tubes and 10 ml of concentrated HCl (Merck 'suprapur') was added. After shaking for 30 min the homogenates were extracted with 10 ml of hexane (pesticide grade) on an automatic shaker for 30 min and then centrifuged. TBT as Sn (TBT-Sn) was measured in the hexane extract after shaking with 3 ml 1 N NaOH for 3 min using a Perkin-Elmer HGA-500 attached to a Perkin-Elmer 5000 AAS with background correction (wavelength 224.6 nm; slit 0.7 nm; injection volume 25 μl). Internal standardisation (standard addition with spiked samples) was employed.

The determination of TBT compounds in all Irish snail samples and in *Littorina littorea* from France was based on the same method with some minor modifications as reported in Bailey (1991). An ICP-MS Elan 5000 (Perkin-Elmer) was used to quantify the TBT.

The TBT analyses in German *Littorina littorea* samples were performed by Dr W. Kalbfus (Bavarian State Administration for Water Resources Management) and were part of the R. & D. project 102 40 303/02 supported by the German Federal Environment Agency (Umweltbundesamt), Berlin. Material and methods are documented in Kalbfus et al. (1996). In short, 5 g of snail tissue was extracted with 60 ml of 0.05% (v/v) tropolone in hexane for at least 8 h and then organotin compounds were derivated with 4 ml 2 M n-pentylmagnesium bromide for 15 min followed by the quench reaction after adding 25 ml water + 10 ml 10% (v/v) HNO_3. The organic phase was cleaned in a florisil column (with 7 g 1% (w/v) deactivated silica) and the column then eluated with 100 ml hexane. Before the samples went on to GC-MS analysis (HP 6890A II and HP 5970A) the eluate was concentrated to a volume of 0.1–1.0 ml.

Quality assurance and quality control steps were employed, including analyses of certified reference material (e.g. PACS-1) and an interlaboratory TBT analysis of identical samples. The analytical methods used for the samples from each country gave very comparable results (variations generally below 8%). All TBT tissue concentrations are given on a dry weight basis as Sn (μg TBT-Sn kg dry wt^{-1}).

Standard statistical analyses (e.g. parametrical correlation, nonparametrical Spearman rank correlation and analyses of covariance (ANCOVA)) of the results were performed according to Weber (1972) and Lozán (1992) using the computer programmes SigmaPlot 5.1 and StatEasy 2.25. A correlation analysis between biological parameters of two different species or between TBT body burdens and biological parameters can only be performed if the conditions for the application of this parametric test, e.g. normal distribution of measuring values are fulfilled. Otherwise, a nonparametric test such as the Spearman rank correlation analysis is more appropriate. But because the latter method fails to visualise the relationship between two tested parameters, not only the Spearman rank correlation (r_s if $n < 30$, t if $n \geqslant 30$) is calculated in these cases but also the parametric correlation data are given. In some of the figures, measures of central tendency like the VDS index and ISI as means of arbitrarily-defined imposex or intersex stages in a sample, are used in spite of the fact that this procedure is not valid from a strictly statistical point of view. But because the information they convey is important and seems to be the most reasonable they are nevertheless presented.

The complete data, especially the exact location of sampling stations with analysed TBT concentrations, intersex and imposex intensities are documented for Ireland in Minchin et al. (1995, 1996, 1997), for France in Oehlmann et al. (1993) and Oehlmann (1994), and for Germany in Bauer et al. (1995) and Oehlmann et al. (1996b).

Results

Intersex and imposex measurements

A common feature of intersex and imposex is the gradual TBT-induced increase of virilisation intensities in females with the final result of female sterility. While in imposex-affected species the entire female genital system is conserved but superimposed (thus the term 'imposex') by male organs like a penis and/or vas deferens, in *Littorina littorea*, an exhibitor of intersex, the female pallial organs are modified towards a male morphological structure in lower intersex stages (1, 2) and then supplanted by the corresponding male formation, a prostate gland (stages 3 and 4). Both phenomena, intersex and imposex, can be described by using evolutive schemes which were first introduced by Bauer et al. (1995) for intersex with 4 different stages (Figure 1) and Gibbs et al. (1987) for imposex development in British *Nucella lapillus* with six stages. The original imposex scheme was modified by Fioroni et al. (1991) and Oehlmann et al. (1991) for the description of imposex in French dogwhelk populations which exhibit a greater variability of phenotypic imposex expression, as well as in other prosobranch species, by considering three different evolutive paths (a, b, and c) in stages 1–3 (Figure 2).

The final result of intersex and imposex development is the sterilisation of females. *Littorina littorea* is definitively sterilised in stages 2, 3, and 4, either because the oocytes and capsular material leak into the mantle cavity (stage 2) or the glands responsible for the formation of the planktonic egg capsule are at least partially missing. Imposex-affected prosobranchs are sterilised in stages 5 and 6. In stage 5 the oviduct is blocked by proliferating vas deferens tissue (stage 5a) or by the formation of a prostate (stage 5b) resulting in the accumulation of abortive egg capsules in the pallial oviduct (stages 6a, b). Another possibility is the inhibition of the ontogenetic closure of the pallial oviduct section which normally infolds from the mantle epithelium. Therefore, this oviduct section is split ventrally in adult stage 5c females with a comparable appearance and reproductive consequences to the intersex stage 2 of the periwinkle.

Intersex and imposex in prosobranch populations can be assessed by a number of different parameters and indices. For *Littorina littorea* the intersex index (ISI) was first introduced by Bauer et al. (1995) as the average intersex stage in a given sample (Σ intersex stages in a sample $\div$ no. of females). In addition to the

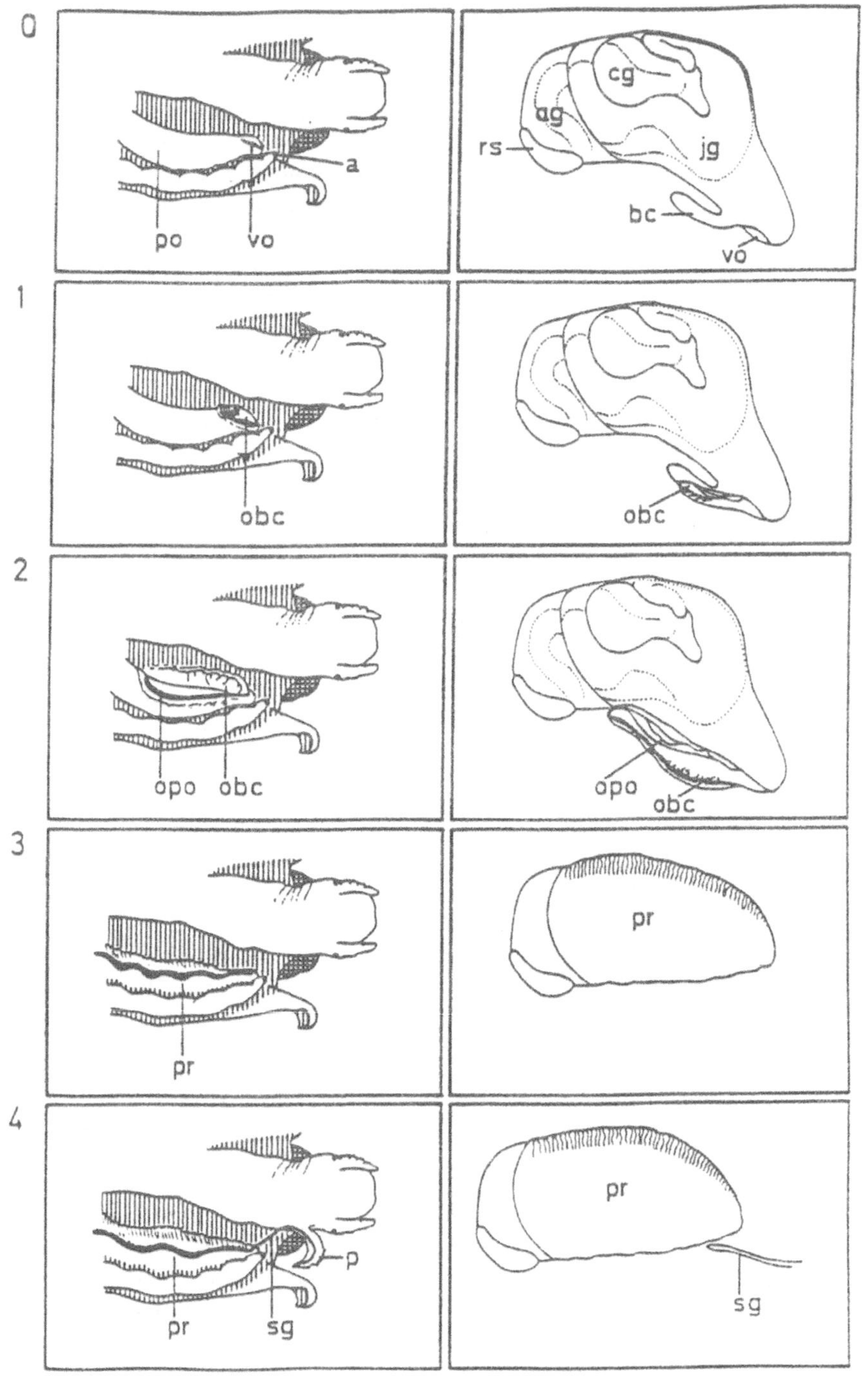

Figure 1. Littorina littorea (female). Intersex development scheme with four different stages. Dorsal views with opened mantle cavity (left) and lateral views of the pallial section of the genital tract (right). Abbreviations: a – anus; ag – albumen gland; bc – bursa copulatrix; cg – capsule gland; jg – jelly gland; obc – open bursa copulatrix; opo – open pallial oviduct; p – penis; po – pallial oviduct; pr – prostate; rs – receptaculum seminis; sg – sperm groove; vo – vaginal opening.

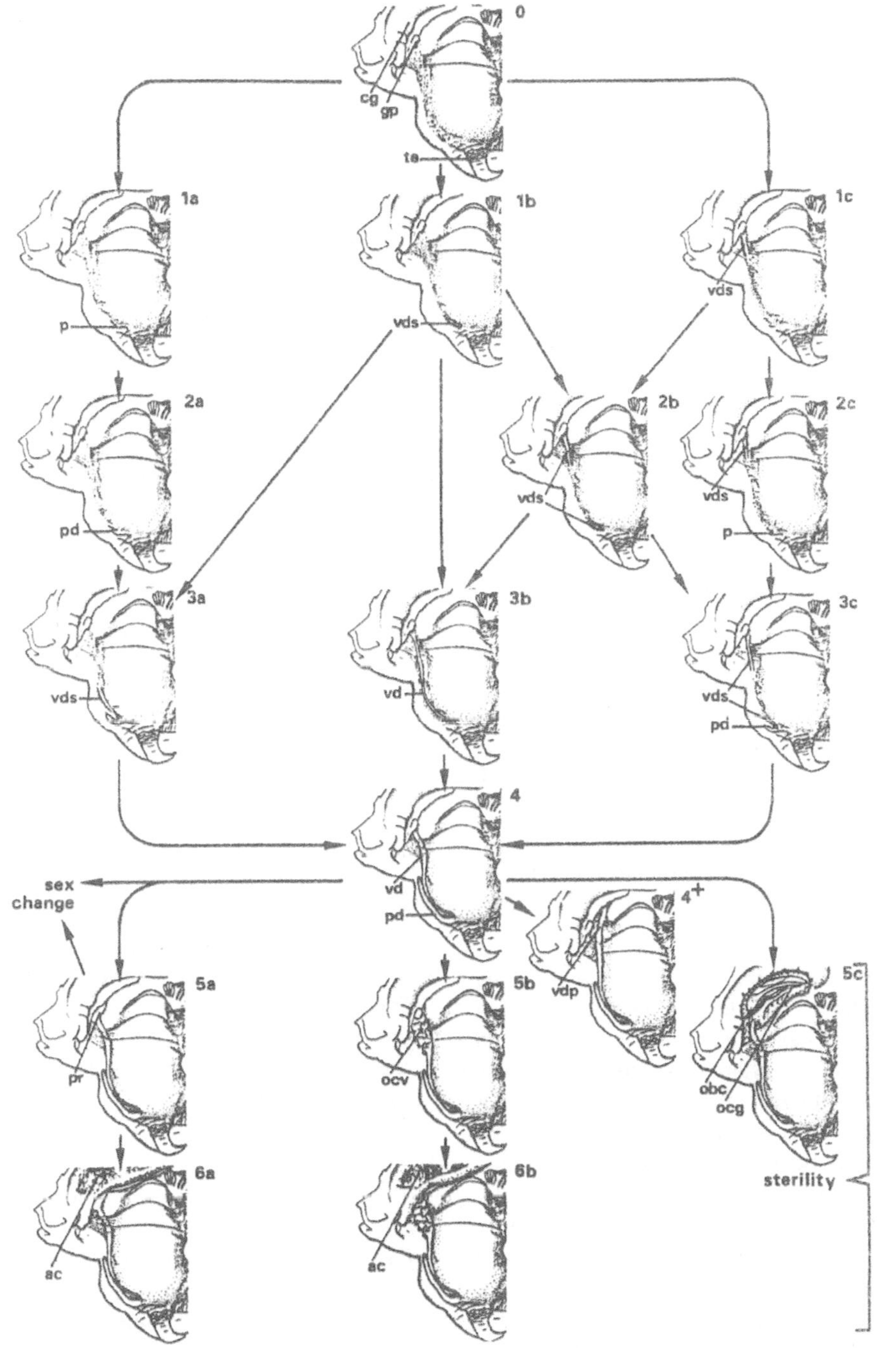

Figure 2. General scheme of imposex development in female prosobranch snails with six different stages. Dorsal views with opened mantle cavity. In *Nucella lapillus* all stages with the exception of stages 4^{+} and 5c occur. Abbreviations: ac – abortive capsule mass; cg – capsule gland; gp – genital papilla; obc – open bursa copulatrix; ocg – open capsule gland; ocv – occlusion of vaginal opening; p – penis without duct; pd – penis with duct; pr – prostate; te – tentacle; vd – vas deferens; vdp – vas deferens passage; vds – vas deferens section.

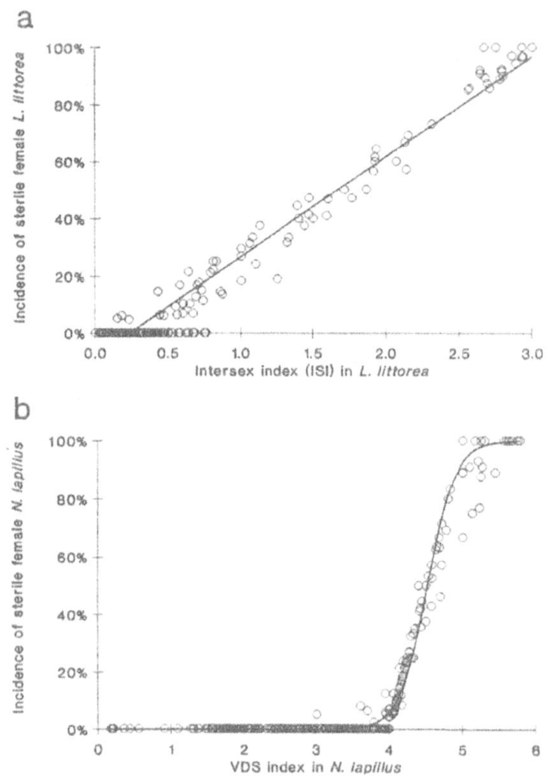

Figure 3. Relationship between the incidence of sterile females and the intersex index in *Littorina littorea* (a) or the vas deferens sequence (VDS) in *Nucella lapillus* (b) from France and Ireland with calculated correlations. (a) $y = 35.0 \cdot x - 8.36$, $n = 237$ samples from 158 stations, $r = 0.983$, $p < 0.0005$. (b) $y = 100 \div (1 + e^{(-5.29 \cdot (x-4.52))}) + 0.0018$, $n = 438$ samples from 178 stations, $r = 0.922$, $p < 0.0005$.

ISI, the incidence of sterile females and the mean value of the prostate length in females can be calculated but these two parameters have some disadvantages when compared with the ISI. They are less sensitive because the values only go above 0 if the intersex stages 2 (percentage of sterile females) or 3 (average female prostate length) can be found in a sample. Consequently, lower TBT exposure levels can by detected by the calculation of ISI values (Oehlmann et al., 1996b).

For the same reason the vas deferens sequence (VDS) index (Σ imposex stages in a sample ÷ no. of females) has proved to be the best suited parameter for the assessment of imposex intensities. Furthermore, the VDS index does not exhibit seasonal variations in contrast to the relative penis size (RPS) and relative penis length (RPL) indices (Oehlmann 1994; Stroben 1994; Stroben et al., 1996). But the main reason for favouring the ISI and VDS index is that both give a very good estimate of the reproductive capability of females in an analysed population (Figure 3). The loss of female fecundity is by far the most important effect during intersex and imposex development and has the greatest ecological relevance.

The first sterile female periwinkles may be found in a given population when the ISI reaches a value of 0.2; above values of 0.8 at least some females will be definitively sterilized and virtually all females have ceased breeding at values above 2.5 (Figure 3a). Because female sterility occurs only in the two highest imposex stages (5 and 6), the relationship between the percentage of sterile females and the VDS index in *Nucella lapillus* can be described by a sigmoid curve (Figure 3b). All females in the imposex stages 5 and 6 are characterised by a blockage of the pallial oviduct and therefore it is impossible for them to deliberate their egg capsules (Gibbs et al., 1987; Oehlmann et al., 1991). Normally no or only very few sterile female dogwhelks can be found in populations with VDS values below 4.0. Above this value the percentage of sterile females increases rapidly and all, or at least the overwhelming majority of females, have ceased breeding in populations with VDS values of 5.0 or more.

The quantification of intersex and imposex intensities by means of ISI and VDS calculations in the populations of both species allows a direct comparison of the TBT sensitivity of *Littorina littorea* and *Nucella lapillus*, a prerequisite for a geographical large-scale biological effect monitoring. Under these circumstances the effects of an assessed TBT exposure level at a given station where only one of both species occurs can be predicted for the absent species. Therefore it is also possible to analyse whether or not the species is missing due to TBT contamination should the habitats be suitable (see *Interspecific comparison of intersex and imposex*).

Geographical uniformity of responses

TBT surveys covering the coastline of several countries or even the complete European Atlantic and North Sea coast (as planned by OSPARCOM; Oslo and Paris Commissions, 1996) are based on the notion that there is a geographical uniformity of the responses in the monitoring species. Those used in monitoring programmes should not only exhibit the same biological effects when exposed to identical xenobiotics over their geographical distribution but should also

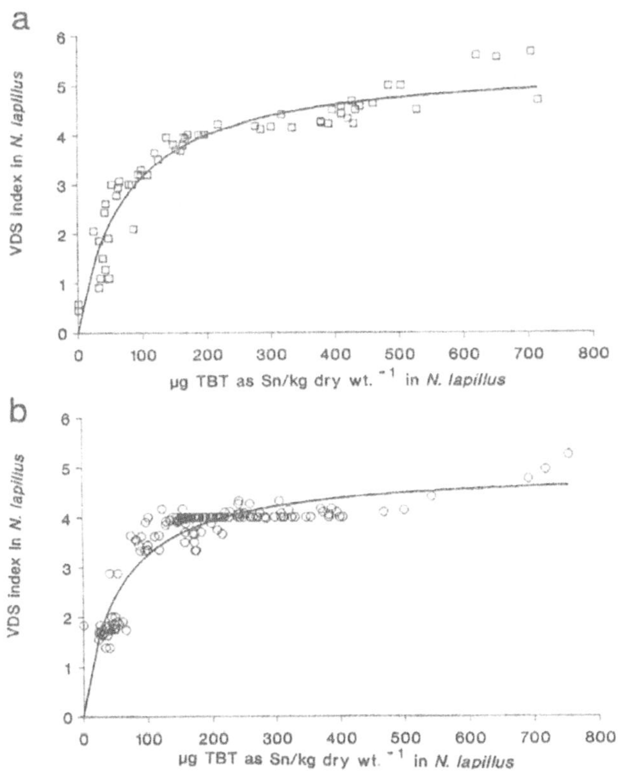

Figure 4. Nucella lapillus. Relationship between the vas deferens sequence (VDS) index and the TBT body burden for Ireland (a) and France (b) with calculated regressions. (a) $y=(5.45\cdot x) \div (72.2+x)$, $n=61$ samples from 53 stations, $r=0.936$, $p<0.0005$; Spearman rank correlation analysis: $t=40.8$, $p<0.0005$. (b) $y=(4.97\cdot x) \div (52.8+x)$, $n=149$ samples from 68 stations, $r=0.899$, $p<0.0005$; Spearman rank correlation analysis: $t=18.0$, $p<0.0005$.

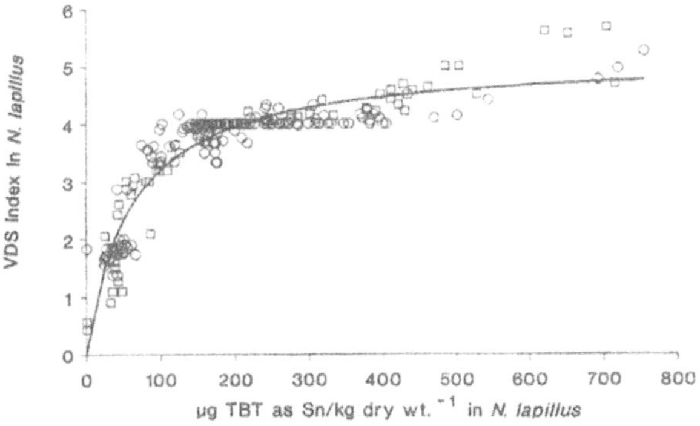

Figure 5. Nucella lapillus. Relationship between the vas deferens sequence (VDS) index and the TBT body burden with results from Ireland (□) and France (○) with calculated regression: $y=(5.14\cdot x) \div (59.6+x)$, $n=210$ samples from 121 stations, $r=0.913$, $p<0.0005$; Spearman rank correlation analysis: $t=31.3$, $p<0.0005$.

Table 1. Nucella lapillus. Analysis of covariance (ANCOVA) for the data from Figure 4. Abbreviations: df, degrees of freedom; MS, mean squares; SS, sum of squares

Variability	SS	df	MS	*F*
Between groups	0.5827	1	0.5827	1.37
Within groups	87.8266	206	0.4263	
Total	88.4094	207		

The regressions are not significantly different from each other ($p<0.001$).

be characterised by identical or at least comparable concentration-effect relationships.

For *Nucella lapillus* a direct comparison of the relationship between imposex intensities and TBT body burden in females is possible for Ireland (Figure 4a) and France (Figure 4b). Dogwhelks, although once present in some areas of Germany, were not available for comparison. Not only does the nonparametric Spearman rank analysis show a significant correlation between the analysed variables, but the parametric analysis also results in highly significant correlation equations which are very similar for both countries. The ANCOVA test applied for both TBT body burden and VDS index relationships reveals that there are no statistically significant differences between the two countries (Table 1) and that therefore the calculation of a common equation is appropriate. This equation is presented in Figure 5.

Data from Ireland, France and Germany were analysed for comparability of ISI in *Littorina littorea*. The calculated regressions which are shown in Figure 6 exhibit greater differences between countries than was found for *Nucella lapillus* in Figures 4 and 5. In France and Germany a greater range of TBT contamination levels was found than in Ireland with maximum TBT body burdens of up to 800 μg TBT-Sn kg^{-1} (dry wt) in France (Figure 6b) and 2 200 μg TBT-Sn kg^{-1} (dry wt) in Germany (Oehlmann et al., 1996b). In Figure 6c only the contamination range up to 800 μg TBT-Sn kg^{-1} (dry wt) is considered. The highest ISI values which were found in France and Germany were 3.0 whereas in Ireland values were generally below 2.0 and the highest TBT concentration in periwinkle tissue did not exceed 500 μg TBT-Sn kg^{-1} (dry wt). These differences seem to be due to a lower TBT contamination level in Irish coastal waters

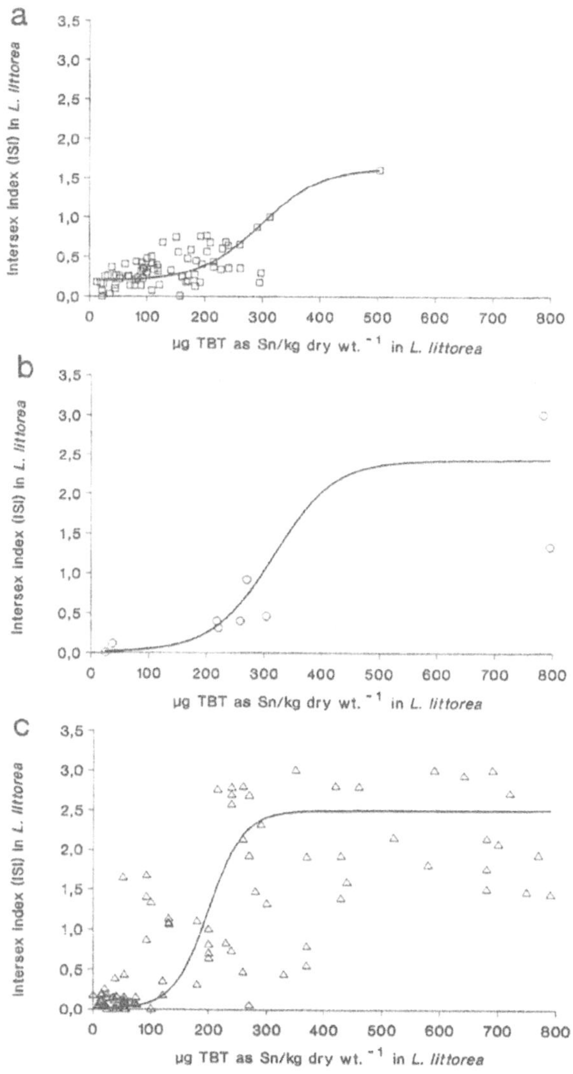

Figure 6. Littorina littorea. Relationship between the intersex index and the TBT body burden for Ireland (a), France (b) and Germany (c) with calculated regressions. (a) $y = 1.43 \div (1 + e^{(-0.019\cdot(x-298))}) + 0.198$, $n = 72$ samples from 59 stations, $r = 0.729$, $p < 0.0005$; Spearman rank correlation analysis: $t = 5.85$, $p < 0.0005$. (b) $y = 2.42 \div (1 + e^{(-0.019\cdot(x-316))}) + 0.011$, $n = 9$ samples from 9 stations, $r = 0.867$, $p < 0.0025$; Spearman rank correlation analysis: $r_s = 0.937$, $p < 0.001$. (c) $y = 2.49 \div (1 + e^{(-0.033\cdot(x-200))}) + 0.009$, $n = 83$ samples from 19 stations, $r = 0.792$, $p < 0.0005$; Spearman rank correlation analysis: $t = 9.44$, $p < 0.0005$.

Table 2. Littorina littorea. Analysis of covariance (ANCOVA) for the data from Figure 6. Abbreviations: df – degrees of freedom; MS – mean squares; SS – sum of squares

Variability	SS	df	MS	*F*
Between groups	0.9622	2	0.4811	2.32
Within groups	32.7750	158	0.2074	
total	33.7372	160		

The regressions are not significantly different from each other ($p < 0.001$).

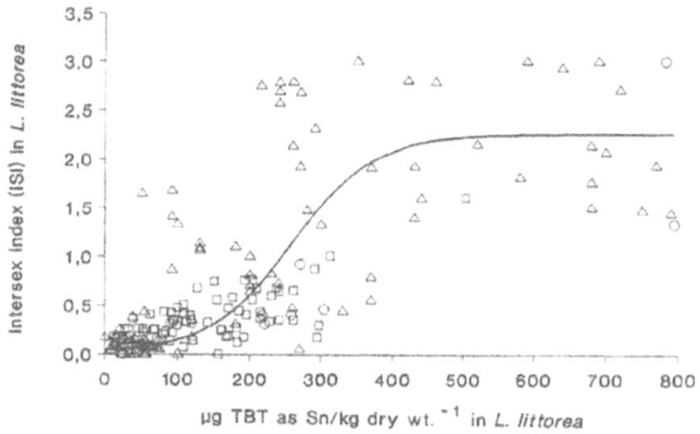

Figure 7. Littorina littorea. Relationship between the intersex index and the TBT body burden with results from Ireland (□), France (o) and Germany (Δ) with calculated regression: $y = 2.24 \div (1 + e^{(-0.018\cdot(x-259))}) + 0.020$, $n = 164$ samples from 87 stations, $r = 0.753$, $p < 0.0005$; Spearman rank correlation analysis: $t = 15.4$, $p < 0.0005$.

and do not necessarily affect the concentration-effect relationship.

The ANCOVA test applied to TBT body burden vs. intersex indices for *Littorina littorea* shows no statistically significant differences between the three countries (Table 2) and so a common equation is permissible (Figure 7).

On the basis of these results and their statistical analysis the geographical uniformity of the imposex and intersex response has been presumed. Another point of interest is a direct comparison of both TBT-induced effects and a possible explanation for the greater deviation of the data points from the calculated concentration-effect regressions in *Littorina littorea* (Figures 6 and 7) compared with *Nucella lapillus* (Figures 4 and 5).

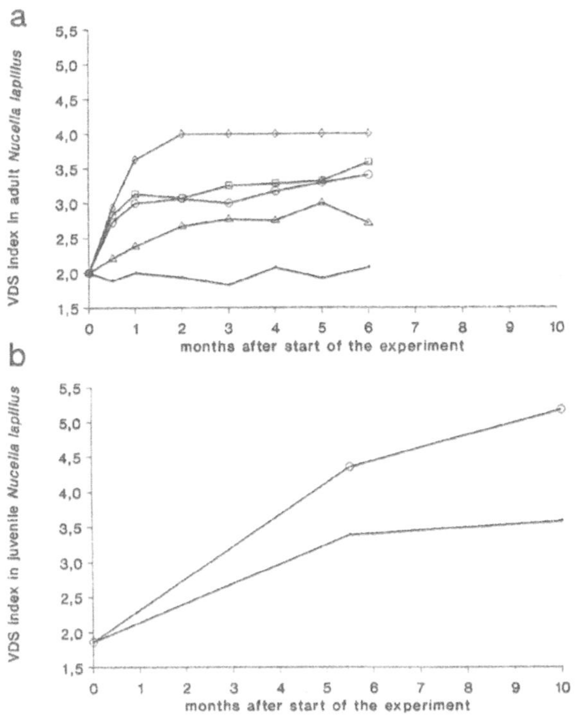

Figure 8. Nucella lapillus. Development of the VDS index under TBT exposure in the laboratory in adult (a) and juvenile specimens (b) (· control, Δ 5 ng TBT-Sn l^{-1}, ∘ 50 ng TBT-Sn l^{-1}, 100 ng TBT-Sn l^{-1}, ◇ injection of 1 μg TBT-Sn specimen^{-1} at the beginning of the experiment). Sample sizes: ≥ 30 specimens.

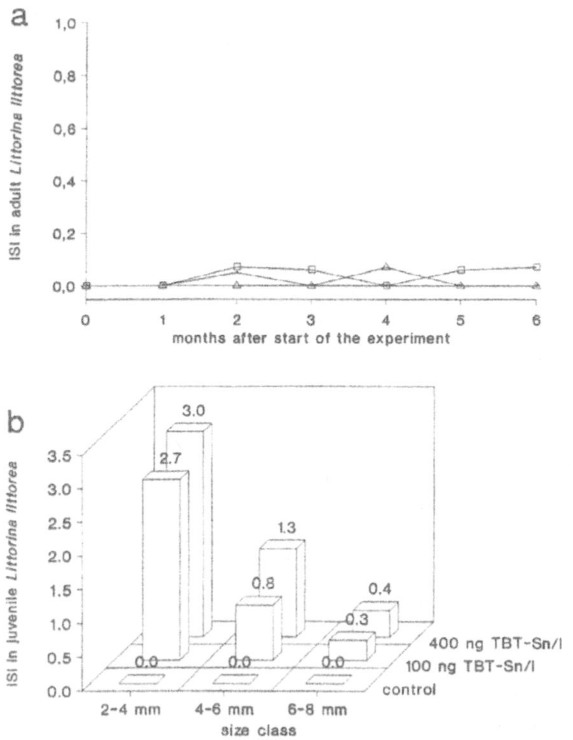

Figure 9. Littorina littorea. Development of the intersex index under TBT exposure in the laboratory in adult (a) and juvenile specimens (b) (· control, Δ 200 ng TBT-Sn l^{-1}, ∘ 400 ng TBT-Sn l^{-1}). Sample sizes: ≥ 30 specimens.

Interspecific comparison of intersex and imposex

Intersex in *Littorina littorea* and imposex in *Nucella lapillus* are both induced by TBT and are irreversible, i.e. an intersex or imposex stage which has developed in a single female as a result of individual TBT exposure cannot be reversed to the normo-morphological situation or reduced to a lower stage even if the animals are kept under TBT-free conditions (see discussion). As shown in Figure 6, for periwinkles there is a greater deviation of data points from calculated concentration-effect equations than for dogwhelks. An explanation for this seems to be an ontogenetic shift in the TBT sensitivity which is more distinct in *L. littorea*.

It has been demonstrated in laboratory experiments with adult and also with sexually immature juvenile dogwhelks that both develop higher imposex stages following a TBT exposure. In Figure 8a the results of an experiment with adult *Nucella lapillus* testing 3 different aqueous TBT concentrations (5, 50, and 100 ng TBT-Sn l^{-1}) and a direct injection of 1 μg TBT-Sn per specimen are shown. Though the VDS index increases in a time- and concentration-dependent manner the maximum observable VDS values of 4.0 – indicating that no sterile females occur – are considerably lower than in the field under comparable exposure conditions. If young hatchlings of the same species are raised in the laboratory at an ambient TBT concentration of 50 ng as Sn l^{-1} the VDS index attains values above 5.0 after 10 months, showing that all females have been sterilised owing to a blockage of the pallial oviduct before they reach sexual maturity (Figure 8b). The increase of the VDS index in the control group during the experiment was owing to the TBT uptake from mussels supplied as food. The same mussels were also used for feeding of the adult snails but because of their lower TBT sensitivity they did not develop higher imposex stages.

In *Littorina littorea* the loss of TBT sensitivity during the ontogenetic development is even more pronounced than in *Nucella lapillus*. If sexually mature periwinkles are exposed to high aqueous TBT concentrations of up to 400 ng as Sn l^{-1} the TBT body burden increases during the experiment in a time- and concentration-dependent manner. However, the same snails taken from Neuharlingersiel (German North Sea coast; ISI range 0–0.08) showed no further increase in intersex intensities (measured as ISI), irrespective of the additional TBT exposure in the laboratory (Figure 9a). In a second experiment, juvenile (i.e. sexually immature) periwinkles were collected at the same field location and separated into three size classes with shell heights of 2–4 mm, 4–6 mm, and 6–8 mm. Each class was divided into groups and exposed to aqueous concentrations of 0 (control), 100, and 400 ng TBT-Sn l^{-1}. Under these exposure conditions the snails were reared for 12 months, when they reached sexual maturity. The results in Figure 9b show that ISI values in the experiments are negatively correlated with the shell height which relates to the age of the young snails at the beginning of the experiment.

These results show that further increases of intersex intensities could not be induced in adult *Littorina littorea* even with high TBT concentrations in contrast to adult *Nucella lapillus* which still develop additional imposex characteristics should environmental TBT concentrations increase, although there is a reduced sensitivity to TBT sensitivity for adult dogwhelks. For this reason imposex intensities in dogwhelks may represent the current exposure level during sampling more exactly than intersex intensities in periwinkles should there be temporal increases in TBT contamination. Intersex intensities in single specimens of *L. littorea* reflect the contamination level during their first year of development, and this may not necessarily be representative for coastal TBT concentrations at the time of sampling and morphometrical analysis. These findings have substantial implications for the sampling strategy in biological effect monitoring programmes (see discussion).

For the interspecific comparison between *Littorina littorea* and *Nucella lapillus* in the field TBT body burdens and biological effects should be considered for populations in Ireland and France which live sympatrically. In Figure 10 the TBT concentrations in periwinkles and dogwhelks of these two countries are compared.

The ANCOVA analysis reveals also in this case that the differences between the two countries are not statistically significant (Table 3), and therefore calculation of a common equation for both countries is appropriate (Figure 11).

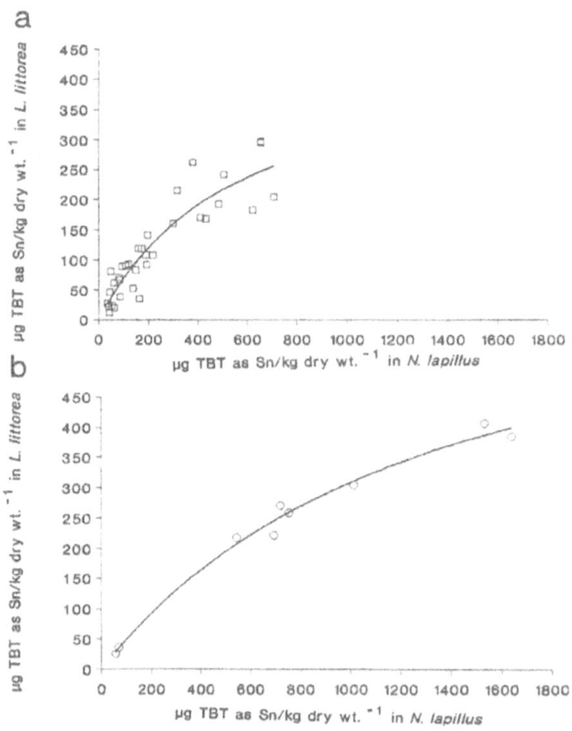

Figure 10. Relationship between the TBT body burden in sympatrically living populations of *Littorina littorea* and *Nucella lapillus* in Ireland (a) and France (b) with calculated correlations. (a) $y = (464 \cdot x) \div (571 + x)$, $n = 34$ samples, $r = 0.893$, $p < 0.0005$; Spearman rank correlation analysis: $t = 12.6$, $p < 0.0005$. (b) $y = (734 \cdot x) \div (1376 + x)$, $n = 9$ samples, $r = 0.991$, $p < 0.0005$; Spearman rank correlation analysis: $r_S = 0.967$, $p < 0.001$.

Table 3. Analysis of covariance (ANCOVA) for the data from Figure 10. Abbreviations: df, degrees of freedom; MS, mean squares; SS, sum of squares

Variability	SS	df	MS	*F*
Between groups	6406.6	1	6406.6	4.86
Within groups	51370.7	39	1317.2	
Total	57777.3	40		

The regressions are not significantly different from each other ($p < 0.001$).

In general, *Nucella lapillus* individuals accumulate more TBT than periwinkles living at the same station. These differences are even more distinct in highly contaminated areas and might reflect the higher TBT concentrations which are found in the prey of dogwhelks when compared with the food of *Littorina littorea* (see discussion).

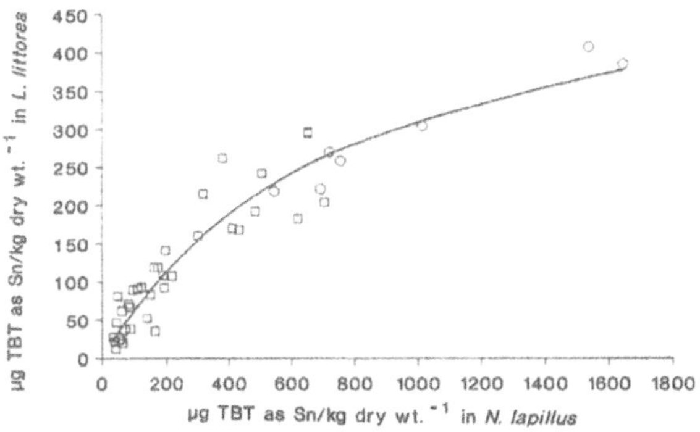

Figure 11. Relationship between the TBT body burden in sympatrically living populations of *Littorina littorea* and *Nucella lapillus* in Ireland (□) and France (○) with calculated correlation: $y = (557 \cdot x) \div (781 + x)$, $n = 43$ samples, $r = 0.932$, $p < 0.0005$; Spearman rank correlation analysis: $t = 17.2$, $p < 0.0005$.

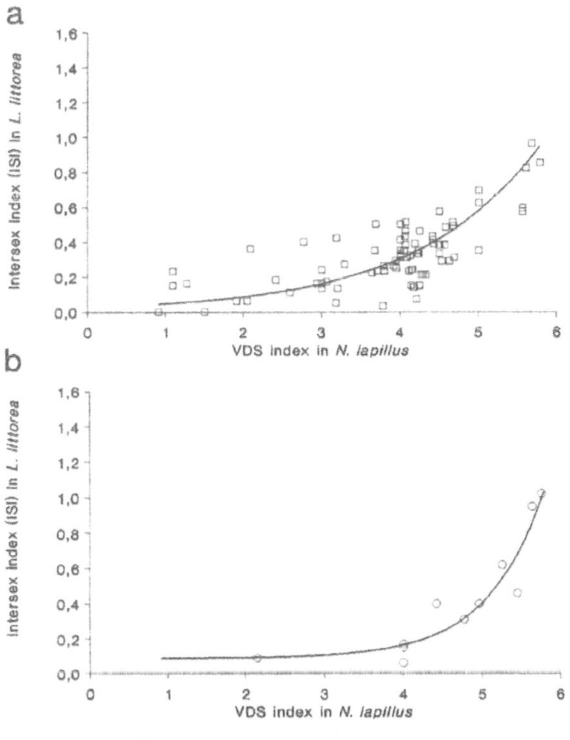

Figure 12. Relationship between the intersex index in *Littorina littorea* populations and VDS indices in sympatrically living populations of *Nucella lapillus* in Ireland (a) and France (b) with calculated correlations. (a) $y = 25.3 \div (1 + e^{(-0.662 \cdot (x - 10.7))}) + 0.0067$, $n = 81$ samples from 80 stations, $r = 0.776$, $p < 0.0005$; Spearman rank correlation analysis: $t = 6.62$, $p < 0.0005$. (b) $y = 62.2 \div (1 + e^{(-1.42 \cdot (x - 8.72))}) + 0.089$, $n = 11$ samples from 11 stations, $r = 0.946$, $p < 0.0005$; Spearman rank correlation analysis: $r_S = 0.947$, $p < 0.0005$.

Table 4. Analysis of covariance (ANCOVA) for the data from Figure 12. Abbreviations: df, degrees of freedom; MS, mean squares; SS, sum of squares

Variability	SS	df	MS	*F*
Between groups	0.1184	1	0.1184	6.67
Within groups	1.5627	88	0.0178	
Total	1.6811	89		

The regressions are not significantly different from each other ($p < 0.001$).

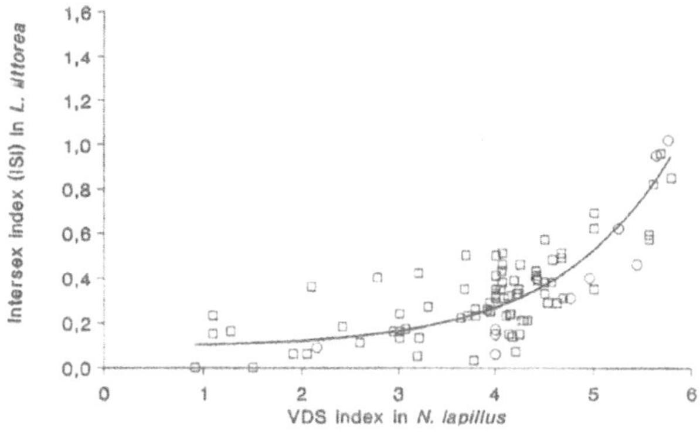

Figure 13. Relationship between the intersex index in *Littorina littorea* populations and VDS indices in sympatrically living populations of *Nucella lapillus* in Ireland (□) and France (○) with calculated correlation: $y = 53.2 \div (1 + e^{(-0.890 \cdot (x - 10.4))}) + 0.092$, $n = 92$ samples from 91 stations, $r = 0.812$, $p < 0.0005$; Spearman rank correlation analysis: $t = 8.10$, $p < 0.0005$.

The interspecific comparison of ISI values in *Littorina littorea* populations with VDS indices in sympatric living dogwhelk populations for Ireland and France shows again only slight differences between these countries (Figure 12).

Again, the ANCOVA analysis reveals that the data and calculated correlations for Ireland and France are not statistically significant from each other (Table 4) and a calculation of a common equation is permissible (Figure 13).

The common relationship between the ISI and VDS index for both countries shows that the use of intersex and imposex has specific advantages over different ranges of contamination. Imposex in *Nucella lapillus* should be used for the assessment of lower TBT exposure levels in only slightly and moderately contaminated areas (ambient TBT concentrations ≤ 2.0 ng as Sn l^{-1}) whereas ISI values in *Littorina littorea* exhibit a kind of background noise with values between 0 and 0.4. At highly contaminated sites with ambient

TBT concentrations above 2.0 ng TBT-Sn l^{-1}, VDS indices in dogwhelks exceed 4.0 and the population becomes endangered due to progressive female sterilization. It is in this contamination range that ISI values begin to increase. Should periwinkle populations exhibit an ISI of 0.5, or more, all females in a dogwhelk population (which lives sympatrically) will have become sterile and the *Nucella* population will soon expire. During the entire investigation period we have not found a single *N. lapillus* population which has survived at a station where periwinkles exhibit ISI values above 1.0. Therefore assessment of coastal TBT pollution in highly polluted areas should be performed using the intersex phenomenon in *L. littorea* whereas imposex in dogwhelks has some advantages in the lower contamination range.

Discussion

Since the first description of intersex in *Littorina littorea* by Bauer et al. (1993) and the more detailed work of Oehlmann et al. (1994) and Bauer et al. (1995) this phenomenon has been used in several surveys in Ireland (e.g. Minchin et al., 1996, 1997) and Germany (Oehlmann et al., 1996b; Bauer et al., 1997) and is now being considered, alongside the imposex response in *Nucella lapillus*, for the biological TBT effect monitoring in coastal waters by OSPARCOM (Oslo and Paris Commissions, 1996). Imposex has been successfully applied in biological TBT effect monitoring studies in Scotland (Bailey & Davies, 1988), England (Gibbs et al., 1987, 1991), Ireland (Schulte-Oehlmann et al., 1993; Minchin et al., 1995, 1996), France (Oehlmann et al., 1993) and outside Europe in the United States (Short et al., 1989; Saavedra Alvarez & Ellis, 1990), Canada (Bright & Ellis, 1990), southeast Asia (Ellis & Pattisina, 1990), New Zealand (Stewart et al., 1992) and Australia (Kohn & Almasi, 1993). Within these programmes different parameters have been used for the assessment of intersex and especially imposex intensities in the analysed populations. The use of the VDS index as the most suited index for various reasons (highest ecological relevance as a direct indicator of female sterility, highest sensitivity, no seasonal variations) was proposed by OSPARCOM for surveys covering the entire convention area (Oslo and Paris Commissions, 1996). The advantages of VDS measurements in direct comparison with other possible imposex parameters are documented and discussed in detail by Stroben et al. (1996).

Recently, the TBT-specifity of the imposex response in neogastropods has been questioned by Nias et al. (1993) and Evans et al. (1995). In laboratory experiments female *Lepsiella vinosa* from an imposex-free population developed imposex characteristics not only when exposed to TBT, as could be expected, but also in the control groups and those which were treated with a copper-based antifoulant. The authors conclude that not only TBT, but also other factors, including environmental stress and copper, may also induce the virilisation of females. Unfortunately, no analytical data are presented and even the aqueous copper and TBT concentrations in the tanks are recalculated from leaching rates of the applied antifouling paints during the experiment. Therefore it is not possible to exclude that TBT was also present in the control and copper exposed groups because either the applied sea water, the mussels used to feed the animals or the copper paint may have been contaminated with the organotin compound. In contrast to the findings of Nias et al. (1993) it has been shown in field experiments by Oehlmann (1994) that *Nucella lapillus* develops imposex only if the shells are marked with a TBT-based antifoulant but not if they are treated with a copper-based and definitively TBT-free paint. More recently, Horiguchi et al. (1995) have reported that also triphenyltin is able to induce imposex in Japanese neo- and mesogastropods. But Bryan et al. (1988), who had tested the effects of five different organotin compounds (including triphenyltin) on *N. lapillus* after an aqueous exposure and direct injection into the foot muscle, have found only for TBT and tetrabutyltin (which is debutylated to TBT by the snails) an imposex inducing activity. The value of imposex in *N. lapillus* and intersex in *Littorina littorea* as biological indicators of TBT contamination depends on the specificity of the responses to TBT. There is only poor evidence that next to TBT other xenobiotics may exhibit an imposex inducing effect with the exception of synthetic and natural androgens (e.g. testosterone, androstenedione) (Oehlmann & Bettin, 1996). The strongest evidence for the responsibility of TBT for the development of imposex in the field is the fact that in all areas where coastal TBT concentrations could be reduced by legislative actions a recovery of prosobranch populations formerly suffering from high imposex levels was found (e.g. Evans et al., 1991; Minchin et al., 1995, 1996, 1997).

Intersex intensities in periwinkle populations should be calculated on the basis of the intersex index (ISI) for comparable reasons. The ISI, as the VDS index in imposex affected species, also gives a direct measurement of the reproductive capability of females and their possible restrictions. Furthermore the ISI allows the detection of lower environmental TBT concentrations by comparison with the average female prostate length and the incidence of sterile females (Oehlmann et al., 1996b).

Though the ultimate effect of intersex and imposex development is sterilisation of females, there are differences between dogwhelk and periwinkle female sterility at the population and ecosystem levels. *Nucella lapillus* produces benthic egg capsules and the offspring develop within the capsule, hatch and crawl onto the shore without having a pelagic larval stage. Because of this absence of a planktonic dispersal stage dogwhelk populations will die out if all females in a given population become sterilised. Additionally, sterilised dogwhelk females suffer from a higher mortality and populations at highly contaminated stations are thus characterised by a shift of the sex ratio in favour of males (Gibbs et al., 1987; Fioroni et al., 1991; Oehlmann et al., 1991). Contrary to the imposex phenomenon in dogwhelks and other muricids, sterilisation due to intersex development in *Littorina littorea* evokes poor recruitment of juveniles but not necessarily the extinction of the population. Because of the planktonic egg capsule, less affected periwinkles from other populations can contribute to the survival of the species, even in highly contaminated areas. Intersex development does not result in increased female mortality because the oviduct does not become blocked. For this reason the sex ratio remains balanced even in highly contaminated areas.

The geographical uniformity of the imposex response in *Nucella lapillus* has already been addressed by Gibbs et al. (1991). Dogwhelks sampled from four different locations around the British Isles developed comparable imposex intensities and TBT body burdens when exposed to identical TBT concentrations in the laboratory. Gibbs et al. (1991) found also some evidence for the geographical uniformity in the field because dogwhelk populations in the southwest of Brittany and the U.K. exhibited the same effects. Our own analysis goes yet a step further. It has not only been shown that *N. lapillus* and *Littorina littorea* can principally develop signs of imposex or intersex but also that the concentration-effect relationship for both species is identical over an even more extended geographical range.

The interspecific comparison of indices and TBT body burdens in this study demonstrates that imposex is more sensitive and is best suited for use in regions where ambient TBT concentrations are below 2.0 ng as Sn l^{-1}. The assessment of intersex intensities should be used in more polluted areas or where there are unsuitable habitats for dogwhelks (Oehlmann et al., 1996b). One of the possible reasons for the higher TBT sensitivity of the imposex phenomenon in *Nucella lapillus* in comparison to intersex in *Littorina littorea* are the differences in TBT body burdens. As predators, dogwhelks probably accumulate more TBT because of the higher contamination of their diet when compared with snails such as buccinids as scavengers or periwinkles as typical herbivores (Bryan et al., 1989; Bailey et al., 1991; Stroben et al., 1992b, c; Stroben, 1994).

Intersex and imposex are irreversible pathomorphological developments in prosobranch snails. Even if affected dogwhelk females are kept under TBT-free conditions for longer periods (Bryan et al., 1987; Bryan & Gibbs, 1989) a once attained imposex stage cannot be reduced to a lower stage or to the normal imposex-free condition. This has been noted in other species (Stroben et al., 1992b, c; Stroben, 1994) as well as in laboratory experiments with *Littorina littorea* (Bauer et al., 1997).

The ontogenetic shift resulting in a relative loss of TBT sensitivity in adults of imposex affected species has already been described by Oehlmann (1994) for muricid and Stroben (1994) for buccinid gastropods. The differences in TBT sensitivity between sexually mature and immature specimens is far more distinct in *Littorina littorea*. The explanation is that a genital tract which once has been formed according to normogenesis, as in an intersex stage 0 female, can no longer be transformed by developing male characteristics. Therefore it seems to be highly improbable that a normal pallial oviduct in such a female should develop a slit-like ventral aperture or that female pallial organs should be reduced completely with a consequent formation of new prostate tissue under the influence of increasing ambient TBT concentrations. Contrary to the situation in periwinkles, sexually mature females of imposex-affected species are still able to develop additional male organs, like a penis and/or vas deferens.

The irreversibility of intersex and imposex development and especially the marked ontogenetic shift in the TBT sensitivity of *Littorina littorea* has major implications for biological effect monitoring surveys. Because it should be the main objective of these programmes to assess the current degree of environmental

TBT contamination and their biological effects, only relatively young specimens which have just reached sexual maturity should be considered in the sampling strategy. This can be achieved for dogwhelks and periwinkles by the exclusion of larger size classes and of specimens with eroded, obviously worn shells which are often characterised by settlement of barnacles and spirorbid polychaetes (Oehlmann et al., 1996b). Another interesting aspect is that a single simultaneous analysis of sexually immature, sexually mature and senile *Littorina littorea* at a given station opens the possibility for a temporal TBT monitoring programme covering a period of several years. This cannot be achieved using *Nucella lapillus* or other imposex-exhibiting species.

Acknowledgements

The financial support of the German Federal Environment Agency (Umweltbundesamt), Berlin for the German survey (R. & D. project 102 40 303/01) is greatly appreciated. We would also like to thank Dr W. Kalbfus, Dipl.-Ing. (FH) A. Zellner, Dipl.-Ing. (FH) S. Frey and T. Knorr (Bavarian State Administration for Water Resources Management, Institute for Water Research) for the performance of the TBT analyses of German periwinkles and Dipl.-Ing. (FH) U.-S. Leffler (IHI Zittau) for her excellent assistance during the TBT analyses of Irish snails.

References

Bailey, S. K. & I. M. Davies, 1988. Tributyltin contamination in the Firth of Forth (1975–1987). Sci. Total Envir. 76: 158–192.

Bailey, S. K., 1991. Tributyltin (TBT) contamination of Scottish coastal waters. Unpublished Ph.D. thesis, Napier University, Edinburgh. 379 pp.

Bailey, S. K., I. M. Davies, M. J. C. Harding & A. M. Shanks, 1991. Effects of tributyltin oxide on the dogwhelk *Nucella lapillus* (L.). Proc. of the Tenth World Meeting of the Organotin Environmental Programme (ORTEP): 7–66.

Bauer, B., M. Brumm-Scholz, U. Deutsch, P. Fioroni, I. Ide, S. Liebe, J. Oehlmann, E. Stroben & B. Watermann, 1993. Basisstudie zur Erfassung pathologischer Effekte von Unterwasseranstrichen bei Schnecken der niedersächsischen Nordseeküste. Final report for 'Aktion Seeklar' e.V. (Hamburg) and 'WWF Germany' (Frankfurt/M.), Ahrensburg and Münster, 105 pp.

Bauer, B., P. Fioroni, I. Ide, S. Liebe, J. Oehlmann, E. Stroben & B. Watermann, 1995. TBT effects on the female genital system of *Littorina littorea*: a possible indicator of tributyltin pollution. Hydrobiologia 309: 15–27.

Bauer, B., P. Fioroni, U. Schulte-Oehlmann, J. Oehlmann & W. Kalbfus, (1997). The use of *Littorina littorea* for tributyltin (TBT) effect monitoring – results from the German TBT survey 1994/95 and laboratory experiments. Envir. Pollut. 96: 299–309.

Bright, D. A. & D. V. Ellis, 1990. A comparative survey of imposex in northeast Pacific neogastropods (Prosobranchia) related to tributyltin contamination, and a choice of a suitable bioindicator. Can. J. Zool. 68: 1915–1924.

Bryan, G. W. & P. E. Gibbs, 1989. Water pollution. In Yearbook of science and technology 1990, McGraw-Hill, New York: 357–359.

Bryan, G. W. & P. E. Gibbs, 1991. Impact of low concentrations of tributyltin (TBT) on marine organisms: a review. In M. C. Newman & A. W. McIntosh (eds), Metal Ecotoxicology: Concepts and Application. Lewis Publisher, Ann Arbor, 361 pp.

Bryan, G. W., P. E. Gibbs & G. R. Burt, 1988. A comparison of the effectiveness of tri-n-butyltin chloride and five other organotin compounds in promoting the development of imposex in the dogwhelk, *Nucella lapillus*. J. mar. biol. Ass. U.K. 68: 733–744.

Bryan, G. W., P. E. Gibbs, G. R. Burt & L. G. Hummerstone, 1987. The effects of tributyltin (TBT) accumulation on adult dog-whelks, *Nucella lapillus*: long-term field and laboratory experiments. J. mar. biol. Ass. U.K. 67: 525–544.

Bryan, G. W., P. E. Gibbs, L. G. Hummerstone & G. R. Burt, 1989. Uptake and transformation of ^{14}C-labelled tributyltin chloride by the dog-whelk, *Nucella lapillus*: importance of absorption from the diet. Mar. Envir. Res. 28: 247–251.

Ellis, D. V. & L. A. Pattisina, 1990. Widespread neogastropod imposex: a biological indicator of global TBT contamination? Mar. Pollut. Bull. 21: 248–253.

Evans, S. M., A. Hutton, M. A. Kendall & A. M. Samosir, 1991. Recovery in populations of dogwhelks *Nucella lapillus* (L.) suffering from imposex. Mar. Pollut. Bull. 22: 331–333.

Evans, S. M., T. Leksono & P. D. McKinnell, 1995. Tributyltin pollution – a diminishing problem following legislation limiting the use of TBT-based anti-fouling paints. Mar. Pollut. Bull. 30: 14–21.

Fioroni, P., J. Oehlmann & E. Stroben, 1991. The pseudohermaphroditism of prosobranchs; morphological aspects. Zool. Anz. 226: 1–26.

Gibbs, P. E., G. W. Bryan & P. L. Pascoe, 1991. TBT-induced imposex in the dogwhelk, *Nucella lapillus*: geographical uniformity of the response and effects. Mar. Envir. Res. 32: 79–87.

Gibbs, P. E., G. W. Bryan, P. L. Pascoe & G. R. Burt, 1987. The use of the dog-whelk, *Nucella lapillus*, as an indicator of tributyltin (TBT) contamination. J. mar. biol. Ass. U.K. 67: 507–523.

Harding, M. J. C., S. K. Bailey & I. M. Davies, 1992. TBT imposex survey of the North Sea. UK Department of the Environment, Contract PECD 7/8/214, Annex 4: Germany (= Scottish Fisheries Working Paper No. 13/92), 7 pp.

Horiguchi, T., H. Shiraishi, M. Shimizu, S. Yamazaki & M. Morita, 1985. Imposex in Japanese gastropods (Neogastropoda and Mesogastropoda): Effects of tributyltin and triphenyltin from antifouling paints. Mar. Pollut. Bull. 31: 402–405.

Kalbfus, W., A. Zellner, S. Frey & T. Knorr, 1996. Analytik von Oberflächenwasser, Sediment und Mollusken zur Validierung des biologischen Effektmonitorings. Final report of R. & D. project 102 40 303/02, Umweltbundesamt, Berlin, 108 pp.

Kohn, A. J. & K. N. Almasi, 1993. Imposex in Australian *Conus*. J. mar. biol. Ass. U.K. 73: 241–244.

Lozán, J. L., 1992. Angewandte Statistik für Naturwissenschaftler, Parey, Berlin, Hamburg, 237 pp.

Minchin, D., J. Oehlmann, C. B. Duggan, E. Stroben & M. Keatinge, 1995. Marine TBT antifouling contamination in

Ireland, following legislation in 1987. Mar. Pollut. Bull. 30: 633–639.

Minchin, D., E. Stroben, J. Oehlmann, B. Bauer, C. B. Duggan & M. Keatinge, 1996. Biological indicators used to map organotin contamination in Cork Harbour, Ireland, Mar. Pollut. Bull. 32: 188–195.

Minchin, D., J. Oehlmann, U. Schulte-Oehlmann, B. Bauer & C. B. Duggan, 1997. Biological indicators used to map organotin contamination from a fishing port, Killybegs, Ireland. Mar. Pollut. Bull. 34: 235–243.

Müller, M. D., L. Renberg & G. Rippen, 1989. Tributyltin in the environment-sources, fate and determination: an assessment of present status and research needs, Chemosphere 18: 2015–2042.

Nias, D. J., S. C. McKillup & K. S. Edyvane, 1993. Imposex in *Lepsiella vinosa* from Southern Australia. Mar. Pollut. Bull. 26: 380–384.

Oehlmann, J., 1994. Imposex bei Muriciden (Gastropoda, Prosobranchia). Eine ökotoxikologische Untersuchung zu TBT-Effekten, Cuvillier Verlag, Göttingen, 167 pp.

Oehlmann, J. & C. Bettin, 1996. TBT-induced imposex and the role of steroids in marine snails. Malacol. Rev. Suppl. 6: 157–161.

Oehlmann, J., P. Fioroni, E. Stroben & B. Markert, 1996a. Tributyltin (TBT) effects on *Ocinebrina aciculata* (Gastropoda: Muricidae): imposex development, sterilization, sex change and population decline. Sci. Total Envir. 188: 205–223.

Oehlmann, J., I. Ide, B. Bauer, B. Watermann, U. Schulte-Oehlmann, S. Liebe & P. Fioroni, 1996b. Erfassung morpho- und histopathologischer Effekte von Organozinnverbindungen auf marine Mollusken und Prüfung ihrer Verwendbarkeit für ein zukünftiges biologisches Effektmonitoring. Final report of R. & D. project 102 40 303/01, Umweltbundesamt, Berlin, 194 pp.

Oehlmann, J., S. Liebe, B. Watermann, E. Stroben, P. Fioroni & U. Deutsch, 1994. New perspectives of sensitivity of littorinids to TBT pollution. Cah. Biol. mar. 35: 254–255.

Oehlmann, J., E. Stroben & P. Fioroni, 1991. The morphological expression of imposex in *Nucella lapillus* (Linnaeus) Gastropoda: Muricidae). J. moll. Stud. 57: 375–390.

Oehlmann, J., E. Stroben & P. Fioroni, 1993. Fréquence et degré d'expression du pseudohermaphrodisme chez quelques Prosobranches Sténoglosses des côtes françaises (surtout de la Baie de Morlaix et de la Manche). 2. Situation jusqu'au printemps de 1992. Cah. Biol. mar. 34: 343–362.

Oslo & Paris Commissions (ed.), 1996. Agenda Item 9 of the Environmental Assessment and Monitoring Committee (ASMO). OSPARCOM, Vila Franca do Campo (= ASMO 96/9/8-E), 23 pp.

Saavedra Alvarez, M. M. & D. V. Ellis, 1990. Widespread neogastropod imposex in the northeast Pacific: implications for TBT contamination surveys. Mar. Pollut. Bull. 21: 244–247.

Schulte-Oehlmann, U., J. Oehlmann & E. Stroben, 1993. TBT-induced imposex of *Nucella lapillus* on Irish coasts in 1990. In J. C. Aldrich (ed.), Quantified Phenotypic Responses in Morphology and Physiology, Proc. 27th European Marine Biology Symposium, Dublin, Ireland, 7–11th September, 1992. JAPAGA, Ashford: 307–311.

Short, J. W., S. D. Rice, C. C. Brodersen & W. B. Stickle, 1989. Occurrence of tri-n-butyltin-caused imposex in the North Pacific marine snail *Nucella lima* in Auke Bay, Alaska. Mar. Biol. 102: 291–297.

Stewart, C., S. J. de Mora, M. R. L. Jones & M. C. Miller, 1992. Imposex in New Zealand neogastropods. Mar. Pollut. Bull. 24: 204–209.

Stroben, E., 1994. Imposex und weitere Effekte von chronischer TBT-Intoxikation bei einigen Mesogastropoden und Bucciniden (Gastropoda, Prosobranchia). Cuvillier Verlag, Göttingen, 193 pp.

Stroben, E., C. Brömmel, J. Oehlmann & P. Fioroni, 1992a. The genital systems of *Trivia arctica* and *T. monacha* (Prosobranchia, Mesogastropoda) and tributyltin induced imposex. Zool. Beitr. 34: 349–374.

Stroben, E., J. Oehlmann & P. Fioroni, 1992b. The morphological expression of imposex in *Hinia reticulata* (Gastropoda: Buccinidae): a potential indicator of tributyltin pollution. Mar. Biol. 113: 625–636.

Stroben, E., J. Oehlmann & P. Fioroni, 1992c. *Hinia reticulata* and *Nucella lapillus*. Comparison of two gastropod tributyltin bioindicators. Mar. Biol. 114: 289–296.

Stroben, E., J. Oehlmann, U. Schulte-Oehlmann & P. Fioroni, 1996. Seasonal variations in the genital ducts of normal and imposex-affected prosobranchs and its influence on biomonitoring indices. Malacol. Rev. Suppl. 6: 173–184.

Ten Hallers-Tjabbes, C. C., J. F. Kemp & J. P. Boon, 1994. Imposex in whelks (*Buccinum undatum*) from the open North Sea: relation to shipping traffic intensities. Mar. Pollut. Bull. 28: 311–313.

Weber, E., 1972. Grundriß der biologischen Statistik, 7th edition. Fischer, Stuttgart, 422 pp.

Hydrobiologia **378:** 215–225, 1998.
R. M. O'Riordan, G. M. Burnell, M. S. Davies & N. F. Ramsay (eds), Aspects of Littorinid Biology.

Toxico-kinetic and -dynamic aspects of TBT-induced imposex in *Hydrobia ulvae* compared with intersex in *Littorina littorea* (Gastropoda, Prosobranchia)

U. Schulte-Oehlmann[1], J. Oehlmann[1], B. Bauer[2], P. Fioroni[2] & U.-S. Leffler[1]
[1] *Internationales Hochschulinstitut Zittau, Lehrstuhl für Umweltverfahrenstechnik, Markt 23, D-02763 Zittau, Germany*
[2] *Universität Münster, Institut für Spezielle Zoologie und Vergleichende Embryologie, Hüfferstr. 1, D-48149 Münster, Germany*

Key words: Hydrobia ulvae, Littorina littorea, imposex, intersex, reproductive failure, TBT biomonitoring

Abstract

A short description of the imposex condition in *Hydrobia ulvae* is presented. In this species four different imposex stages with two types of development were identified. In laboratory experiments the mud snail accumulates TBT in a time and concentration-dependent manner. Increasing biological indices e.g. VDSI (vas deferens sequence index), female penis length (FPL) and the imposex incidence are significantly correlated with increasing TBT body burden. The percentage of sterile females depends on the TBT contamination of sediments. An interspecific comparison of the mud snail *H. ulvae* and the periwinkle *Littorina littorea* elucidates that the mud snail is the more sensitive species at lower environmental TBT concentrations.

Introduction

The colonisation of underwater surfaces (fouling) by epizoa (e.g. Phycophyta, Anthozoa, Hydroidea, Bivalvia, Bryozoa, Cirripedia, Ascidiaceae) leads to an increase in frictional resistance of merchant ships and other boats. Consequently this results in a steep rise in fuel consumption. Antifouling paints or biocides were developed to prevent organisms from settlement. Their mode of operation is largely based on the continuous emission of extremely toxic substances e.g. copper, tributyltin (TBT) and further organic compounds used as cotoxicants. TBT compounds belong to the most toxic xenobiotics ever produced and released into the environment (Müller et al., 1989; Stewart et al., 1992). Their use is not restricted to the marine antifouling sector; organotins are furthermore applied in the preservation and conservation of outdoor facilities. The biocide is added to a variety of materials (e.g. textiles, dispersion paints, PVC etc.). Therefore leaching is also responsible for an increasing TBT contamination of freshwater resources (Kalbfus et al., 1991; Schebek et al., 1991; Oehlmann et al., 1996c; Schulte-Oehlmann et al., 1996b). TBT is known to produce a variety of malformations in aquatic animals. One of them is termed imposex (occurrence of male parts in addition to the female genital system in dioecious prosobranchs), another one intersex (disturbance of the congruity between gonad and genital tract in the periwinkle *Littorina littorea*). Both phenomena describe gradual morphological reactions which correlate with an increasing TBT exposure and originate from the androgenic activity of the biocide (Oehlmann & Bettin, 1996). Until now imposex and intersex of prosobranch species are the only substance specific effect-monitoring tools known to assess the degree of TBT pollution of the aquatic environment. Both phenomena have already been used in several TBT surveys in different countries (e.g. Germany, Ireland, Great Britain, France and The Netherlands) to determine the actual TBT contamination of coastal ecosystems (Bryan et al., 1986; Gibbs et al.,

1987; Ten Hallers-Tjabbes et al., 1994; Oehlmann, 1994; Minchin et al., 1997). In general these studies focused on neogastropod species (*Nucella lapillus, Hinia reticulata, Buccinum undatum*) which are absent in brackish estuaries or in the Baltic Sea, where *Hydrobia ulvae*, as a typical mud flat inhabitant, can be found in high population densities. For biomonitoring purposes the use of indigenous indicator organisms is desirable. Therefore imposex in *H. ulvae* offers a suitable alternative in areas of variable salinities (Schulte-Oehlmann et al., 1996a, 1997).

Materials and methods

In 1994 and 1995 samples of *Hydrobia ulvae* and *Littorina littorea* from 16 different stations along the German North Sea and Baltic coast were collected; more than 1700 mud snails and 3000 periwinkles were examined (the sample size was a minimum of 30 individuals per station). First the specimens were analysed morphometrically and then narcotised in 2–7% $mgCl_2$ (according to salinity at sampling stations) dissolved in distilled water. Shell and aperture height were measured, and after cracking the shell with a vice, the external properties of the genital system were analysed with a stereo-dissecting microscope. All organs were measured to an accuracy of 0.05 mm using an eyepiece. For imposex and intersex measurements the snails were sexed. For *H. ulvae* the penis length was analysed and the imposex stage (according to Schulte-Oehlmann et al., 1997) determined. As imposex indices, the VDS (Gibbs et al., 1987; Fioroni et al., 1991) (vas deferens sequence = average imposex stage of a population with values of 0 to 4 in *H. ulvae*, compare also Results) and the average female penis length (FPL) of a sample were calculated and imposex incidences determined. The intersex determination in *L. littorea* was carried out following Bauer et al. (1995); for the assessment of intersex intensities in periwinkle populations, the ISI (average intersex stage of a sample with values of 0–4 in *L. littorea*) was calculated.

The toxico-kinetic results of TBT on laboratory experiments were assessed with more than 870 specimens of *Hydrobia ulvae*. Mud snails were kept in 10 litre glass aquaria with a 48 h static renewal system and provided with air filters. Water temperature was 14 ± 1 °C. The snails were exposed to different nominal aqueous TBT concentrations (0, 200, 400 and 800 ng TBT as Sn l^{-1} (TBT-Sn), stock solutions in 96% acetic acid) for seven months. The specimens of the solvent control group received only acetic acid. The animals were analysed at monthly intervals and the complete tissues of 30–100 individuals were deep frozen for TBT analyses.

Chemical analyses were performed in order to determine the relationship between biological measurements and TBT contaminations of tissues and sediments. The determination of TBT compounds was based on Ward et al. (1981) and modified after Bailey (1991). The analysis includes an extraction of TBT and DBT compounds with hexane, the elimination of DBT (dibutyltin) by washing the hexane extract with NaOH and quantification using ICP-MS Elan 5000 (Perkin-Elmer). Internal standardisation (standard addition with spiked samples) after Welz (1983) was employed. Certified reference material (CRM: PACS-1, delivered by the National Research Council of Canada; certified concentration of 1.27 ± 0.22 mg TBT-Sn kg^{-1} and 1.16 ± 0.18 mg DBT-Sn kg^{-1}) was analysed additionally. Results were within the standard deviation for the CRM (measured concentration: 1.25 ± 0.08 mg TBT-Sn kg^{-1} and 1.13 ± 0.09 mg DBT-Sn kg^{-1}). The TBT analyses of sediments were performed by Dr W. Kalbfus (Bavarian State Administration for Water Resources Management) and were part of the R. & D. project 102 40 303/02 supported by the Umweltbundesamt, Berlin. Material and methods are documented in Kalbfus et al. (1996).

Results

Imposex in *Hydrobia ulvae* can be described by a classification scheme (Figure 1) with stages 0–4 and an additional stage 5 (sterile females) already at the beginning of the virilisation (Schulte-Oehlmann et al., 1997). It includes two different paths of evolution: the a-types in this scheme are females which first develop a penis and then gradually a vas deferens by stage 3a. Stage 1a represents a female with a penis primordium without a penis duct. In stage 2a a penis duct is formed and the extension of the copulatory organ increases. A distal vas deferens section appears at the base of the penis in stage 3a. It grows toward the genital apertures. In stage 4 a continuous vas deferens connects the female orifices with the penis base. In the b-path the occurrence of male organs is vice versa. The formation of a vas deferens initiates the process which is completed with the appearance of a penis in stage 3b. A gutter-like depression can be found in front of

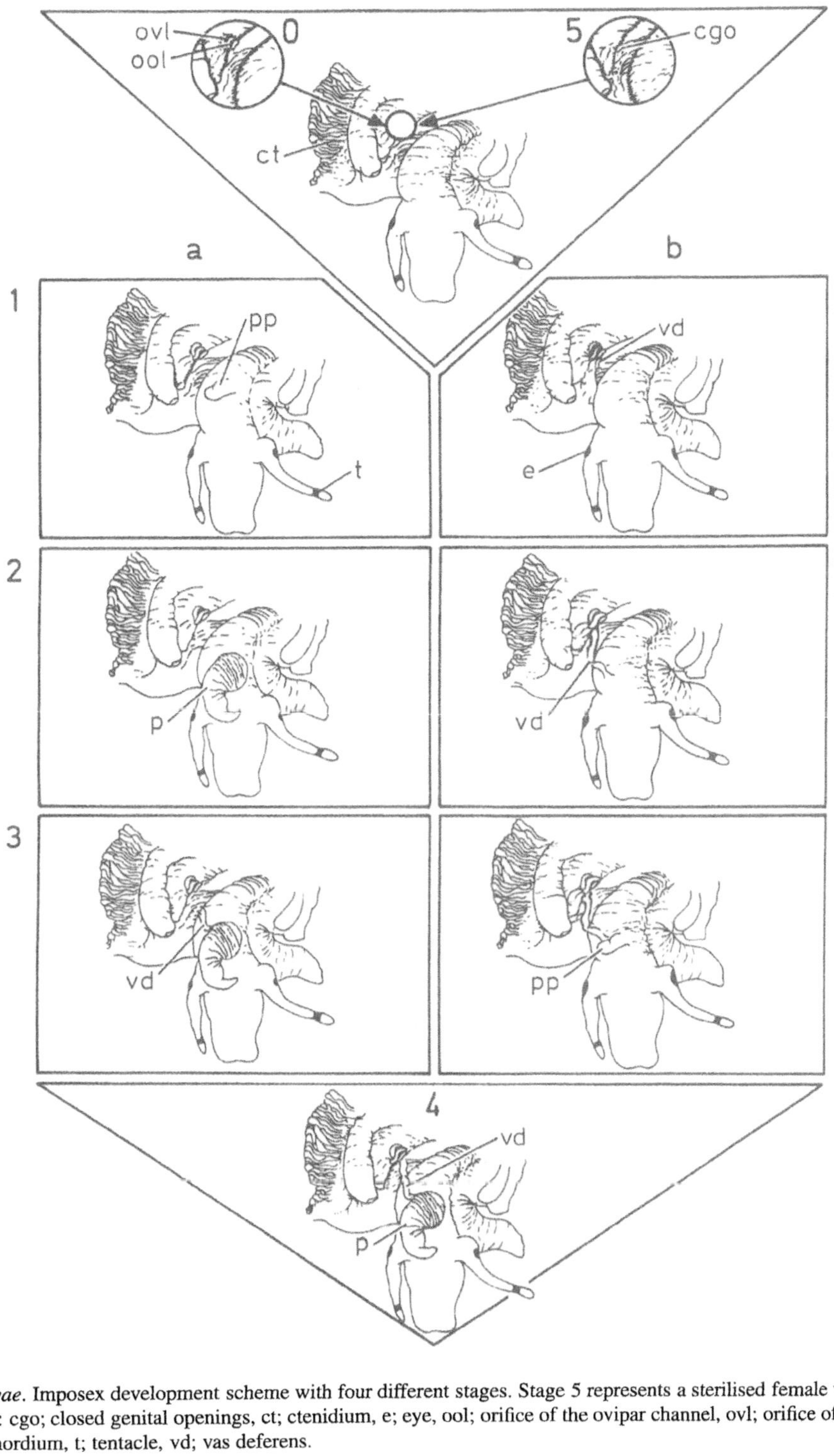

Figure 1. Hydrobia ulvae. Imposex development scheme with four different stages. Stage 5 represents a sterilised female without male characteristics. Abbreviations: cgo; closed genital openings, ct; ctenidium, e; eye, ool; orifice of the ovipar channel, ovl; orifice of the vaginal channel, p; penis, pp; penis primordium, t; tentacle, vd; vas deferens.

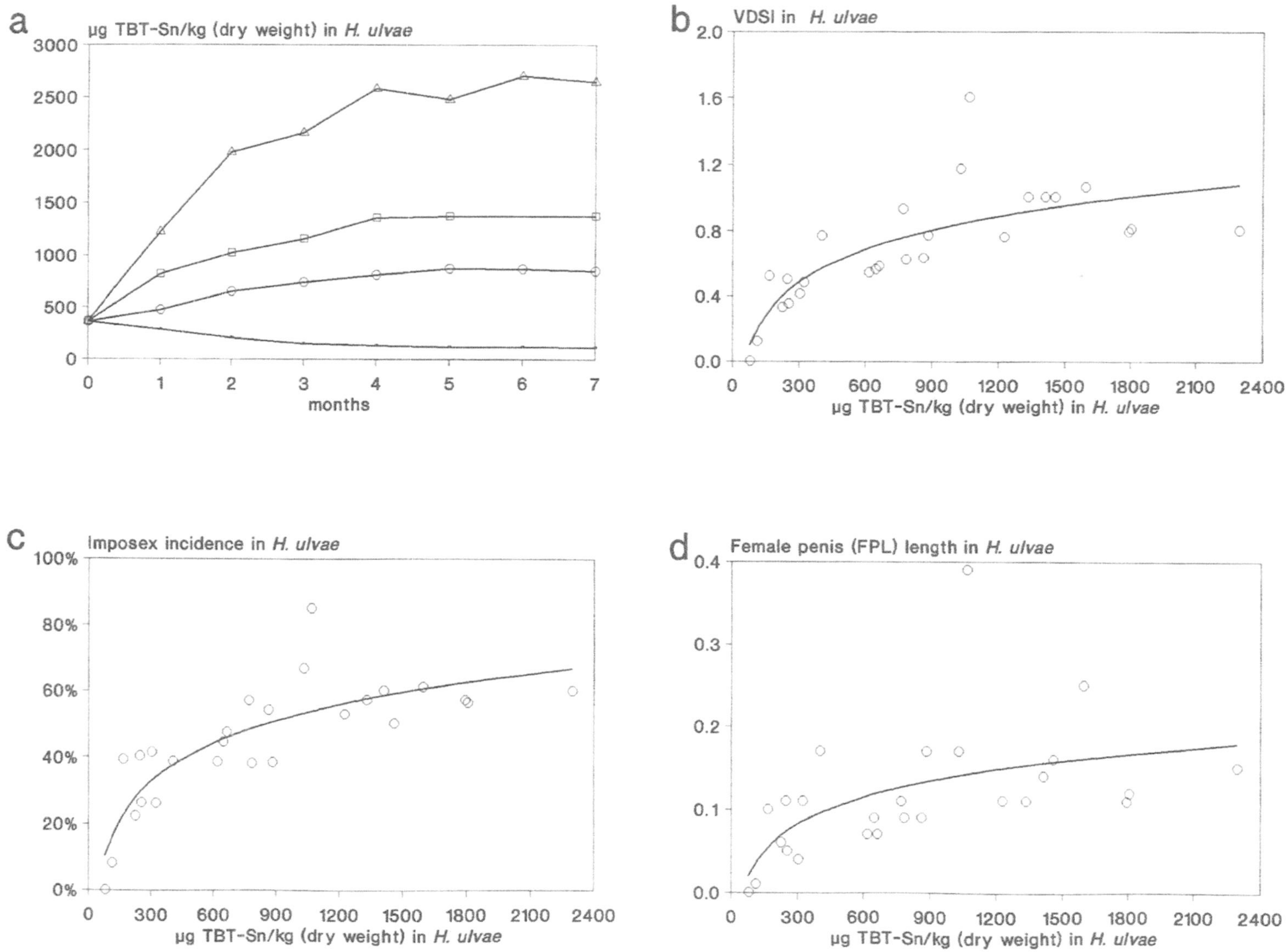

Figure 2. Hydrobia ulvae. Development of TBT body burden in laboratory experiments (a) and relationship between TBT body burden and imposex parameters with calculated regressions in field studies (a–c). (a) Exposure groups: (˙) solvent control group, (○) 200 ng TBT-Sn l^{-1}, (□) 400 ng TBT-Sn l^{-1}, (Δ) 800 ng TBT-Sn l^{-1}. (b) $y = -1.16 + 0.289 \cdot \ln(x)$, $n = 26$ samples, $r = 0.780$, $p < 0.0005$. (c) $y = -63.0 + 16.8 \cdot \ln(x)$, $n = 26$ samples, $r = 0.839$, $p < 0.0005$. (d) $y = -0.186 + 0.047 \cdot \ln(r)$ $n = 26$ samples, $r = 0.559$, $p < 0.0005$.

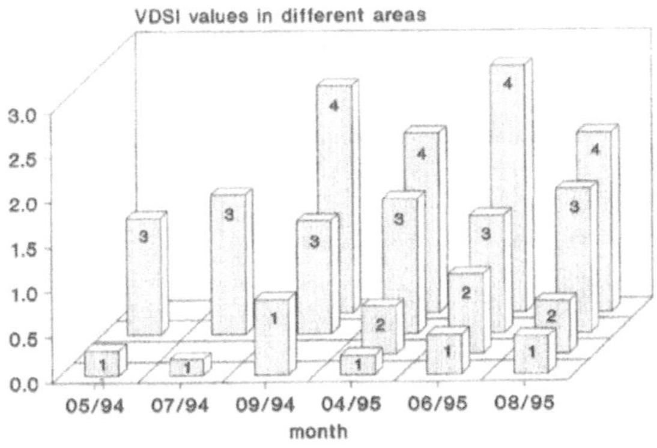

Figure 3. Hydrobia ulvae. VDSI values at different stations during an investigation period of two years. (1) = Neuharlingersiel, (2) = Gollwitz, (3) = Wismar/Wendorf, (4) = Kiel/Schilksee.

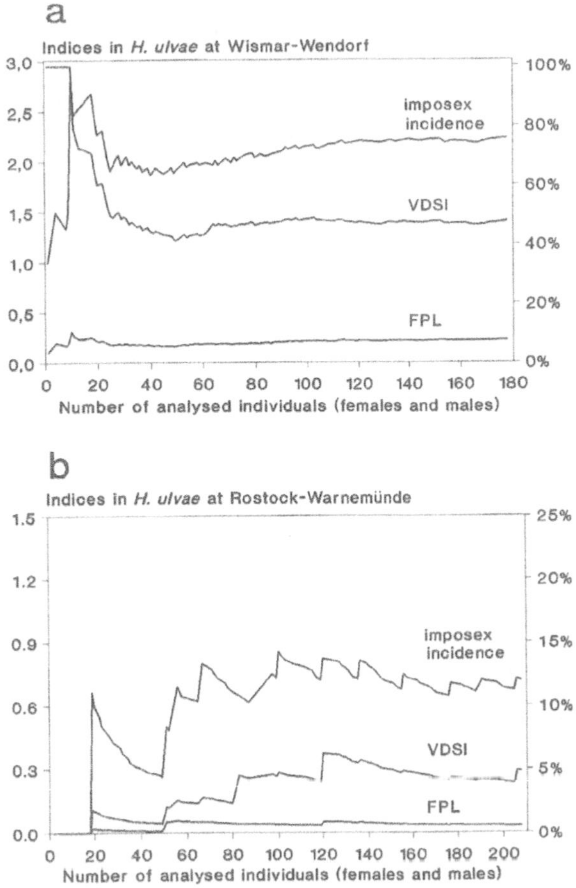

Figure 4. Hydrobia ulvae. Imposex parameters (imposex incidence in%, vas deferens sequence index (VDSI) and average female penis length (FPL)) at stations with different imposex frequencies in relation to the number of analysed individuals. (a) Wismar/Wendorf, (b) Rostock/Warnemünde.

the vulva in stage 1b. Towards stage 2b it becomes a completely closed vas deferens. In stage 3b a penis primordium without a penis duct is developed. This organ increases in size and length throughout stage 3b into stage 4. A penis duct finally appears in stage 4.

Furthermore in *H. ulvae* the morphological appearance of the female genital apertures is variable. Although most of the virilised mud snails exhibit open orifices some also possess closed orifices irrespective of their imposex stage. Although a stage 5 is not explicitly included in the classification scheme, the value 5 should be added to the imposex stage of all females which exhibit closed genital apertures, similar to the original schemes of Gibbs et al. (1987), Fioroni et al. (1991) and Oehlmann et al. (1991), where sterile females have been ranked as stages ≥5. According to the scheme suggested here for *Hydrobia ulvae* a virilised female of stage 2 with closed female genital opening will be given the value 7.

Figure 2a demonstrates that the mud snail accumulates TBT in a time- and concentration-dependent manner under laboratory conditions. The TBT uptake is equilibrated after 3 months (exposure to 200 and 400 ng TBT-Sn l^{-1}) and 4 months (exposure to 800 ng TBT-Sn l^{-1}). Although the individuals of the 800 ng TBT-Sn l^{-1} exposure group had the highest biocide concentrations in their tissues, the bioconcentration factors (BCFs) (compare Table 1) of the different experimental groups are nearly identical and independent of the aqueous TBT-concentration. Only individuals of the 200 ng TBT-Sn l^{-1} exposure group accumulated slightly more TBT (~ 25%) than those of the other groups.

The relationship between biological indices and TBT body burden is shown in Figures 2b–c. All examples reflect highly significant correlations between TBT concentrations in the tissue and the biological parameters VDSI, imposex incidence and FPL. The values of the indices increase sharply with enhancement of the TBT body burden. Threshold concentrations (Table 1) can be determined in the range of 100 μg TBT-Sn kg^{-1} (dry weight). An equilibration level is attained between 1500–2400 μg TBT-Sn kg^{-1} (dry weight). Regarding the BCFs of *Hydrobia ulvae* determined under laboratory conditions, a threshold concentration for biological effects in water can be estimated in the range of 20 ng TBT-Sn l^{-1} in the laboratory. The correlation between the TBT body burden and imposex indices gives preference to the

Table 1. Hydrobia ulvae. Bioconcentration factors (BCF) for TBT and TBT threshold concentrations for biological effects. BCF value are based on laboratory experiments and threshold concentrations from field studies

Nominal TBT concentrations (ng as Sn l^{-1})	TBT body burden (μg as Sn kg^{-1} dry wt)	BCF	Threshold concentrations (μg Sn kg^{-1} dry wt)	
			in tissue	in sediments
200	842.5	4.2×10^3	100 (VDSI)	20–30 (VDSI)
400	1375	3.4×10^3		8000 (sterility)
800	2648	3.3×10^3		

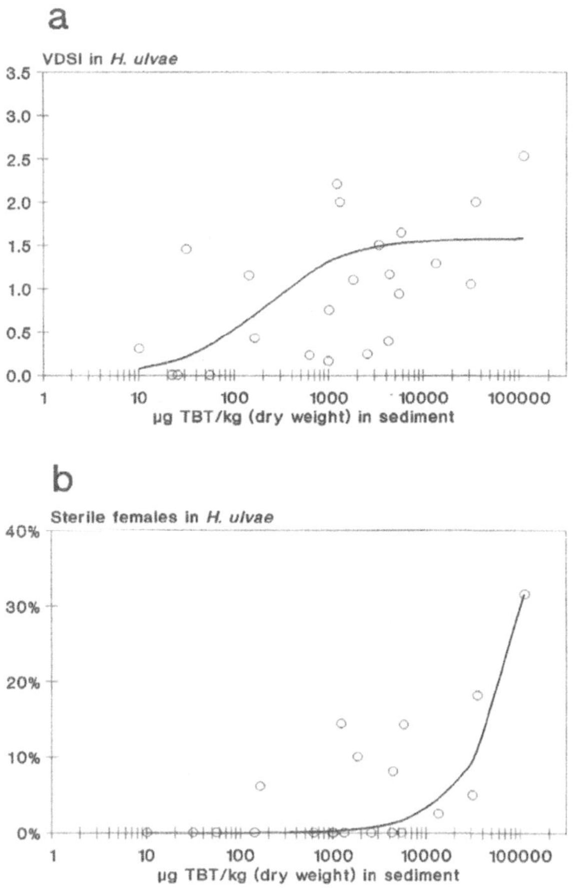

Figure 5. Hydrobia ulvae. Relationship between TBT concentrations in sediments and imposex parameters with calculated regressions in field studies. (a) VDSI: $y = (1.58 \cdot x) - (203 + x)$, $n = 23$ samples, $r = 0.592$, $p < 0.0025$. (b) Percentage of sterile females: $y = 0.0003 \cdot x - 0.008$, $n = 20$ samples, $r = 0.742$, $p < 0.0005$.

VDSI ($r = 0.780$) and imposex incidence ($r = 0.889$) in comparison to the FPL ($r = 0.559$). VDS-Index and imposex incidence can therefore be recommended as the best biological parameters reflecting the actual contamination level in *H. ulvae* (Schulte-Oehlmann et al., 1997).

On the other hand our investigations also made clear that statistical considerations regarding the sampling size have to be taken into account for monitoring purposes (Figures 4a, b). If the performance of only one sampling campaign is planned, a minimum number of 60–80 analysed individuals per station should be considered. This must be explained in the context of the imposex frequency at different sampling stations. At stations where imposex is very common (high imposex incidences) e.g. Wismar/Wendorf (Figure 4a), a sample size of 30 analysed individuals is absolutely sufficient in order to achieve reproducible results. At those stations where the phenomenon is rarer (low imposex incidences) e.g. Rostock/Warnemünde (Figure 4b), the number of analysed specimens has to be much higher to get representative results.

The biological parameters (VDSI, imposex incidence, FPL) were useful monitoring tools not only to assess but also to differentiate between contamination levels of various sampling stations without greater deviations during the investigation period of two years (Figure 3). For example the VDSI at Kiel/Schilksee, a highly frequented marina, exhibited the highest values compared to the 15 other sampling stations; simultaneously sediment samples from this area had the highest TBT contamination of all analysed stations (compare Kalbfus et al., 1996; Oehlmann et al., 1996a). Contamination levels in the centrefield were found at stations at a geographical distance of one to three kilometres from TBT point sources, like at Wismar/Wendorf, whereas at reference or control stations like Gollwitz and Neuharlingersiel the VDSI values were comparatively low.

Correlations between the VDSI (Figure 5a) or the percentage of sterile female *Hydrobia ulvae* (Figure 5b) and TBT concentrations in sediments are statistically significant, too. TBT measurements of sediments documented in Figure 5 were determined as μg TBT kg^{-1} (dry weight). For a direct comparison of TBT residue values in sediments and tissues the sediment concentrations were converted in μg TBT-Sn (μg TBT kg^{-1} $\times$ 0.4 = μg TBT-Sn kg^{-1}). A TBT increase above the threshold concentration of = 20–30 μg TBT-Sn kg^{-1} leads to a marked ascent of VDSI values. A biocide concentration of = 8000 μg TBT-Sn kg^{-1} in sediments (Table 1) causes a steep climb in the percentage of sterile females (Figure 5b). The increase of the VDSI is equilibrated at 3000 μg TBT-Sn kg^{-1} in sediments.

Finally some aspects of an interspecific comparison of *Littorina littorea* and *Hydrobia ulvae* as TBT effect monitoring species are presented. Comparisons of the biological indices determined for both analysed snails in sympatric populations are summarised in Figures 6a–d. The most important biological parameter for the periwinkle, the ISI, is compared to the best biological parameters, the VDSI (Figure 6a), the imposex incidence (Figure 6b) and the female penis length (FPL; Figure 6c) for the mud snail. Additionally the percentage of sterile females in *H. ulvae* is compared to the percentage of infertile female *L. littorea* (Figure 6d). For all compared biological parameters highly significant correlations can be found, but the biological indices of *Hydrobia* increase faster and more sharply than the ISI. To give an example: in the sympatrically living populations which are compared directly in Figure 6b, the imposex incidence of *H. ulvae* is about 40–60% (i.e. maximum values for the species at these selected stations) whereas ISI values of *L. littorea* are in the range of 0–0.2 (i.e. minimum values for the species at these selected stations). Similar observations can be made if the ISI values are compared to the VDSI or female penis length (FPL) of *H. ulvae*. Only if the percentage of sterile females in mud snail and periwinkle populations is compared does this biological parameter of the periwinkle increase faster.

Discussion

Except for slight differences, the imposex development scheme of *Hydrobia ulvae* is in good accordance to the general imposex classification scheme of muricid gastropods (Oehlmann et al., 1991, 1996b, c).

Generally the bioaccumulation of noxious substances depends on different exposure paths. As this is also valid for TBT, this organotin compound is accumulated either directly from ambient water or after mobilisation from contaminated sediments, and also directly from sediments and food (Bryan, 1979). The degree of bioaccumulation is species-specific. In the case of aqueous exposure it can be expressed by the BCF (bioconcentration factor). The latter depends on the exposure concentration and the bioavailability of the compound. BCFs for TBT (3.3 $\times$ 10^3 at 200 ng TBT-Sn l^{-1} and 3.4 $\times$ 10^3 at 400 ng TBT-Sn l^{-1}) in *Hydrobia ulvae* were determined under laboratory conditions. They are in good agreement with those of *Littorina littorea* (7.1 $\times$ 10^3 at 200 ng TBT-Sn l^{-1} and 5.1 $\times$ 10^3 at 400 ng TBT-Sn l^{-1}), *Marisa cornuarietis* (9.1 $\times$ 10^3 at 200 ng TBT-Sn l^{-1}) and *Hinia reticulata* (1.1 $\times$ 10^4 at 200 ng TBT-Sn l^{-1}) (Stroben, 1994; Bauer et al., 1995; Schulte-Oehlmann et al., 1995). A comparison with the data for the other species is possible because the tested TBT concentrations and the employed experimental set-up were identical. Although BCFs of the mud snail were not determined in field studies it can be presumed that they will be much higher under field conditions. Bryan et al. (1987), Bailey et al. (1991) and Oehlmann (1994) reported that the dog whelk *Nucella lapillus* accumulates significantly more TBT under field conditions than in the laboratory. TBT contaminations below 1 μg l^{-1} resulted in relatively low BCFs in laboratory experiments with other mollusc species also whereas identical concentrations under field conditions resulted in much higher BCFs (Salazar, 1989). Furthermore it has to be considered that in our laboratory experiments the only contamination path was from ambient water; other natural contamination sources, e.g. sediments and food, were missing. It has been shown (Figures 5a–b; Oehlmann et al., 1996a) that TBT exposure via sediments also plays an important role in the uptake of this xenobiotic and in inducing imposex development in the mud snail. As *H. ulvae* is a typical mud flat inhabitant, the species lives directly on or partially in the sediment. Sediments exhibit an accumulation factor of 10^3 for TBT (Kram et al., 1989) and thus are known to be an important source for re-

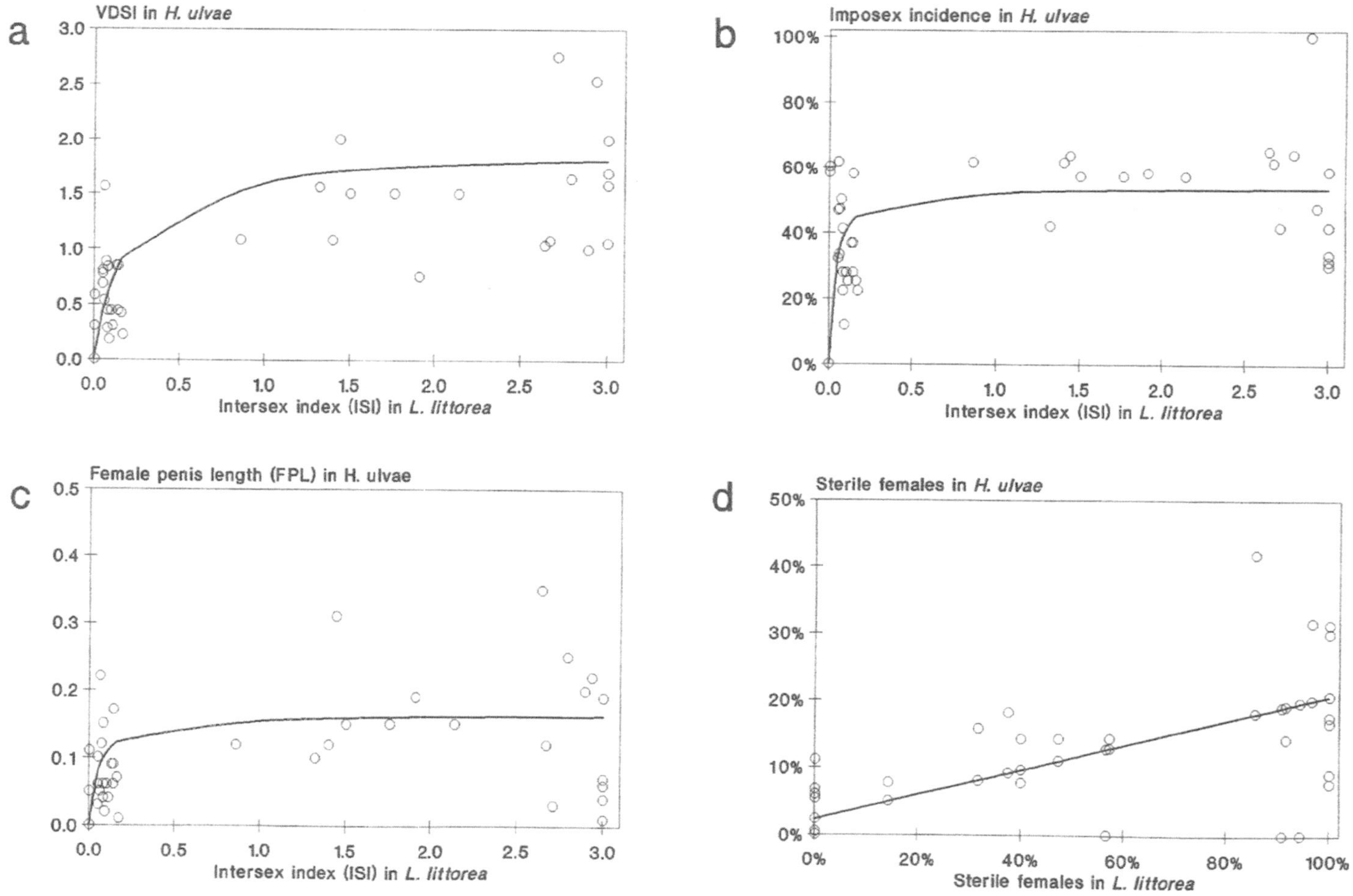

Figure 6. Hydrobia ulvae and *Littorina littorea*. Relationship between biological parameters in *H. ulvae* (VDSI, imposex incidence, female penis length) and *L. littorea* (intersex index) (a–c) and percentage of sterile females in both species (d). (a) $y=(1.92 \cdot x)-(0.181+x)$, $n=44$ samples, $r=0.791$, $p < 0.000$. (b) $y=(54.0 \cdot x)-(0.0336+x)$, $n=44$ samples, $r=0.536$, $p < 0.0005$. (c) $y=(1.166 \cdot x)-(0.0572+x)$, $n=44$ samples, $r=0.531$, $p < 0.0005$. (d) $y=2.27 \cdot x+0.185$, $n=44$ samples, $r=0.0.710$, $p < 0.0005$.

mobilisation of the biocide (Langston & Pope, 1995; Kalbfus et al., 1996). The half-life of TBT especially in anaerobic harbour sediments is up to 15 years (de Mora et al., 1989) and therefore contaminated sediments will have an influence on the masculinisation phenomenon of prosobranchs even if TBT emissions are reduced seriously. A threshold tissue concentration of 100 μg TBT-Sn kg^{-1} dry weight for imposex development in *Hydrobia ulvae* is comparable to those values reported for the nassariid snail *Hinia reticulata* (100 μg TBT-Sn kg^{-1} dry weight) (Stroben et al., 1992) and the periwinkle *Littorina littorea* (150 μg TBT-Sn kg^{-1} dry weight) (Oehlmann et al., 1996a). In muricid snails such as *Nucella lapillus* and *Ocenebra erinacea* threshold concentrations (10 μg TBT-Sn kg^{-1} dry weight for the dog whelk and 50 μg TBT-Sn kg^{-1} dry weight for the rough tingle) for virilisation effects are lower (Oehlmann, 1994) whereas in the limnic ampullariid *Marisa cornuarietis* the value is much higher (1500 μg TBT-Sn kg^{-1} dry weight; Schulte-Oehlmann et al., 1995). The reasons for these differences can be explained by the different ecology of the investigated species. Predatory living snails generally accumulate more TBT at a given station than scavenger species and again considerably more than herbivores and lastly limnic gastropods (Bryan et al., 1986; Stickle et al., 1990; Tsuda et al., 1990; Bailey et al., 1991; Oehlmann, 1994; Schulte-Oehlmann et al., 1995). The reason for this seems to be the possible biomagnification potential of organotin compounds. Muricids are probably the most sensitive group within the prosobranchs, developing pathological alterations corresponding to the environmental TBT pollution at lower concentration ranges than other systematic groups (Oehlmann, 1994). They are followed by some nassariids (Stroben, 1994; Ten Hallers-Tjabbes et al., 1994) and different species of mesogastropods (Bauer et al., 1995; Schulte-Oehlmann et al., 1995, 1997).

It cannot be expected that all biomonitoring parameters possess the same suitability for the assessment of TBT effects. In *Hydrobia ulvae* the mean shell height is between 3–5 mm and therefore the female penis length (FPL) is not easy to measure, especially when it is below 0.05 mm. Furthermore, in areas where female mud snails exhibit exclusively imposex development according to the b-path of the classification scheme (penis development not before stage 3b), the FPL would be useless.

It is not inconsistent that the percentage of sterile females of *Littorina littorea* increases more sharply than the percentage of sterile females in *H. ulvae* although the latter was identified as the more sensitive species on the basis of a direct comparison of ISI values with VDSI, FPL and imposex incidences in sympatrically living populations. This can be explained by the fact that in *L. littorea* female sterility is an obligatory symptom of masculinisation whereas in the mud snail infertility is also observable irrespective of the virilisation phenomenon. At a given TBT contamination level, female mud snails exhibit strong signs of masculinisation (i.e. high imposex stages) while the percentage of sterile females can be comparatively low. Sympatrically living female periwinkles might exhibit weak signs of masculinisation (i.e. low intersex stages) but simultaneously the percentage of infertile females can be very high. As female *L. littorea* are already definitively sterile at intersex stage 2, also low masculinisation stages go along with high percentages of female sterility.

The findings of Oehlmann (1994), concerning the minimum number of snails for a representative sampling strategy, support our own results for *Hydrobia ulvae* (Figure 4). However, the author recommends a minimum number of 30 dog whelks per sample for analysis even at stations of low imposex frequency. For *Nucella lapillus* this seems to be right. With respect to the mud snail the number of analysed snails has to be higher because at stations with low imposex frequency, the variation amplitude of the different parameters of 30 individuals is higher for *Hydrobia* than for *Nucella*.

Finally, it should not be concealed that Figure 3 represents a selection of 16 sampling stations; for two of them the results were not constant and exhibited greater deviations (VDSI between 0.06 and 1.0 and 0.17 and 2.31) (compare Oehlmann et al., 1996a). The reasons for these results are unknown but might be explained by different ecological factors. At sampling stations where those problems arise, ecological studies, for example, including the floating behaviour of hydrobiids, should be considered for future analyses. Decreasing salinity, circadian rhythms or interspecific concurrence in mixed *Hydrobia* populations may secondarily influence the floating behaviour of the species in those biotopes. Therefore ecologists should be asked for their recommendations if areas for biomonitoring purposes also include stations with high mobilities of individuals.

Acknowledgements

This work was supported in part by the Umweltbundesamt, Berlin (R. & D. project 102 40 303/01). We are grateful to Dr W. Kalbfus and colleagues (Bavarian State Administration for Water Resources Management, Institute for Water Research) for the performance of the TBT analyses in sediments.

References

Bailey, S. K., 1991. Tributyltin (TBT) contamination of Scottish coastal waters. Unpublished Ph.D. thesis, Napier University, Edinburgh, 379 pp.

Bailey, S. K., I. M. Davies, M. J. C. Harding, & A. M. Shanks, 1991. Effects of tributyltin oxide on the dogwhelk *Nucella lapillus* (L.). Proc. of the Tenth World Meeting of the Organotin Environmental Programme (ORTEP) Association, Berlin, Germany, September 26/27 1991: 7–66.

Bauer, B., P. Fioroni, L. Ide, S. Liebe, J. Oehlmann, E. Stroben & Watermann, 1995. TBT effects on the female genital system of *Littorina littorea*: a possible indicator of tributyltin pollution. Hydrobiologia 309: 15–27.

Bryan, G. W., 1979. Bioaccumulation of marine pollutants. Phil. Trans. r. Soc., Lond. 286: 483–505.

Bryan, G. W., P. E. Gibbs, L. G. Hummerstone & G. R. Burt, 1986. The decline of the gastropod *Nucella lapillus* around south-west England; evidence for the effect of tributyltin from antifouling paints. J. mar. biol. Ass. U.K. 66: 611–640.

Bryan, G. W., P. E. Gibbs, G. R. Burt & L. G. Hummerstone, 1987. The effects of tributyltin (TBT) accumulation on adult dog-whelks, *Nucella lapillus*: long-term field and laboratory experiments. J. mar. biol. Ass. U.K. 67: 525–544.

de Mora, S. J., N. G. King & M. C. Miller, 1989. Tributyltin and total tin in marine sediments: profiles and the apparent rate of TBT degradation. Envir. Technol. Lett. 10: 901–908.

Fioroni, P., J. Oehlmann & E. Stroben, 1991. The pseudohermaphroditism of prosobranchs; morphological aspects. Zool. Anz. 226: 1–26.

Gibbs, P. E., G. W. Bryan, P. L. Pascoe & G. R. Burt, 1987. The use of the dog-whelk, *Nucella lapillus*, as an indicator of tributyltin (TBT) contamination. J. mar. biol. Ass. U.K. 67: 507–523.

Kalbfus, W., A. Zellner, S. Frey & T. Knorr, 1996. Analytik von Oberflächenwasser, Sediment und Mollusken zur Validierung des biologischen Effektmonitorings. Final report of R. & D. project 102 40 303/02, Umweltbundesamt, Berlin, 108 pp.

Kalbfus, W., A. Zellner, S. Frey & E. Stanner, 1991. Gewässergefährdung durch organozinnhaltige Antifouling-Anstriche. Final report of R & D project 126 050 10, Umweltbundesamt, Berlin, 163 pp.

Kram, M. L., P. M. Stang & P. F. Seligman, 1989. Adsorption and desorption of tributyltin in sediments of San Diego Bay and Pearl Harbor. Appl. Organometal. Chem. 3: 523–536.

Langston, W. J. & N. D. Pope, 1995. Determinants of TBT adsorption and desorption in estuarine sediments. Mar. Pollut. Bull. 31: 32–43.

Minchin, D., J. Oehlmann, U. Schulte-Oehlmann, B. Bauer & C. B. Duggan, 1997. Biological indicators used to map organotin contamination from a fishing port, Killybegs, Ireland. Mar. Pollut. Bull. 34: 235–243.

Müller, M. D., L. Renberg & G. Rippen, 1989. Tributyltin in the environment – sources, fate and determination: an assessment of present status and research needs. Chemosphere 18: 2015–2042.

Oehlmann, J., 1994. Imposex bei Muriciden (Gastropoda, Prosobranchia). Eine ökotoxikologische Untersuchung zu TBT-Effekten, Cuvillier Verlag, Göttingen, 167 pp.

Oehlmann, J. & C. Bettin, 1996. Tributyltin-induced imposex and the role of steroids in marine snails. Malacol. Rev. Suppl. 6: 157–161.

Oehlmann, J., I. Ide, B. Bauer, B. Watermann, U. Schulte-Oehlmann, S. Liebe & P. Fioroni, 1996a. Erfassung morpho- und histopathologischer Effekte von Organozinnverbindungen auf marine Mollusken und Prüfung ihrer Verwendbarkeit für ein zukünftiges biologisches Effektmonitoring. Final report of R & D project 102 40 303/01, Umweltbundesamt, Berlin, 194 pp.

Oehlmann, J., P. Fioroni, E. Stroben, B. Markert, 1996b. Tributyltin (TBT) effects on *Ocinebrina aciculata* (Gastropoda: Muricidae): imposex development, sterilization, sex change and population decline. Sci. Total Envir. 188: 205–223.

Oehlmann, J., E. Stroben & P. Fioroni, 1991. The morphological expression of imposex in *Nucella lapillus* (Linnaeus) (Gastropoda: Muricidae). J. moll. Stud. 57: 375–390.

Oehlmann, J., E. Stroben, U. Schulte-Oehlmann, B. Bauer, P. Fioroni & B. Markert, 1996c. Tributyltin biomonitoring using prosobranchs as sentinel organisms. Fresenius J. Anal. Chem. 354: 540–545.

Salazar, M. H., 1989. Mortality, growth and bioaccumulation in mussels exposed to TBT: differences between the laboratory and the field. Oceans '89, Conference Record 2: 530–536.

Schebek, L., M. O. Andreae & H. J. Tobschall, 1991. Methyl- and butyltin compounds in water and sediments of the Rhine River. Envir. Sci. Technol. 25: 871–878.

Schulte-Oehlmann, U., C. Bettin, P. Fioroni, J. Oehlmann & E. Stroben, 1995. *Marisa cornuarietis* (Gastropoda, Prosobranchia): a potential TBT bioindicator for freshwater environments. Ecotoxicology 4: 372–384.

Schulte-Oehlmann, U., J. Oehlmann, B. Watermann, B. Bauer, I. Ide & P. Fioroni, 1996a. Effekte von Tributylzinn (TBT) aus Antifoulinganstrichen auf Schneckenpopulationen in der Ostsee. In J. L. Lozán, R. Lampe, W. Matthäus, E. Rachor, H. Rumohr & H. von Westernhagen (eds), Warnsignale aus der Ostsee: wissenschaftliche Fakten. Parey, Berlin: 176–179.

Schulte-Oehlmann, U., J. Oehlmann, P. Fioroni & B. Bauer, 1997. Imposex and reproductive failure in *Hydrobia ulvae* (Pennant) (Gastropoda, Prosobranchia). Mar. Biol. 128: 257–266.

Schulte-Oehlmann, U., E. Stroben, P. Fioroni & J. Oehlmann, 1996b. Beeinträchtigungen der Reproduktionsfähigkeit limnischer Vorderkiemerschnecken durch das Biozid Tributylzinn (TBT). In J. L. Lozán & H. Kausch (eds), Warnsignale aus Flüssen und Ästuaren: wissenschaftliche Fakten, Blackwell, Berlin, Oxford: 249–255.

Stewart, C., S. J. de Mora, M. R. L. Jones & M. C. Miller, 1992. Imposex in New Zealand neogastropods. Mar. Pollut. Bull. 24: 204–209.

Stickle, W. B., J. L. Sharp-Dahl, S. D. Rice & J. W. Short, 1990. Imposex in *Nucella lima* (Gmelin) via mode of exposure to tributyltin. J. exp. mar. Biol. Ecol. 143: 165–180.

Stroben, E., 1994. Imposex und weitere Effekte von chronischer TBT-Intoxikation bei einigen Mesogastropoden und Bucciniden (Gastropoda, Prosobranchia). Cuvillier Verlag, Göttingen, 193 pp.

Stroben, E., J. Oehlmann & P. Fioroni, 1992. The morphological expression of imposex in *Hinia reticulata* (Gastropoda: Buc-

cinidae): a potential indicator of tributyltin pollution. Mar. Biol. 113: 625–636.
Ten Hallers-Tjabbes, C. C., J. F. Kemp & J. P. Boon, 1994. Imposex in whelks (*Buccinum undatum*) from the open North Sea: relation to shipping traffic intensities. Mar. Pollut. Bull. 28: 311–313.
Tsuda, T., S. Aoki, M. Kojima & H. Harada, 1990. Differences between freshwater and seawater-acclimated guppies in the accumulation and excretion of tri-n-butyltin chloride and triphenyltin chloride. Wat. Res. 24: 1373–1376.
Ward, G. S., G. C. Cramm, P. R. Parrish, H. Trachman & A. Slesinger, 1981. Bioaccumulation and chronic toxicity of bis(tributyltin)oxide (TBTO): test with a saltwater fish. In D. R. Branson & K. L. Dickson (eds), Aquatic Toxicology and Hazard Assessment: Fourth Conference. ASTM STP 737, American Society for Testing and Materials, Philadelphia: 183–200.
Welz, B., 1983. Atomabsorptionsspektrometrie, Verlag Chemie, Weinheim, 349 pp.

Hydrobiologia **378:** 227–233, 1998.
R. M. O'Riordan, G. M. Burnell, M. S. Davies & N. F. Ramsay (eds), Aspects of Littorinid Biology.

Ultrastructure of prostate gland tissue in males and females with intersex phenomena of *Littorina littorea* L.

G. Sundermann[1], B. Bauer[1] & J. Oehlmann[2]
[1] *Institut für Spezielle Zoologie und Vergleichende Embryologie, Hüfferstrasse 1, D-48149 Münster, Germany*
[2] *Internationales Hochschulinstitut, Markt 23, D-02763 Zittau, Germany*

Key words: Littorina littorea, prostate gland, histology, intersex, female stage 3

Abstract

In the prosobranch snail *Littorina littorea* (L., 1758) the ultrastructure of the prostate gland cells (pgc) in males and the altered glandular epithelium of the pallial oviduct of females in intersex stage 3 is compared. Regarding form, structure, organelles and secretory products the pgc in males are identical with the corresponding gland cells of the females. Consequently these results demonstrate that in females of intersex stage 3 the epithelium of the pallial oviduct, which originally consists of several (3) glandular parts, is transformed into a male prostate gland.

Introduction

Malformations in prosobranchs as a reaction to Tributyltin (TBT) contamination include, among others, imposex and intersex phenomena (Short et al., 1989; Bright & Ellis, 1990; Oehlmann et al., 1991; Stroben et al., 1992a & b; Bauer et al., 1995). Imposex is a superimposition of male sex organs on females while intersex is defined in *L. littorea* as any sex determination disturbance between gonad and genital tract. During intersex development no superimposition of male characters occurs but the organs of the pallial oviduct are first modified towards and then supplanted by the corresponding male morphological structures (Bauer et al., 1995). These authors classify the transformation of the pallial oviduct during intersex development in 5 stages (0–4). The present study compares the ultrastructure of the prostate gland in males and females of intersex stage 3 in the course of investigations into the suitability of *L. littorea* for TBT effect monitoring.

Materials and methods

Male *L. littorea* as well as females of intersex stage 3 were collected in September/October on the German North Sea Coast.

For semi-thin sections and transmission electron microscopy (TEM) the prostate gland of males and the pallial oviduct of females were dissected. Tissue fixation started according to Eisenman and Alfert (1982) with a pre-fixation in 0.5 ml post-fixative added to 10 ml main fixativc (for 5–10 min). Main-fixation was carried out with 4% glutaraldehyde in 0.2 m Na cacodylate (4.28 g 100 ml^{-1}), 0.1 m NaCl (0.58 g 100 ml^{-1}), and 0.35 m sucrose (11.98 g 100 ml^{-1}), (pH 7.2), followed by post-fixation in 1% OsO_4 in 0.2 m Na cacodylate (4.28 g 100 ml^{-1}), and 0.3 m NaCl (1.75 g 100 ml^{-1}), (pH 7.2). Fixed tissue was embedded in epoxy resin (Durcupan, Fluka, Switzerland). Semi-thin and ultra-thin sections were cut on an LKB Ultrotome III with glass knives. Semi-thin sections were stained with 1% toluidine blue in 1% borax (aqueous solution), ultra-thin sections with lead citrate. The transmission electron microscope Zeiss EM 900 used for the study, is on loan from the Stiftung Volkswagenwerk.

Results

Males

In *L. littorea* the pallial section of the gonoduct ontogenetically originates from an infolding of the mantle

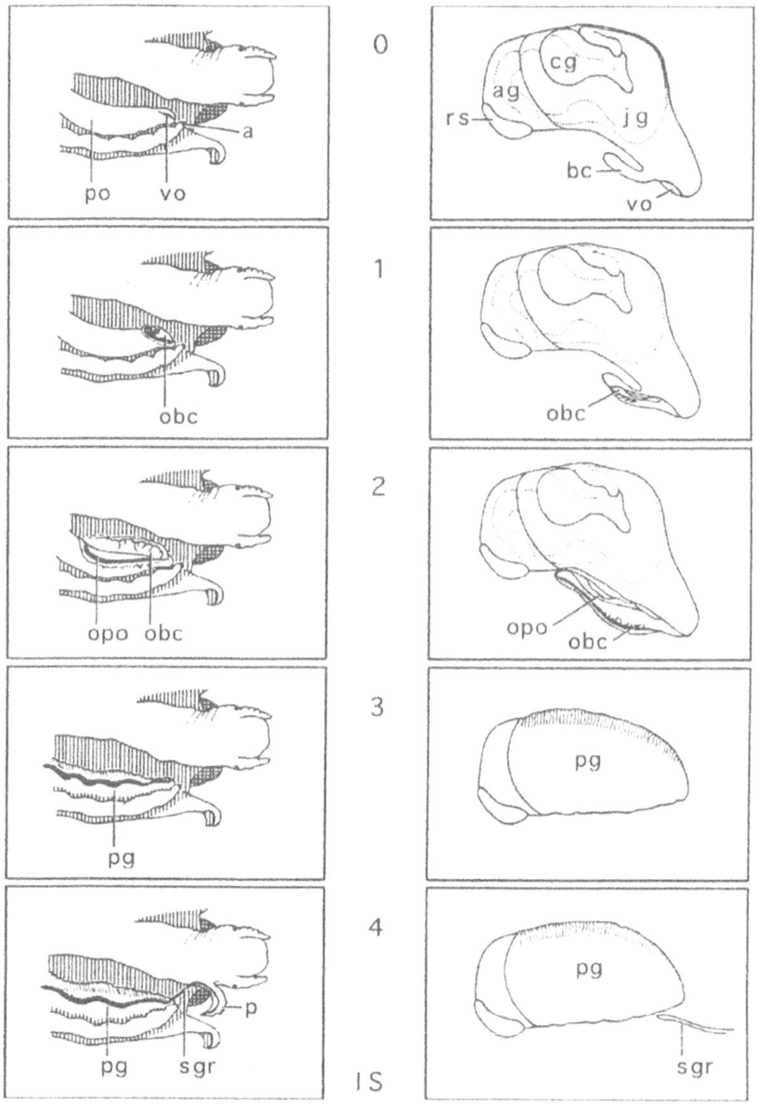

Figure 1. Female, scheme of intersex development. Left side: stage 0 to 4, dorsal view with opened mantle cavity. Right side: lateral view of pallial oviduct. Abbreviations: a, anus; ag, albumen gland; bc, bursa copulatrix; cg, capsule gland; IS, intersex stage; jg, jelly gland; obc, open bursa copulatrix; opo, open pallial oviduct; p, penis; pg, prostate gland; po, pallial oviduct; rs, receptaculum seminis; sgr, seminal groove; vo, vaginal opening.

epithelium. In males this open structure remains and provided with intraepithelial gland cells serves as a prostate in adult specimens (Figure 5). There are at least five clearly distinguishable types of gland cells. Their nuclei lie in the basal region of the granule-filled cells. Most abundant is the prostate gland cell, here abbreviated pgc. Pgc's are slender, spindle-shaped cells which mainly build up the prostate gland; their plasm, occasionally even in subnuclear position, is filled with spherical secretion granules which stain deep-blue with toluidine blue in semi-thin sections. Because of the staining properties the pgc's are easy to distinguish in histological sections (Figure 8). On TEM pictures (Figures 10, 11) the secretion granules all have the same homogeneous electron-dense structure, are membrane bounded (Figure 13) and of different size (diameters from ca. 2.5 μm to smaller). The dictyosome region, with small golgi complexes is

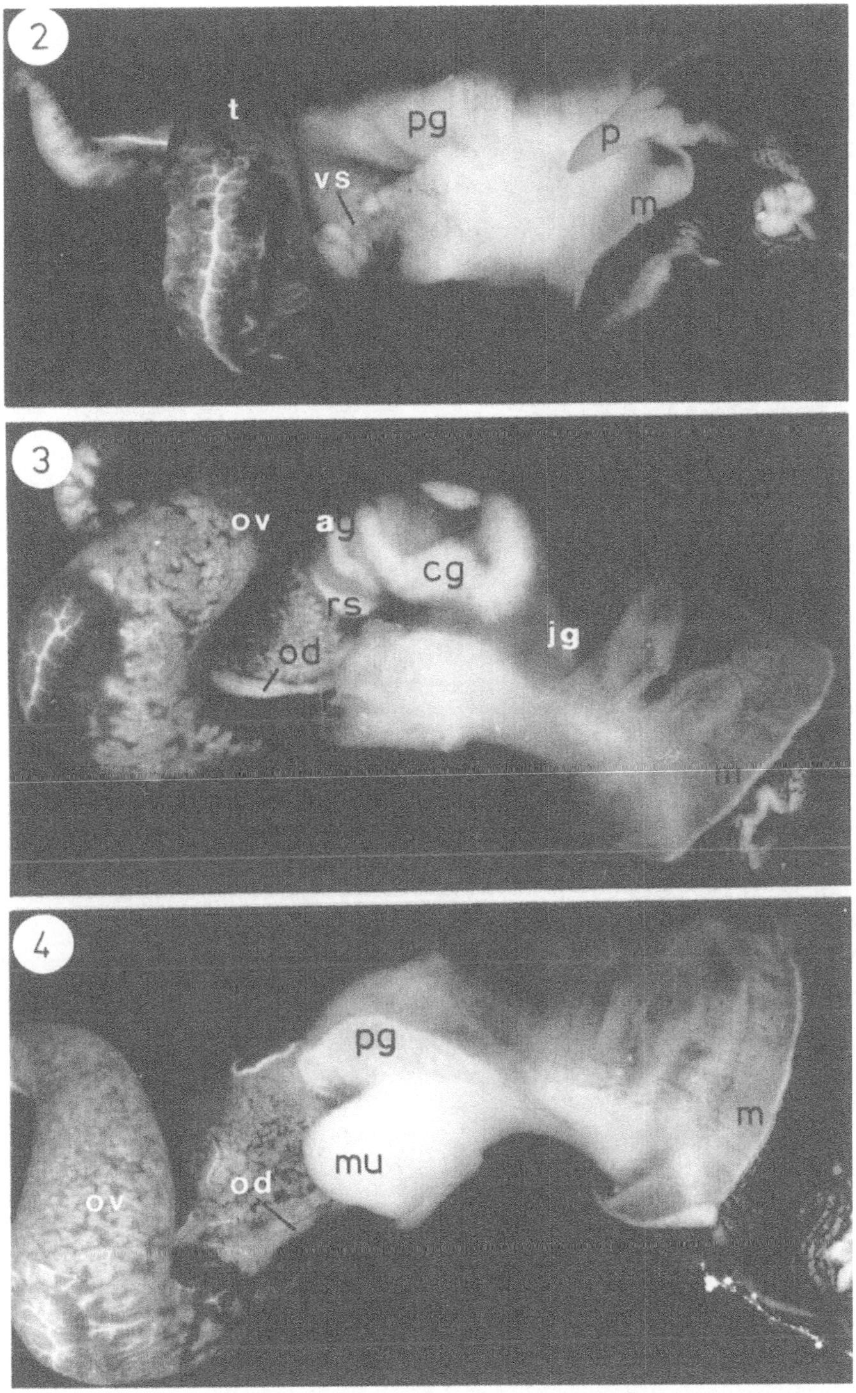

Figures 2–4. Figure 2. Male, shell removed, lateral view on right side. Figure 3. Female without intersex phenomena (intersex stage 0), shell removed, lateral view on right side. Figure 4. Female, intersex stage 3, shell removed; lateral view on right side. Abbreviations: ag, albumen gland; cg, capsule gland; jg, jelly gland; m, mantle; mu, musculature; od, gonadial and renal oviduct; ov, ovary; p, penis; pg, prostate gland; rs, receptaculum seminis; t, testis; vs, vesicula seminalis.

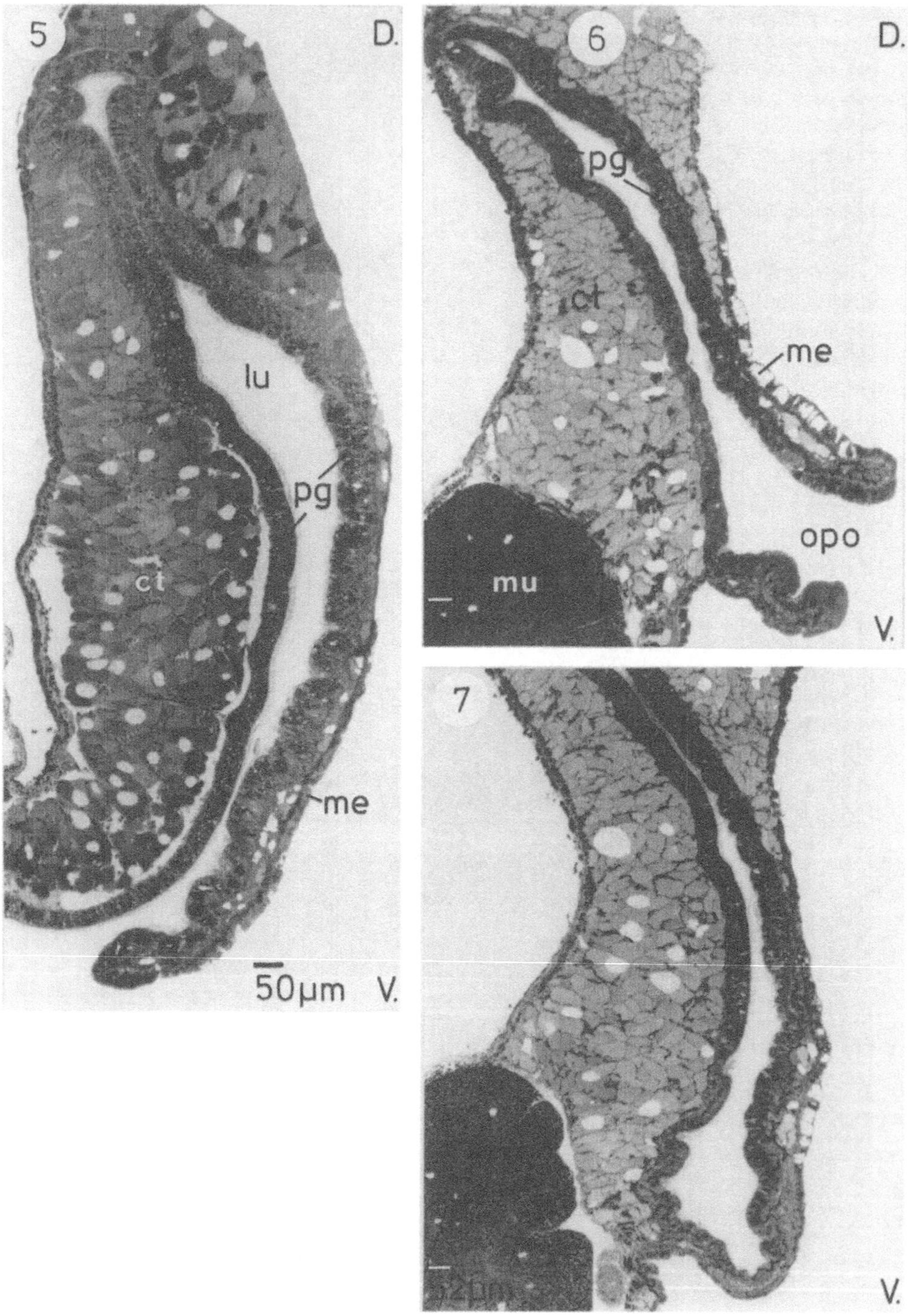

Figures 5–7. Figure 5. Male, transverse semi-thin section through prostate region. Figure 6. Female, intersex stage 3, transverse semi-thin section through distal region of transformed pallial oviduct; prostate gland ventrally open. Figure 7. Female, same animal as in Figure 6, semi-thin section in a more proximal region; prostate gland ventrally closed. Abbreviations: D., dorsal, ct, connective tissue; lu, lumen of prostate gland; me, mantle epithelium; mu, musculature; opo, open pallial oviduct; pg, prostate gland; V., ventral.

situated distal of the nucleus. In this region mitochondria are numerous and big membranous vesicles are found regularly. The latter show an unhomogeneous, electron-light content consisting of granular accumulations and occasional small membranous stacks; lysosomes are present in this region, too. REr extends in the lateral cell body and the spaces between the secretion granules. In the apical cell region microtubules are arranged parallel to the long axis of the cell. The apical surface bears some microvilli (Figure 12).

Unlike the adjoining mantle epithelium, in the prostate gland ciliated supporting cells intersperse the gland cells. Their nuclei lie in the distal cell region where the cytoplasm has its greatest extent (Figure 10). A thin stalk which is connected to the basal lamina represents the middle and basal region of the cell body. The cilia show the characteristic axoneme of $9\times2+2$ microtubules and possess small ciliary rootlets at their bases. In the distal cell region mitochondria are accumulated and lysosomes commonly found.

The basal region of the prostate gland epithelium is well provided with nerves and accompanying glial cells; they are frequently cut (Figure 11).

Intersex affected females

All examined specimens have been identified undoubtedly as females by the presence of an ovary (compare Figure 4).

In stage 3 the pallial oviduct is in total or partially supplanted by a prostate gland (Figures 1, 4, 6, 7); compared to stage 4 only the penis is missing (Figures 1, 2). The pallial oviduct of the examined females is split in the basal parts, only the most proximal region is a closed tube. Its epithelium is totally transformed into a prostate (Figure 4). No vestiges of the female bursa copulatrix, the jelly gland, the capsule gland, the albumen gland, and the seminal receptacle, which are clearly visible in non affected females, (Figure 3) are found. In the female prostate the histological structure of the epithelium as well as the ultrastructure of the glandular cells are essentially the same as the corresponding structures in the male (Figures 5, 6, 7, 8, 9, 10, 11, 14).

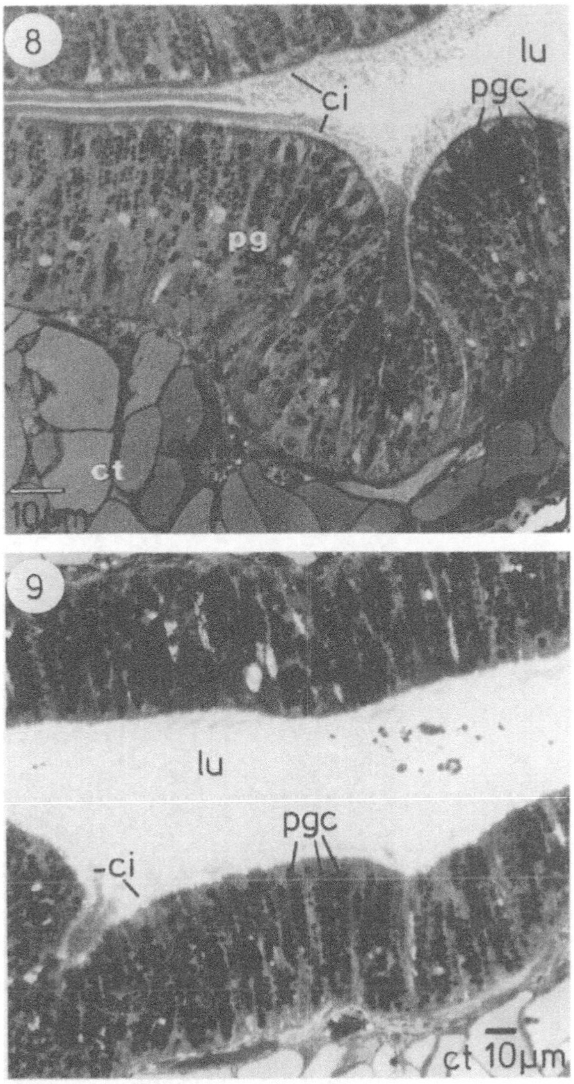

Figures 8–9. Figure 8. Male, longitudinal semi-thin section through prostate gland epithelium. Figure 9. Female, same animal as in Figure 6, longitudinal semi-thin section through prostate gland epithelium in proximal region. Abbreviations: ci, cilia; ct, connective tissue; lu, lumen of prostate gland; pg, prostate gland; pgc, prostate gland cells.

Discussion

In females affected by intersex the distal pallial oviduct transforms gradually into a prostate gland from stage 0 to stage 3. From the exterior this conversion is recognisable by a progressive ventral splitting of the ventral wall of the tube. In stage 3 the inner epithelium of the pallial oviduct is supplanted by a prostate gland (Bauer et al., 1995). The histological investigations prove that the female prostate gland is in accordance with the male prostate gland concerning structure and ultrastructure. The present results confirm the assumption that intersex phenomena caused by TBT pollution lead to substitution of the numerous

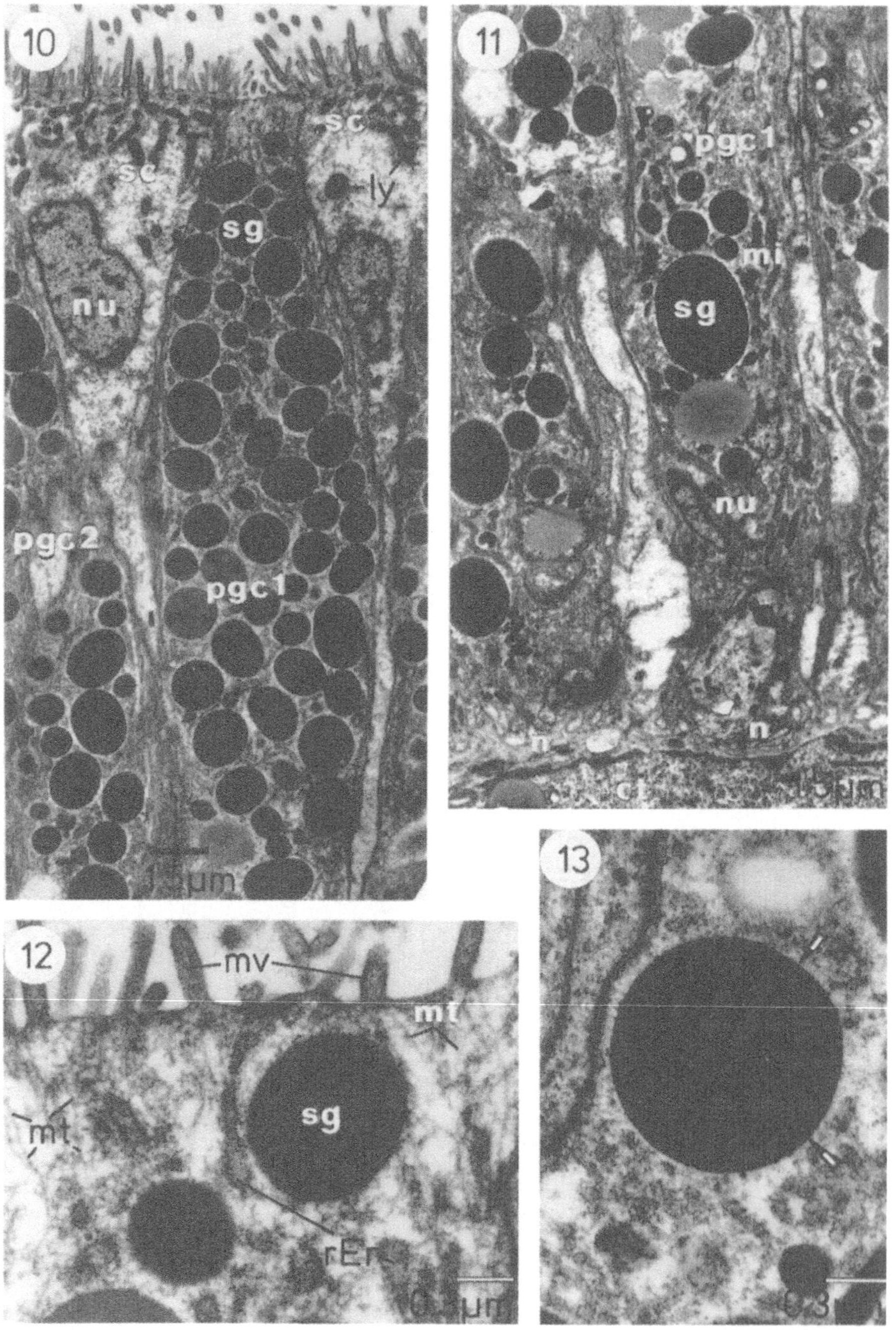

Figures 10–13. Figure 10. Male, prostate gland, TEM, upper half of longitudinally cut prostate gland cells and ciliated supporting cells. Figure 11. Lower half of Figure 10 cells. Figure 12. Male, prostate gland, TEM, longitudinal section through apical surface of prostate gland cell. Figure 13. Male, prostate gland, TEM, membrane-bound secretory granule of prostate gland cell. Abbreviations: ct, connective tissue; ly, lysosomes; mi, mitochondria; mt, microtubules; mv, microvilli; n, nerve; nu, nucleus; pgc, prostate gland cell; rEr, rough endoplasmatic reticulum; sg, secretion granule; sc, supporting cell; arrowheads, membrane of secretion granule.

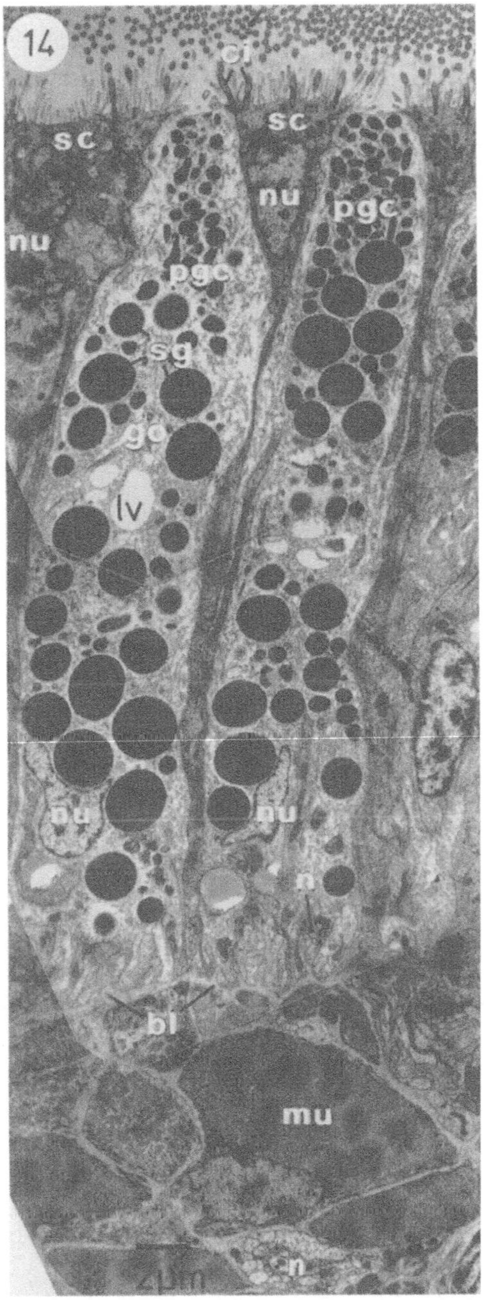

Figure 14. Female, same animal as in Figure 6, prostate gland, TEM, longitudinal section through prostate gland cells and ciliated supporting cells. Abbreviations: bl, basal lamina; ci, cilia; go, golgi complex; lv, electron-light vesicle; mu, musculature; n, nerve; nu, nucleus; pgc, prostate gland cell; sc, supporting cell; sg, secretion granule.

different female pallial oviduct cells by male prostate gland cells.

Apart from the morphological malformations (Bauer et al., 1995) the cellular alterations will make reproduction impossible.

Acknowledgement

The authors thank very much Christel Mehlis and Barbara Hasert for technical assistance.

References

Bauer, B., P. Fioroni, I. Ide, S. Liebe, J. Oehlmann, E. Stroben & B. Watermann, 1995. TBT effects on the female genital system of *Littorina littorea*: a possible indicator of tributyltin pollution. Hydrobiologia 309: 15–27.

Bright, D. A. & D. V. Ellis, 1990. A comparative survey of imposex in northeast Pacific neogastropods (Prosobranchia) related to tributyltin contamination, and a choice of a suitable bioindicator. Can. J. Zool. 68: 1915–1924.

Eisenman, E. A. & M. Alfert, 1982. A new fixation procedure for preserving the ultrastructure of marine invertebrate tissues. J. Microscopy 125: 117–120.

Oehlmann, J., E. Stroben & P. Fioroni, 1991. The morphological expression of imposex in *Nucella lapillus* (Linnaeus) (Gastropoda: Muricidae). J. moll. Stud. 57: 375–390.

Oehlmann, J., E. Stroben & P. Fioroni, 1992. The rough tingle *Ocenebra erinacea* (Gastropoda: Muricidae): an exhibitor of imposex in comparison to *Nucella lapillus*. Helgoländer wiss. Meeresunters. 46: 311–328.

Short, J. W., S. D. Rice, C. C. Brodersen & W. B. Stickle, 1989. Occurrence of tri-*n*-butyltin-caused imposex in the North Pacific marine snail *Nucella lima* in Auke Bay. Alaska. Mar. Biol. 102: 291–297.

Stroben, E., C. Brömmel, J. Oehlmann & P. Fioroni, 1992a. The genital systems of *Trivia arctica* and *Trivia monacha* (Prosobranchia: Mesogastropoda) and tribultyltin induced imposex. Zool. Beitr. N. F. 34: 349–374.

Stroben, E., J. Oehlmann & P. Fioroni, 1992b. The morphological expression of imposex in *Hinia reticulata* (Gastropoda: Buccinidae): a potential biological indicator of tributyltin pollution. Mar. Biol. 113: 625–636.

Hydrobiologia **378:** 235–236, 1998.
R. M. O'Riordan, G. M. Burnell, M. S. Davies & N. F. Ramsay (eds), *Aspects of Littorinid Biology.*

Diet in *Littoraria*

J. T. Christensen
Department of Marine Ecology, Institute of Biological Sciences, Aarhus University, Denmark

Key words: Diet, *Littoraria*, mangroves, fungi

Mangroves of the Indo-Pacific region harbour a varied and usually abundant fauna of littorinids of the genus *Littoraria*, the taxonomy, geographical distribution and anatomy of which have been described by Reid (1986, 1989). Mangrove-dwelling species of the genus spend all of their adult lives on stems, prop-roots and leaves of the trees, but very little is known of their diets.

As an initial study of the food of Indo-Pacific *Littoraria,* gut contents of the closely related *L. pallescens* (Philippi, 1846), *L. intermedia* (Philippi, 1846) and *L. scabra* (L., 1758) living on *Rhizophora apiculata* were analysed. *L. pallescens* lives on the leaves of *Rhizophora* while *L. intermedia* spends most of its time on stems and prop-roots. *L. scabra* is also found on stems, and is usually found lower than *L. intermedia. L. pallescens* feeds on the leaf surface apparently without damaging the leaf, though Ohgaki (1990) reported radular marks on leaves of *R. stylosa*.

Snails were collected from mangrove trees on Phuket, Thailand, and killed by dropping into 70% alcohol. Stomach contents and faeces were removed under a dissecting microscope, smeared on to microscope slides and imbedded in glycerol gelatin. Some smears were heat-fixed, flushed with aniline blue, rinsed with water and imbedded in glycerol gelatin before microscopic examination.

For Scanning Electron Microscopy of *Rhizophora* leaf surfaces, leaves were picked from trees and immediately fixed in alcohol (70% v/v) or formalin (4% v/v in seawater). For identification of mucous trails, leaves were dipped in a solution of toluidine blue and stained sections cut out and fixed.

Output of faecal material was determined by placing snails individually in small vials which contained seawater for one hour. Faecal pellets were counted and their size (volume) determined as the mean of ten pellets. The height of the shell was also measured. Additional microscopic preparations were made from the faeces collected in this manner.

Stomachs of *L. pallescens* contained much fungal hyphae, hyphal fragments and a variety of fungal spores. Spores from 9 genera were found in addition to mycelium and spores of *Cladosporium cladosporioides* (Figure 1) (identification by K. D. Hyde, Hong Kong University). Fragments of leaf epidermis were also found. Spores were present in the faeces, but hyphae were only found as small fragments here. Hyphae stained blue with aniline blue. In *L. intermedia* large amounts of plant material, mainly cork cells, were found in stomachs in addition to fungal hyphae and spores. Some of the cork cells contained fungal mycelium. Faeces also contained large amounts of cork cells, but faecal material was more fragmented. Stomach contents of *L. scabra* were similar to that of *L. intermedia*, but fragments were larger. Diatoms were occasionally found. Scanning electron microscopy of leaf surfaces revealed fungi as the most conspicuous biotic element (Figure 2), and in mucous trails diatoms and foraminiferans could also be found.

The size of the faecal pellets was correlated with the size of the snail ([log(pellet vol.) = 0.05 (shell height) − 2.07]; $r = 0.76$, $p < 0.0001$) but was independent of species. Within species the number of pellets produced was independent of snail size. Thus, the number of pellets produced can be used as a relative measure of faecal output in interspecies comparisons. There was a highly significant difference between species with *L. intermedia* having a higher output than *L. pallescens* (Mann-Whitney $U = 582.5$, $p < 0.001$). A similar difference in faecal output was found between *L. scabra* and *L. pallescens* collected on *Avicennia* on Bunaken Island, Indonesia ($U = 102.5$, $p < 0.001$; data supplied by F. Boneka) (Figure 3).

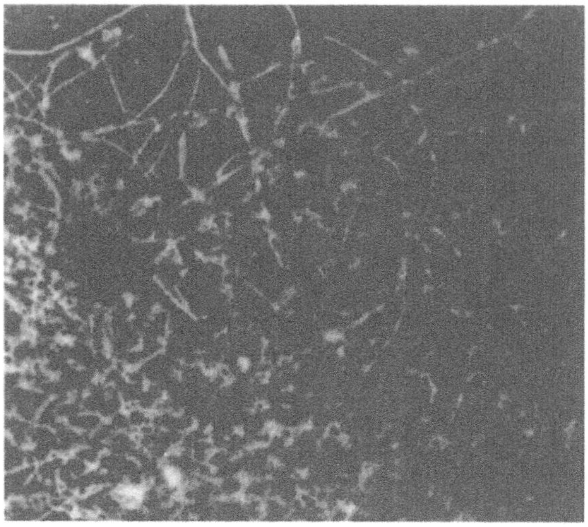

Figure 1. Fungal hyphae in stomach of *Littoraria pallescens.*

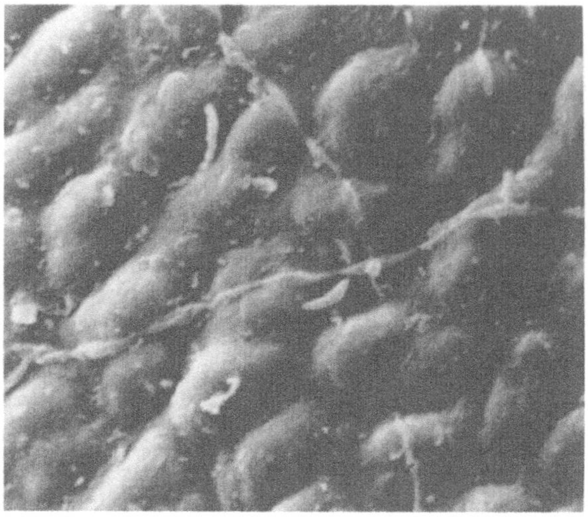

Figure 2. Fungal hyphae and spores on the leaf surface of *Rhizophora apiculata.*

Littorinid snails feed on a wide range of plant material from macro- and microalgae to lichens (Norton et al., 1990). The ingestion of fungi and fungus-infested plant material has been found in *L. angulifera* in Central America (Kohlmeyer & Bebout, 1986) and in *L. irrorata* in salt marshes in U.S.A. (Newell & Bärlocher, 1993). The latter species was able to grow on a diet solely consisting of fungi, though growth was better on fungus-infested standing dead cordgrass (Bärlocher & Newell, 1994). The finding of fungi in the guts of Indo-Pacific *Littoraria* may imply that fungivory is widespread in the genus. However, Ohgaki (1990 and pers. comm.) reported detritus, diatoms and foliose green algae in diets of *L. pallescens* and *L. intermedia* on Ishigaki Island, Okinawa.

It appears that *L. pallescens* has a lower rate of food intake (as measured by faecal output) than *L. intermedia* and *L. scabra*, possibly owing to a higher digestibility of its diet. The role of fungi in the energy budget of the snails remains to be established, but fungi may aid in the breakdown of plant structural material.

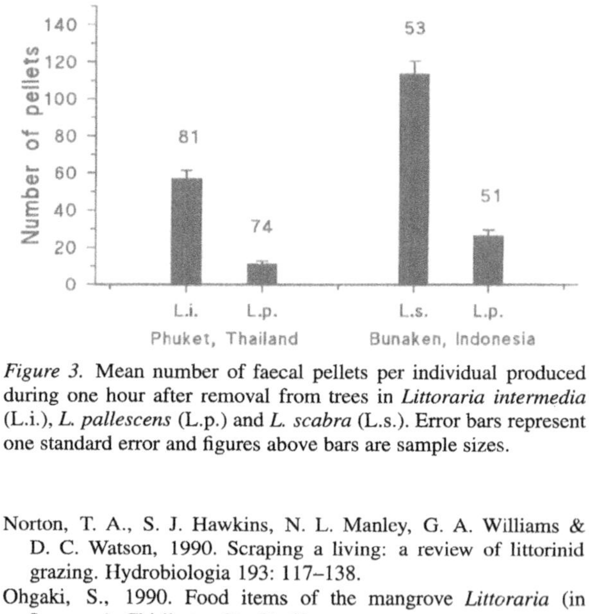

Figure 3. Mean number of faecal pellets per individual produced during one hour after removal from trees in *Littoraria intermedia* (L.i.), *L. pallescens* (L.p.) and *L. scabra* (L.s.). Error bars represent one standard error and figures above bars are sample sizes.

References

Bärlocher, F. & S. Y. Newell, 1994. Growth of the salt marsh periwinkle *Littoraria irrorata* on fungal and cordgrass diets. Mar. Biol. 118: 109–114.

Kohlmeyer, J. & B. Bebout, 1986. On the occurrence of marine fungi in the diet of *Littorina angulifera* and observations on the behaviour of the periwinkle. P. S. Z. N. I. mar. Ecol. 7: 333–343.

Newell, S. Y. & F. Bärlocher, 1993. Removal of fungal and total organic matter from decaying cordgrass leaves by shredder snails. J. exp. mar. Biol. Ecol. 171: 39–49.

Norton, T. A., S. J. Hawkins, N. L. Manley, G. A. Williams & D. C. Watson, 1990. Scraping a living: a review of littorinid grazing. Hydrobiologia 193: 117–138.

Ohgaki, S., 1990. Food items of the mangrove *Littoraria* (in Japanese). Chiribotan 21: 51–53.

Reid, D. G., 1986. The littorinid molluscs of mangrove forests in the Indo-Pacific region. British Museum (Natural History), London. 227 pp.

Reid, D. G., 1989. The comparative morphology, phylogeny and evolution of the gastropod family Littorinidae. Phil. Trans. r. Soc. Lond. B 324: 1–110.

Hydrobiologia **378:** 237–239, 1998.
R. M. O'Riordan, G. M. Burnell, M. S. Davies & N. F. Ramsay (eds), *Aspects of Littorinid Biology.*

A novel method to assess the cercariae emission intensity of littorinid populations

A. J. Knowles & M. S. Davies
Ecology Centre, University of Sunderland, Sunderland SR1 3SD, U.K.

Key words: Parasite, *Cryptocotyle lingua*, *Littorina littorea*, infection, rocky shore, north-east England

The ecological implications of parasitism of *Littorina littorea* (L.) (Gastropoda: Prosobranchia) and other gastropods by larval digenean trematodes has attracted considerable research interest (e.g. Lambert & Farley, 1968; Williams & Ellis, 1975; Lauckner, 1987; Sokolova, 1995). Such implications hinge on an accurate quantification of parasite loading within an infected population. There are several techniques available for determining parasite load or an index of it. Dissection, usually of the digestive gland, is the most commonly used destructive technique (e.g. Williams & Brailsford, 1990; Granovitch, 1992; Siddal et al., 1993), which can give an accurate estimate of parasite load in terms of the number of cercariae present. The subjective 'foot colour' technique correlates the intensity of orange colour of the foot of *L. littorea* to the degree of infection by *Cryptocotyle lingua* (Creplin) (Digenea: Heterophyidae), (e.g. Willey & Gross, 1957; Huxham et al., 1993), however, dissection should be employed to investigate the presence of other trematode cercariae. The non-destructive technique most usually used is the isolation technique (e.g. Bourns, 1963; Boerner Brodsky, 1971; Robson & Williams, 1971; Hellmann Craig, 1975; Haas et al., 1994; Wood & Matthews, 1996), which quantifies the rate of cercariae emission, and allows for their identification, from the molluscan host. This requires isolating the mollusc in a small volume of water, and allowing cercariae to emerge. One drawback, however, is the ability of some species of cercariae to re-encyst into the mollusc, resulting in an underestimate of parasite load, and we have observed the re-encystment of cercariae into *L. littorea*. Presented here is an inexpensive adaptation to the isolation technique concept, which removes cercariae from the vicinity of the host as they emerge, providing an accurate quantification of relative cercariae emission rates whilst reducing the possibility of re-encystment.

A tower (height 2.0 m, width 0.32 m, breadth 0.53 m), was constructed to house 5 plastic tanks (base area 0.3 m × 0.45 m), stacked vertically above a reservoir tank containing 7–8 l of filtered (20 μm) seawater. Excluding the reservoir tank, the base of each of the stacked tanks was perforated with 20 evenly spaced holes, each housing an isolation chamber, constructed from a polyethylene cup (volume 30 ml) which had 4 holes (diameter 2–3 mm) equally spaced around the edge of the base. Suspended below each of the 5 tanks was a wire mesh cradle holding 20 similar cups, each positioned below an isolation chamber, whose bases were removed and replaced with 20 μm mesh (Normesh, UK). A quick-setting epoxy resin was used to stick the mesh to the cups. All cups were allowed to cure for $>$ 24 h and were soaked in seawater for $>$ 24 h prior to use to remove any leachates. One *Littorina littorea* was introduced into each isolation chamber after washing the shell with filtered (20 μm) seawater to remove particulate matter. A layer of plastic mesh (aperture $\sim$ 10 mm) was fixed to the internal base of the tanks to prevent any *L. littorea* escaping from the isolation chambers during the procedure.

Seawater was pumped from the reservoir tank to the uppermost tank, where it was distributed through a horizontal pipe. Seawater flowed over the *L. littorea* and through the filter cups, and emerged cercariae were trapped on the (20 μm) filter. In this way a total of 100 animals could be analysed at once.

L. littorea ($n = 500$ per shore) were collected between July and September 1996, from four rocky shores on the north-east coast of England: Whitburn (National Grid Reference NZ416628), Tynemouth (NZ375691), Whitley Bay (NZ354751) and Cresswell

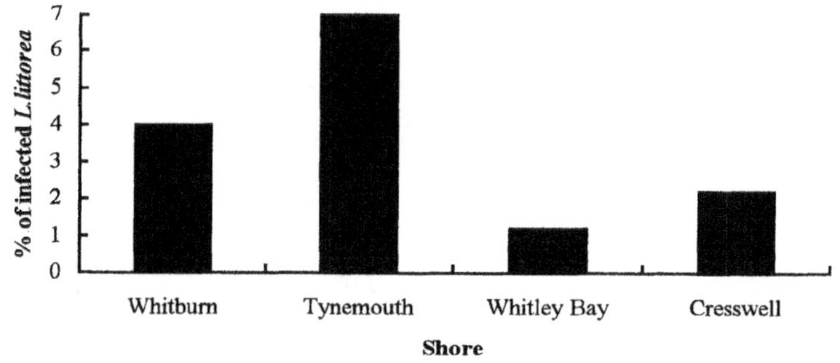

Figure 1. Percentage of *L. littorea* releasing cercariae over 15 h ($n = 500$ for each shore).

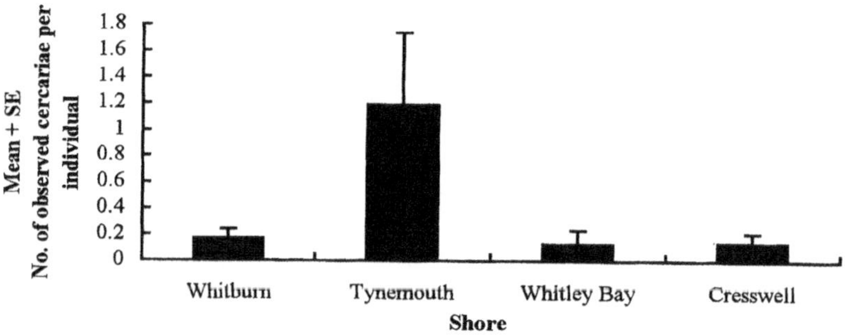

Figure 2. Mean number of cercariae released by individual *L. littorea* over 15 h ($n = 500$ for each shore).

(NZ298934). Individual littorinids were placed into separate plastic bags to prevent cross infection occurring during transit. All isolations were performed in a constant temperature room (25 °C) for 15 h overnight, after which all *L. littorea* were removed from the system and their shell heights recorded (±0.01 mm). The resultant biological material trapped on the filters was fixed in 10% w/w formalin and examined for the number and species of cercariae present using the key of James (1968).

Some *L. littorea* from each rocky shore released cercariae (Figure 1). Only one species of trematode was found to infect *L. littorea*: *Cryptocotyle lingua.* The population at Tynemouth had the greatest proportion of infected *L. littorea* (7%), followed by Whitburn (4%), Cresswell (2.2%) and Whitley Bay (1.2%, Figure 1). The population at Tynemouth also showed the greatest number of cercariae per individual (mean $\pm$ SE $= 1.2 \pm 0.5$) and although this value was much higher than that of the populations from the other three shores (Figure 2), there was no significant difference between populations in the mean number of cercariae released per individual (Kruskal-Wallis test, $H = 2.9$, df $= 3$, $p > 0.05$). The population at Tynemouth also showed the greatest number of cercariae per infected individual (mean $\pm$ SE $= 16.9 \pm 2.8$, Figure 3), but again there was no significant difference between the populations (Kruskal-Wallis test, $H = 3.23$, df $= 3$, $p > 0.05$). Although the littorinids from the rocky shore at Whitburn had the second largest number of infected *L. littorea* (Figure 1), the mean number of cercariae per infected individual was the lowest of the populations investigated.

There was no significant correlation between shell height (a measure of body size) and the number of released cercariae for any of the rocky shores sampled (Whitburn, $r = 0.148$ $p > 0.05$; Tynemouth, $r = 0.300$ $p > 0.05$; Whitley Bay, $r = 0.281$ $p > 0.05$; Cresswell, $r = 0.185$ $p > 0.05$).

The technique presented is no more time-consuming than current isolation techniques and allows large numbers of animals to be analysed at the same time. The equipment can be constructed with little expense and flowing seawater is likely to reduce the potential for re-encystment over isolation techniques that employ still seawater. The presented technique, of course, does not detect all the cercariae within a single *L. littorea*, but merely offers a comparative index of loading. Our technique might be useful where an accurate index of parasite emission rate is required from a small population which could then be re-introduced to its natural environment. The filters which trap the

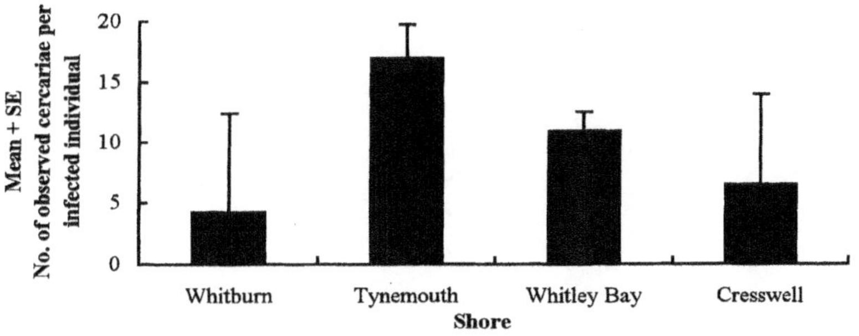

Figure 3. Mean number of cercariae released from infected *L. littorea* over 15 h.

biological material released by *L. littorea* also allow other aspects of littorinid biology to be examined, e.g. analysis of egestion rates and faecal contents or egg capsule release.

The lack of variation in parasite loading between the populations of *Littorina littorea* studied here might indicate similar levels of intermediate and final hosts (fish and seabirds, respectively) at the shores, but this, and the variation in parasite loading with time of year are yet to be analysed.

Acknowledgements

Thanks to L. McMahon, C. Knowles, K. Ball and T. Horan for their valuable assistance with field work and to two anonymous referees for useful comments.

References

Boerner Brodsky, D. K., 1971. Selected aspects of the parasitism of the gastropod *Littorina littorea* by *Cryptocotyle lingua* (Trematoda). Unpublished Ph.D. thesis, Northwestern University, Illinois.

Bourns, T. K. R., 1963. Larval trematodes parasitizing *Lymnaea stagnalis appressa* Say in Ontario with emphasis on multiple infections. Can. J. Zool. 41: 937–941.

Curtis, L. A. & K. M. Hubbard, 1990. Trematode infections in a gastropod host misrepresented by observing shed cercariae. J. exp. mar. Biol. Ecol. 143: 131–137.

Granovitch, A. I., 1992. The effect of trematode infection on the population structure of *Littorina saxatilis* (Olivi) in the White Sea. In J. Grahame, P. Mill & D. G. Reid (eds), Proceedings of the Third International Symposium on Littorinid Biology. The Malacological Society of London, London: 255–263.

Haas, W., M. Körner, E. Hutterer, M. Wegner & B. Haberl, 1995. Finding and recognition of the snail intermediate hosts by 3 species of echinostome cercariae. Parasitology 110: 133–142.

Hellmann Craig, L., 1975. *Hismasthla quissetensis* and *Lepocreadium setiferoides*: emergence patterns from their molluscan host, *Nassarius obsoletus*. Expl Parasit. 38: 56–63.

Huxham, M., D. Raffaelli & A. Pike, 1993. The influence of *Cryptocotyle lingua* (Digenea: Platyhelminthes) infections on the survival and fecundity of *Littorina littorea* (Gastropoda: Prosobranchia); an ecological approach. J. exp. mar. Biol. Ecol. 168: 223–238.

James, B. L., 1968. The distribution and keys of species in the family Littorinidae and of their digenean parasites, in the region of Dale, Pembrokeshire. Field Stud. 2: 615–650.

Lambert, T. C. & J. Farley, 1968. The effect of parasitism by the trematode *Cryptocotyle lingua* (Creplin) on zonation and winter migration of the common periwinkle, *Littorina littorea* (L.). Can. J. Zool. 46: 1139–1147.

Lauckner, G., 1987. Ecological effects of larval trematode infestation on littoral marine invertebrate populations. Int. J. Parasitol. 17: 391–398.

Matthews, P. M., 1986. Ecological and taxonomic studies on larval Digenea infecting *Littorina* spp. on the Co. Down coast. Unpublished Ph.D. thesis, The Queen's University, Belfast.

Robson, E. M. & I. C. Williams, 1971. Relationships of some species of digenea with the marine prosobranch *Littorina littorea* (L.) II. The effect of larval digenea on the reproductive biology of *L. littorea*. J. Helminth. 45: 145–159.

Siddal, R., A. W. Pike & A. H. McVicar, 1993. Parasites of *Buccinum undatum* (Mollusca: Prosobranchia) as biological indicators of sewage-sludge dispersal. J. mar. biol. Ass. U.K. 73: 931–948.

Sokolova, I. M., 1995. Influence of trematodes on the demography of *Littorina saxatilis* (Gastropoda: Prosobranchia: Littorinidae) in the White Sea. Dis. Aquat. Org. 21: 91–101.

Willey, C. H. & P. R. Gross, 1957. Pigmentation in the foot of *Littorina littorea* as a means of recognition of infection with trematode larvae. J. Parasitol. 43: 324–327.

Williams, G. A. & T. J. Brailsford, 1990. The relationship between flat periwinkle life histories and digenean infections. Hydrobiologia 193: 183–190.

Williams, I. C. & C. Ellis, 1975. Movements of the common periwinkle, *Littorina littorea* (L.), on the Yorkshire coast in winter and the influence of infection with larval digenea. J. exp. mar. Biol. Ecol. 17: 47–58.

Wood, B. P. & R. A. Matthews, 1996. Reactions of sera from immunised and experimentally infected thick-lipped grey mullet, *Chelon labrosus* (Risso 1826), with isolated proteins of *Cryptocotyle lingua* (Creplin, 1825). Fish Shellfish Immun. 6: 483–492.

Hydrobiologia **378:** 241–242, 1998.
R. M. O'Riordan, G. M. Burnell, M. S. Davies & N. F. Ramsay (eds), Aspects of Littorinid Biology.

Additional records of rough periwinkles in northwest Scotland

D. C. Smith
Dept. of Biological Sciences, State University of New York, Brockport, NY 14420, U.S.A.

Key words: Scotland, rough periwinkle distribution

The status of the species of rough periwinkles has become relatively stable since the elevation of *Littorina nigrolineata* (now *L. compressa* (Reid, 1996)) to species status (Heller, 1975; Sacchi, 1975) and by the recognition of *L. arcana* as a species (Hannaford Ellis, 1978). Because of these revisions of the rough periwinkle species earlier reports on their distribution are no longer useful. The most comprehensive paper on their distribution since the current definitions of the species is that of Mill and Grahame (1990) based on numerous sites in England and Wales, and a few in Scotland.

In order to fill in some of the unknown areas, I visited twelve locations on the Isle of Skye and on the mainland coast of Scotland north of Skye as far as Point of Stoer. The sites chosen were all rocky shores consisting of cliff faces and/or large boulder fields. All but one location (Inverewe, 5/7/92/1) were exposed shores, and the collections were made in the lower half of the zone occupied by the periwinkles; the selection of exposed shores and the lower region of the shore was to increase the likelihood of finding *L. arcana* and *L. compressa* should they be present.

About 100 individuals were gathered from each site. The animals were boiled and then frozen for transport to the laboratory where the thawed soft parts were removed for species identification. Identification of *L. compressa* was based upon shell morphology (flattened ridges wider than the intervening grooves which usually contain a dark line; secondary grooves are frequently present on some of the wider ridges and may also contain a dark line) as well as distinctive penis morphology in the males (Sacchi, 1975). Separation of female *L. saxatilis* and *L. arcana* was based on the presence of a brood pouch (*L. saxatilis*) or a jelly gland (*L. arcana*). No attempt was made to separate the males of these two species.

Table 1. Location and number of littorinids collected

Collection	Location	OS reference	No. of animals	*Ls*	*Lc*	*La*
5/2/92	Vaternish pen.	NG222613	91	X	X[1]	X[2]
5/3/92/1	Trotternish pen.	NG405732	109	X	X[3]	
5/3/92/2	Staffin	NG496683	125	X		
5/4/92/1	Neist pen. W	NG126471	131	X		
5/4/92/2	Neist pen. SE	NG130470	53	X		
5/5/92	Elgol	NG515136	99	X	X[4]	
5/6/92/1	Laide E	NG905921	102	X		
5/6/92/2	Point of Stoer W	NC003328	103	X		
5/6/92/3	Point of Stoer E	NC043331	96	X		
5/7/92/1	Inverewe	NG855821	79	X		
5/7/92/2	Melvaig	NG739862	121	X		
5/7/92/3	Opinan	NG735709	109	X		

Ls = *Littorina saxatilis; Lc* = *L. compressa; La* = *L. arcana*
1. 27 individuals of *L. compressa* in this collection
2. 14 females of *L. arcana* (males not distinguished from *L. saxatilis*)
3. 10 individuals
4. 10 individuals

Littorina saxatilis was found at all 12 sites; *L. arcana* was found at only one of the sites (Vaternish peninsula, Skye, 5/2/92), and *L. compressa* was found at 3 of the 12 (Vaternish peninsula, 5/2/92; Trotternish peninsula, 5/3/92/1; and Elgol, 5/5/92, all on Skye). The 6 mainland sites only yielded *L. saxatilis* to my collecting efforts (Figure 1, Table 1).

Littorina saxatilis appears to be ubiquitous on British shores where there is a suitable firm rocky foundation. *L. arcana* is considerably more limited. According to Mill & Grahame (1990) there are extensive gaps in the distribution of *L. arcana* along the south and southeast coasts of England and along the west coast of Britain north of Anglesey, Wales. Only three reports of *L. arcana* along the northwest coast of Scotland were previously made; this present report only extends that by one site. The distribution of *L.*

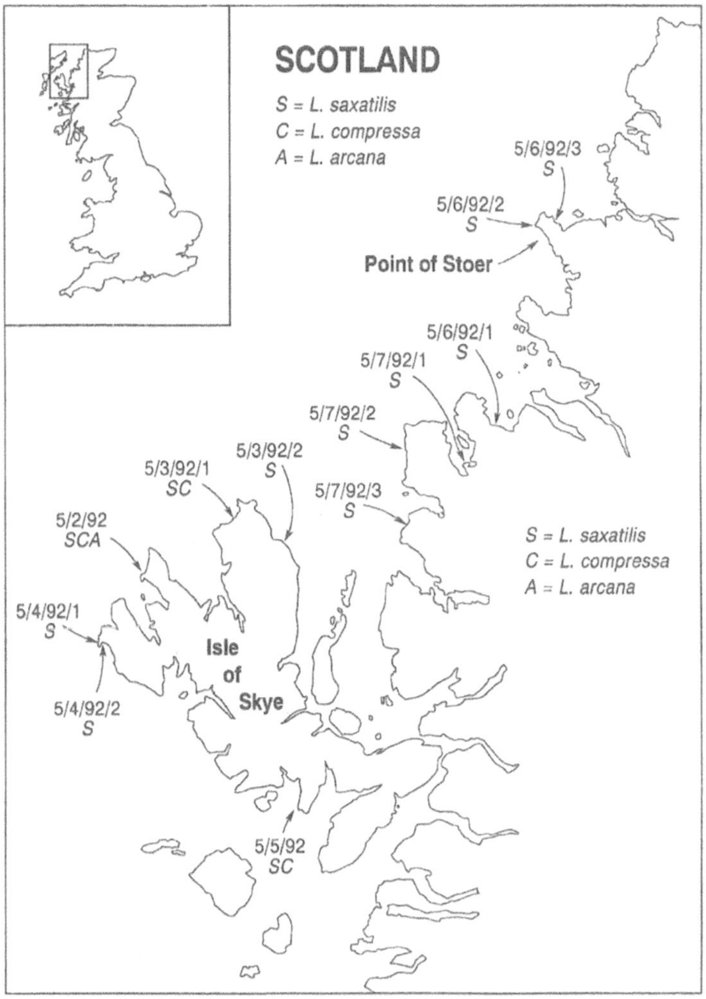

Figure 1. Distribution of species of rough periwinkles in northwest Scotland.

compressa is even more limited; Mill and Grahame reported its presence in Cornwall, several locations in Wales, and a few scattered locations in Scotland. In this report I extend that distribution by 3 locations, all on the Isle of Skye.

References

Hannaford Ellis, C. J., 1978. *Littorina arcana* sp. nov.: a new species of winkle (Gastropoda: Prosobranchia: Littorinidae). J. Conch. 29: 304.

Heller, J., 1975. The taxonomy of some British *Littorina* species, with notes on their reproduction (Mollusca: Prosobranchia). J. linn. Soc., Zool. 56: 131–151.

Mill, P. J. & J. Grahame, 1990. Distribution of the species of rough periwinkle (*Littorina*) in Great Britain. Hydrobiologia 193: 21–27.

Reid, D. G., 1996. Systematics and evolution of *Littorina*. Ray Society, London. 463 pp.

Sacchi, C. F., 1975. *Littorina nigrolineata* (Gray), (Gastropoda, Prosobranchia). Cah. Biol. mar. 16: 111–120.

Hydrobiologia **378:** 243–246, 1998.
R. M. O'Riordan, G. M. Burnell, M. S. Davies & N. F. Ramsay (eds), Aspects of Littorinid Biology.

Epilogue

Mark S. Davies[1] & Gray A. Williams[2]
[1] *Ecology Centre, University of Sunderland, Sunderland SR1 3SD, U.K.*
[2] *The Department of Ecology & Biodiversity and The Swire Institute of Marine Science, The University of Hong Kong, Hong Kong*

Introduction

It is rare to find a family with sufficient a scientific following to merit its own series of symposia. The Littorinidae are cosmopolitan, accessible, extremely abundant and exhibit wide ranges of morphs and colours; properties that are probably the source of their interest to amateur naturalists and scientists alike. The comparative wealth of knowledge of this family is due in no small part to the co-operation and collaboration that exists between littorinid biologists and the Fifth International Symposium on Littorinid Biology at Cork provided a further opportunity for discussion and exchange of current littorinid research. Attendance has grown with every symposium in this series and now, it would seem, there are more littorinid biologists than ever before.

This communication highlights themes arising from the papers presented at Cork, most of which are contained within this volume, and attempts to identify future research directions. It is eight years since this was last attempted as an epilogue (Raffaelli, 1990) and an update is, therefore, timely. Recently both Reid (1996) and McQuaid (1996a, b) have produced authoritative texts reviewing the systematics and evolution of the genus *Littorina* and the ecology and evolution of the Littorinidae, respectively. These reviews provide a framework to stimulate further research and the papers presented at Cork illustrated some of these directions. Littorinid research spans diverse disciplines, from molecular biology to elegant anatomical studies and the papers in this Proceedings' volume reflect this, and fall into three broad themes: systematics, ecology, and pollution studies.

Systematics

Raffaelli (1990) reported that the taxonomic problems with the European littorinids had been largely resolved, yet the Proceedings published since 1990 (Grahame et al., 1992; Mill & McQuaid, 1995), including this volume, contain various interpretations of littorinid taxonomy, especially with regard to the *Littorina saxatilis* ('rough periwinkle') complex. It is now generally accepted that the complex comprises three species, *L. saxatilis, L. arcana* and *L. compressa* (formerly known as *L. nigrolineata*) (see Reid, 1996), although there is still debate over the status of the 'neglectoid' morphs. Hull (this volume) provides evidence of assortative mating in the zone of overlap between two mostly allopatric populations of *L. saxatilis* (a thin-shelled, wide-apertured, small form and a thick-shelled, tall form) on the North Sea coast of the UK. Where crosses between the forms occur, the hybrids are typically less fit than the parents and Wilding et al. (this volume) provide RAPD evidence for the separation of these two forms. A similar system also appears to exist in north-west Spain (Cruz et al., this volume). *L. neglecta* remains a contentious species, though Black et al. (this volume) present further morphological evidence for its separation from *L. saxatilis*.

In the light of these studies it appears that a complex pattern of incipient speciation is occurring. Such a state of flux should not prevent valuable physiological and ecological work on these organisms being carried out, providing that the individual animals used are described in sufficient detail to make their status clear with respect to future revisions of the group (see, for example, Smith, this volume). This situation presents a valuable and rare opportunity for study of speciation processes, and it is to be expected that there will be disagreements over classification in such a complex system, where the term 'species' is difficult to apply (see Gosling, 1994). We should recognise that the system is dynamic and formal nomenclature should have low priority until patterns of gene flow are clarified.

Raffaelli (1990) noted that ecological work often precedes systematic investigations. Perhaps there has been a shift in emphasis, particularly with the advent of new techniques or the re-application of old methods (Medeiros et al., Mikhailova & Johannesson, Olabarria et al., all this volume), so that we now often lack firm ecological data as a context for systematic data. De Wolf et al. (this volume) use myoglobin and iso-electric focusing of esterases to show that *L. striata* from sites as much as 2000 km apart show little genetic variation despite marked differences in morphology. Variation in dispersal strategies within the Littorinidae and the influence of local environmental conditions at different sites may result in great intraspecific variability and this has been investigated in a number of species (in this volume: *Bembicium vittatum*, Johnson & Black; *Littorina saxatilis*, Cruz et al.; the flat winkles, Olabarria et al.; *Nodilittorina lineolata*, De Magalhães).

Ecology

In the 1980s and early 1990s much attention was directed towards radular structure, food preference and the complex subject of plant-herbivore interactions, as evident from the contents of earlier Proceedings volumes. In the present volume such subjects are conspicuous by their almost total absence, the exception being Christensen's informative account of the fungal diet of mangrove-dwelling *Littoraria*. This absence is surprising since the relationship between grazer and grazed is still largely unclear. For example, we know precious little of the impact of littorinids in arresting or driving patterns of succession on both a macro- and micro-scale, or of the impact of most littorinids on their algal resources, although McQuaid (1996b) reviews our current knowledge. Ecologists may, unfortunately, have been discouraged from such studies by perceived taxonomic confusion.

Boulding & Harper (this volume) demonstrate that we have much to learn about the basic ecology of littorinids. In an elegant series of experiments they investigated the relationship between littorinid distribution and the topographic features presented by a substratum covered with barnacles. Approximately half the variation in snail abundance could be accounted for by the number of barnacles and the number of small and very small 'pockets' (spaces between the mats of barnacles). Boulding & Harper suggest that the pockets represent refuges that could act to regulate the population size of littorinids. Studies such as this are important in understanding basic distribution patterns and their aetiologies, and have been extensively performed in Australia (see, for example Chapman & Underwood, 1996).

Much of the ecological work presented at Cork provided evidence to aid the systematists. Size in relation to habitat was a recurring theme; both Johnson & Black (this volume) and De Magalhães (this volume) relate size to wave exposure in Australian *Bembicium vittatum* and Brazilian *Nodilittorina lineolata*, respectively.

Studies on reproduction clearly have much to offer the systematists and this may explain the popularity of such studies within this volume. Studies on mating behaviour yield valuable, though often complex, information for both the systematist and the ecologist. Erlandsson & Rolan-Alvarez (this volume) suggest that mate choice is size-based in some Swedish *L. saxatilis* populations and describe how a sheltered shore population preferred large mates while an exposed shore population preferred small mates. Nevertheless, when the two populations were brought together, mating between the morphs was random with respect to size. Hull (this volume, see above) describes similar assortative mating between two forms of *L. saxatilis* in north-east England. Studies on the timing and intensity of reproduction are equally important. Cruz et al. (this volume) describe variations in fecundity with shore height in *L. saxatilis*. Egg production by overlapping populations of three species of *Nodilittorina* was investigated by Mak (this volume) and such baseline studies on thus far little-investigated genera should be encouraged. So far, patterns of dispersion of pelagic eggs and larvae have largely been neglected, despite the vital role played in gene flow by this stage in the life-cycle.

The synergy between ecology and systematics is vital and the Littorinidae give ample opportunity to study selection and dispersal. A note of caution should, however, be added when seeking ecological explanations for reproductive patterns, as evolutionary and phylogenetic constraints are ultimately the basic framework within which any species can vary (see McQuaid, 1996a).

Pollution studies

In recent years (in this series of symposia, from the third) there has been interest in the use of littorinids as sentinel species for the effects of organotin pollutants, and papers in this volume have extended this work. The transformation of female structures to male structures during 'intersex' is further documented by Sundermann et al. (this volume) who report a tributyltin (TBT)-induced transformation of the female pallial oviduct to a male prostate gland in *Littorina littorea,* and the effects of TBT are reported in the hydrobiid *Hydrobia ulvae* which is described by Schulte-Oehlmann et al. (this volume) as being more sensitive to TBT pollution than *L. littorea.* Comparison between intersex in littorinids and 'imposex' in the muricid *Nucella lapillus* is long overdue and is provided by Oehlmann et al. (this volume). Imposex in dogwhelks is more sensitive to TBT, but intersex in *L. littorea* is a preferable measurement at higher TBT concentrations when dogwhelks can be eliminated. Intersex and imposex occurrence were geographically uniform in the locations examined (Ireland, France and Germany). Oehlmann et al. recommend that the 'intersex index' and the 'vas deferens sequence index' are the most valuable parameters to record for monitoring the development of imposex in *L. littorea* populations. Littorinids, therefore, may provide a useful additional group to assess the effects of TBT and we may see their adoption in future monitoring studies.

Conclusions

Reid's (1989, 1990, 1996) development of a phylogenetic analysis of the Littorinidae has helped to focus approaches to variability between species, emphasising the constraints of ancestral lineages, and these studies remain landmarks in our knowledge of the family. There still remains debate on the status of some forms and this area is the subject of intense investigation. The possibility of incipient speciation remains an exciting avenue for further exploration. One aspect which is increasingly recognised as important is that of behaviour and its implications for gene flow. Simple mating experiments involving the monitoring of offspring viability, phenotypic plasticity and success in different environments would yield a wealth of information, as proposed by Raffaelli (1990). The role of mating and mate selection may well be the key to understanding the status of different morphs. It is, however, important to recognise that most work has been restricted to narrow geographic areas and investigations of patterns of gene flow over wider areas are clearly needed.

In contrast, recent studies of the role of littorinids in structuring communities are rare (although see the review by McQuaid 1996b), and given littorinid abundances and their ubiquitous distribution, such studies would be fruitful. In an earlier symposium, the global high-shore dominance of littorinids was discussed in terms of their adaptations to this environment (McMahon, 1990). The Littorinidae provide an excellent study group for such broad-scale, geographic comparisons. An examination of the ecological role of littorinids in the high rocky intertidal around the world offers an excellent opportunity to investigate variation in herbivore effectiveness on a range of scales. Perhaps future meetings will promote further collaborations to realise such an investigation.

There is a definite need for more comparative research, especially over a wider geographic scale. Although studies are progressing over a wider area than ever, littorinid research is still rather fixed in a Eurocentric mode. The convening of the next symposium in Jamaica and the advent of more studies on lesser-known genera, such as *Nodilittorina* and *Littoraria* may go some way to redress the balance and will expand our knowledge of this fascinating family on a truly international scale.

Acknowledgements

We are grateful to C. Little, C. D. McQuaid and D. G. Reid for comments on a draft of this epilogue.

References

Chapman, M. G. & A. J. Underwood, 1996. Influences of tidal conditions, temperature and desiccation on patterns of aggregation of the high-shore periwinkle, *Littorina unifasciata*, in New South Wales, Australia. J. exp. mar. Biol. Ecol. 196: 213–237.

Gosling, E. M., 1994. Speciation and wide-scale genetic differentiation. In A. R. Beaumont (ed.), Genetics and Evolution of Aquatic Organisms. Chapman and Hall, London: 1–15.

Grahame, J., P. J. Mill & D. G. Reid (eds), 1992. Proceedings of the Third International Symposium on Littorinid Biology. The Malacological Society of London, London. 324 pp.

McMahon, R. F., 1990. Thermal tolerance, evaporative water loss, air–water oxygen consumption and zonation of intertidal prosobranchs: a new synthesis. In Johannesson, K., D. G. Raffaelli & C. J. Hannaford Ellis (eds), Progress in Littorinid and Muricid

Biology. Kluwer Academic Publishers, Dordrecht. Hydrobiologia 193: 241–260.

McQuaid, C. D., 1996a. Biology of the family Littorinidae. I. Evolutionary aspects. Oceanogr. Mar. Biol. Ann. Rev. 34: 233–262.

McQuaid, C. D., 1996b. Biology of the family Littorinidae. II. Role in the ecology of intertidal and shallow marine ecosystems. Oceanogr. Mar. Biol. Ann. Rev. 34: 263–302.

Mill, P. J. & C. D. McQuaid (eds), 1995. Advances in Littorinid Biology. Kluwer Academic Publishers, Dordrecht. Hydrobiologia 309: 193 pp.

Raffaelli, D., 1990. Epilogue. In Johannesson, K., D. G. Raffaelli & C. J. Hannaford Ellis (eds), Progress in Littorinid and Muricid Biology. Kluwer Academic Publishers, Dordrecht. Hydrobiologia 193: 271–273.

Reid, D. G., 1989. The comparative morphology, phylogeny and evolution of the gastropod family Littorinidae. Phil. Trans. Roy Soc., Lond. B 324: 1–110.

Reid, D. G., 1990. A cladistic phylogeny of the genus *Littorina* (Gastropoda): implications for evolution of reproductive strategies and for classification. In Johannesson, K., D. G. Raffaelli & C. J. Hannaford Ellis (eds), Progress in Littorinid and Muricid Biology. Kluwer Academic Publishers, Dordrecht. Hydrobiologia 193: 1–20.

Reid, D. G., 1996. Systematics and evolution of *Littorina*. Ray Society, London. 463 pp.

Hydrobiologia **378:** 247–248, 1998.
R. M. O'Riordan, G. M. Burnell, M. S. Davies & N. F. Ramsay (eds), Aspects of Littorinid Biology.

Index

GPSR Compliance

The European Union's (EU) General Product Safety Regulation (GPSR) is a set of rules that requires consumer products to be safe and our obligations to ensure this.

If you have any concerns about our products, you can contact us on ProductSafety@springernature.com

In case Publisher is established outside the EU, the EU authorized representative is:

Springer Nature Customer Service Center GmbH
Europaplatz 3
69115 Heidelberg, Germany

Batch number: 10762189

Printed by Printforce, the Netherlands